AF531299

TEXT BOOK
OF
BIOSTATISTICS
I

DPH MATHEMATICS SERIES

TEXT BOOK OF BIOSTATISTICS I

By

A.K. Sharma

DISCOVERY PUBLISHING HOUSE
NEW DELHI-110002

Published by:

DISCOVERY PUBLISHING HOUSE PVT. LTD.
4383/4B, Ansari Road, Darya Ganj
New Delhi-110 002 (India)
Phone : +91-11-23279245; 23253475; 43596065
E-mail : discoverybooksindia@gmail.com
discoverypublishinghouse@gmail.com
namitwasan9@gmail.com
web : www.discoverypublishinggroup.com

First Published: **2005**

Reprinted: **2022**

ISBN: 978-81-8356-030-6

Text Book of Biostatistics - I

Printed at:
Infinity Imaging Systems
Delhi

Preface

This book of Biostatistics has been specially written to meet the requirements of B.A./B.Sc. students of all Indian Universities.

The subject matter has been discussed in such a simple way that the student will find no difficulty to understand it. The proof of various theorems and examples have been given with minute details each chapter of this book contains complete theory and large number of solved examples sufficient problems have also been selected from various Indian Universities and competitive examination.

I hope that this book warmly received by the student and teacher.

I wish to thank my colleague and friends for countless helpful remarks and suggestion. The author wish to express his thanks to the publisher M/s Discovery Publishing House, New Delhi, for bringing out their book in present nice form.

A.K. Sharma

Contents

Pages

Preface

1. **Introduction to Biostatistics** 1

Introduction, Definition of Biostatistics, Brief History of Biostatistics, What Can be Done with Biostatistics?, Some Definitions and Concepts, Applications of biostatistics, Statistical Symbols, Scales of Measurement.

2. **Population and Samples** 14

Introduction, Advantages of Using a Sample, Use of Tables of Random Numbers for Sampling, Other Sampling Methods, Sampling Techniques, Sampling and Non-Sampling Errors.

3. **Describing the Data (*Tabular and Graphical Approaches*)** 27

Introduction, Tabular Representation, Graphical Representation, Properties of Graphically Represented Frequency Distribution.

4. **Measures of Central Location** 54

Various Measures of Central Tendency, Arithmetic Mean, Mathematical Properties of Arithmetic Mean, Weighted Arithmetic Mean, Median, Relationship Among Mean, Median and Mode, Geometric Mean, Compound Interest Formula, Relationship Among the Averages.

5. **Hypothesis Testing** 166

Introduction, Significance of Measuring Variation, Properties of Good Measure of Variation, The Mean Elevation, The Interquartile Range or the Quartile Deviation, The Standard Deviation, Difference Between Mean Deviation and Standard Deviation, Calculation of Standard Deviation, Mathematical

Properties of Standard Deviation, Variance and Standard Deviation Compared.

6. **The Chi-Square (χ^2) Test** 260

Introduction, χ^2 Defined, The Chi-Square Distribution, Conditions for Applying χ^2 Test, Yates' Corrections, Uses of χ^2 Test, Additive Property of χ^2, Chi-Square Test for Specified Value of Population Variance, Misuse of Chi-Square Test, Limitations on the Use of χ^2 Test.

7. **Partial and Multiple Correlation** 317

Introduction, Partial Correlation, Multiple Correlation, Multiple Regression Analysis, Reliability of Estimates.

8. **Sampling and Designs** 353

Sample Method of Census, Theoretical Basis of Sampling Methods of Sampling, Random Sampling Methods, Size of Samples, Merits and Limitations of Sampling, Sampling and Non-Sampling Errors.

9. **Tests of Significance** 373

Introduction, Standard Error and Sampling Distribution, Estimation.

1

Introduction to Biostatistics

INTRODUCTION

Application of statics in biological problems is called *Biostatistics.* Statistical knowledge plays an important role in many aspects of science and education. Some knowledge of statistics is essential for the students of Zoology, Botany, Medicine, Psychology, Microbiology and many other disciplines for the reasons given below :

This text introduces the students of biology to the concepts and methods used in the biostatistical treatment of data. Broadly speaking, the application of statistical methods to biological problems is called biostatistics. The biological data which result from the conduct of experiments of various kinds, are usually collection of measurements and observations. At times such collections may be large and require classification, summary description and rules of evidence for drawing valid inferences. Biostatistics provides a conceptual framework, and practical methodology, whereby this can be done.

(i) Use of Scientific Knowledge

Many Articles, Books and Journal articles report experimental findings in statistical form but also present theories and arguments involving statistical concepts. The statistical knowledge helps to evaluate the results and significance. One may come across a research article with results projected as quite significant but when examined statistically the results may not be significant at all. A person with the knowledge of statistics often evaluates the scientific results and claims by examining the sample size (that is, the number of individuals or items studied), sample collection methods, and analysis procedures. Understanding of statistics helps to attain an attitude to respect the possibilities of statistical proof with a healthy scepticism.

In most of biological and medical experiements the statistics is extensively used to study varibility among individuals and effectiveness of different drugs and treatments. A little understanding of statistics can enable us to appreciate the role of these approaches in different fields. Unfamiliarity with statistical jargon can result in complete lack of understanding of simple publications.

Statistics is often misused or misinterpreted but to those who understand the misinterpretations they tell the truth or they say nothing.

(ii) Design and Conduct of Experiments

Many students as a part of their training are required to design and conduct original investigations. The design of an experiment is inseparable from the treatment of results and conclusions drawn. A simple experiment, for example, must be designed to facilitate the treatment of data in such a way as to permit clear interpretation. The sample, size, sampling method, summarization and treatment of data must be done in correct scientific way as to withstand the evaluation of experts.The experiments designed and conducted by using statistics often lead to interpretation that conforms to known scientific evidence.

If the design is seriously flawed, no amount of statistical manipulation can lead to the drawing of valid conclusion.

(iii) Theoretical Research

The research in the field of animal behaviour has become more quantitative, The development of theories serves to organize our information. Theorists predict what we expect to observe in specified circumstances. The means by which we test the theories are largely statistical.

(iv) Problem Solving

Often research is conducted on a limited scale, not to test a theory, but to uncover information vital to the solution of a practical problem. We may be interested to seek the answers of such questions as : how can a particular test be improved; is this variable, for example, wing-length related to the other, variable like age of the bird; does the new diet has better effect than the common diet? Such questions call for the application of statistical methods to the available data.

(v) Daily Use

Statistics are of immediate and practical utility. They help us get our work done more quickly and efficiently. They aid the teachers in the assignment of grades and in the construction of test. The students can make their results more presentable by using statistical methods. The validity of various types of claims in the advertisements we see daily, can be tested by just examining their experimental design and procedures adopted to draw inferences. A person with the basic knowledge of statistics may not find it easier to accept the false claims.

DEFINITION OF BIOSTATISTICS

Biostatistics deals with the application of statistics to the biological data. By statistics we mean the aggregates affected to a marked extent by multiplicity of causes numerically expressed, enumerated or estimated according to a reasonable standard of accuracy, collected in a systematic manner for predetermined purpose and placed in relation to each other. It involves collection, *arrangements, analysis and interpretation of the numerical facts so as to draw scientific conclusions or to make effective decisions.*

It is branch of applied mathematics and may be regarded as mathematics applied to the observational biological data. The students should not try to make rigid lines of demarcation between biology, biostatistics, statistics and mathematics as it is not possible and probably also not desired to do so. Biostatistics is simply the appli'9'ition of statistical methods to the solution of biotogical problems. It is also called biometry.

There are two types of statistics more commonly used in biological studies. These are *sample statistics* and *test statistics,*

(1) *Sample Statistics :* It is utilised to calculate population parameters, such as mean, standard deviation, etc. It explains the nature of distribution of data. The data in this case are immediately calculated in order to provide biologists the first hand results. It also gives the knowledge of statistically significant or non-significant results.

(2) *Test Statistics :* It is used to test the hypothesis about one or more sample data. In statistical test, one has to choose more data for analysis, obtained by conducting the experimental design. It includes the study of chi-sqaare and t-test, etc.

BiosLatistics can be divided into two subcategories : *descriptive biostatistics and inferential biostatistics.*

(a) Descriptive Biostatistics

The study of bicxtatfetieal procedures which deal with summarization of data to make it more informative and comprehensible is called *descriptive biostatistics.*

The descriptive biostatistics involves graphical and tabular app. oaches to describe, sunmarize and analyze the data. This includes anything done to data which is designed to summarize or describe them without attempting to infer anything that goes beyond the data.

An example of a descriptive biostatistics is the decennial census of India, in whichlall the citizens are required to provide such information as age, sex, religion, marital status, education and occuoation. The data obtained in such a census can then be compiled and arranged into tables and graphs that describe characteristics of the population at a given time.

(b) Inferential Biostatistics

Although descriptive biostatistics is an important branch of statistics, statistical inference usually arises from samples and this means that its analysis will require generalizations which go beyond the data. *The procedures which serve to make generalizations on the basis of the studies of a sample (which is a part of the pupulation of interest) constitute inferential biostatistics.*

Such methods are required, for instance, to predict the mean weight of fishes in a big pond; to compare the effectiveness of two reducing diets (on -the bases of the weight losses of two groups of persons who have been on diet).

A scientist studying the weight gains in pigs after being fed on a new diet, after taking the weights of control and experimental groups may be interested to find out :

(i) If there is any real difference in the weights of pigs fed on normal diet and those fed on the new diet.

(ii) He may also be interested to draw conclusions about the efficacy of the new diet and may like to generalize his findings.

The ability to make generalized conclusions, inferring characteristics of the whole from the characteristics of its parts lies within the realms of inferential statistics. Statistical inference is the process of drawing conclusions about population characterstics based on the information container in a sample.

BRIEF HISTORY OF BIOSTATISTICS

Francis Galton (1822-1911), a cousin of Charles Darwin is called the *"father of biostatistics"* and eugencis. Before ninteenth century statistics was mathematical, political and governmental. In the early nineteenth century, a famous Belgian statistician Quetelet combined the theory and practical methods of statistics and applied them to the problems of biology, education, medicine and sociology. Quetelet is credited with the development of statistical theory as a general method of research applicable to any observational science. Beyond any doubt the individual who had the greatest effect upon the introduction and use of statistics in biology was Francis Gallon. The inadequacy of Darwin's genetic theories stimulated Gallon to try and solve the problems of heredity. In the course of his long life he made notable contributions in the fields of heredity eugenics, psychology, anthropometry, and statistics. Our present understanding of correlation, the measure of agreement between two variables, is credited to him. Pearson collaborated with Gallon in later years and was instrumental in developing many of the correlation and regression formulas that are in use today. Gallon also contributed to the development centiles or percentiles.

The term biometry was coined by *W.F.R. Weldon* (1860-1906). In the twentieth century new techniques and methods were applied to the study of small samples. The major contributions in small sample theory were made by an English statistician R.A. Fischer. He developed most of his methods in an agricultural and biological setting. Fischer's methods are used in almost all the fields of science.

Statistics is a broad and extremely active field whose applications touch almost every aspect of science. Although statistics arose to satisfy the needs of scientific research, the development of its niethodology in turn affected the sciences in which statistics is applied.

WHAT CAN BE DONE WITH BIOSTATISTICS?

Biologists mainly deal with animals in which variability is much more as compared to the other systems. A knowledge ofbiostatistics aids in the correct interpretation of data from animal systems. Suppose a biologist wants to carry out an experiment to study the effect of a new diet on the weights of rats. He can take the help of biostatistics to decide how many rats should be included in the experiment, how to select normal and experimental rats; to make observations more concise and informative, and to draw correct conclusions. What are some of the things that can be done to get the maximum information from the results of such experiments? Here is the list of some of the possibilities :

(1) Averages can be calculated. These averages give a picture of the typical performance of the groups.

(2) The variability of the measurements can be determined. By using the average as a point of reference, one can determine how the observations spread about this central point.

(3) Graphs, tables, and figures can be prepared to portray clearly the nature of the group or groups.

(4) The raw observations can be transformed into a more meaningful form. The most common of these forms are centiles (or percentiles) and standard scores.

(5) The relationship of one variable to another can be determined. These statistics, called correlation coefficients, are among the most useful. For example, it might be of interest to find the relationships between the height and weight; between the dosage and the response.

(6) The reliability of the measurement instruments can be determined. This is done by making two measurements of the same individuals with the same or parallel devices and finding the correlation between the two sets of data. As will be seen, there are also other ways of computing reliability coefficients.

(7) The validity of the measurements can be determined. For example, intelligence tests are often validated by correlating scores on these tests with grade-point averages. If the intelligence tests are valid,

those who obtained the highest scores will also receive the highest academic grades.

(8) One set of measurements, or a combination of variables, can be used to predict future status or behavior. This is probably the major end of all correlational work; in themselves correlation coefficients are of little value.

(9) From the measurement of a sample of individuals, inferences can be made about the larger population from which the sample was drawn. The drawing of statistical inferences is one of the chief activities of modern research. Information about a small sample is usually of limited use. The research worker hopes to generalize and draw conclusions about a larger group.

(10) The performance of one group can be compared with that made by another, and the significance of any difference can be tested. Suppose that a school is trying out a new reading curriculum. We have two groups : one is handled in the usual way in the course of a semester or a year and the other, an experimental group, is treated in a new fashion. That is, the experimental group is exposed to the new reading curriculum. At the end of the school year the same reading test is administered to both groups. The scores made by each group on this test can be averaged. By inspection it cannot be determined whether or not the averages are significantly different. At this point a test of significance can be made to see whether or not the difference is attributable to chance variation. *The processes of drawing inferences, making predictions, and testing significance are examples of sampling or inferential statistics.* The other methods mentioned are included in what is called descriptive statistics. Modern experimental research in fields of knowledge from anthropology to zoology uses sampling statistics.

SOME DEFINITIONS AND CONCEPTS

(a) *Data :* While observing various phenomena, one is usually interested in getting information on specific characteristics, e.g., age, sex, weight, height, marital status, occupation, blood pressure, body temperature, etc. These characteristics are referred to as variables whereas the values of observations recorded for them are referred to as data. Thus, the data are the raw materials of statistics. There are two types of statistical data :

(i) *Primary Data :* It is the data collected by a particular person or organisation for his or her own use from the primary uses.

(ii) *Secondary Data :* It is the data collected by some other person or organisation for his or her own use but the investigator can also use it. It means that a data can be primary for one person and secondary for the other.

(b) *Population : A population is any actual or conceptual collection of individuals defined by stated characteristics.* Every set of statistical data must be interpreted in relation to the popul?tion. Population is a group of individuals about which the scientific inferences are made. All university students in Delhi, all bacteria in a specific bottle of milk, all progeny of a particular rat; all persons who undergo heart surgery in 1998, etc. constitute different populations. On occassions biostatistican may deal with a population of measurements. By this is meant an indifinitely large aggregates of meansurements which hypothetically might be obtained under specific conditions.

(c) *Variable : A variable is any property of an individual (person, tree, bird, rat etc.) that can be/expressed in numerical terms.* For example the height of a 12 year old boy; the number of leaves on a branch, liver weight of a rat, percentage of calcium in the blood of a patient. We are interested in the values of a variable (sometimes of several distinct variables) that relate to the different members of a population. A variable may be readily measured or determined for any member of the population. It may be the product of a complicated analysis or laborious counting.

Variable is a property whereby the members of a group or set differ from one another. Theorirticular values of a variable are referred to as *variate* or *variate values*. A variable may be continuous or discrete. *A continuous variable* may take any value within a defined range of values. The possible values of the variable belong to a continuous series. Between any two values of the variable an (definitely large number of in between values may occur. Height, weight, and chronological time are examples of continuous variables. A *discrete variable* or discontinuous *variable* can lake specific values only. Size of a family, the number of female patients, the number of leaves etc. are discrete variables.

Besides these two categories of *measurement variables* which can be expressed in numerically ordered fashion there are other two types of variables which are often used in biology. These are the *ranked variables* and *attributes. A variable which is expressed as a series of ranks such as 1, 2, 3, 4, 5 is called ranked variable.* For example, in an experiment one may record the rank order of litters born or of emergence or pupae without specifying the exact time. Variables which are expressed qualitatively are called attributes. These are all properties such as red or white, pregnant or not pregnant, dead or alive etc.

(d) *Constant* : The term constant refers to a property whereby the members of a group do not differ from one another. In a sense constant is a particular type of variable. It is a variable which does not vary from one member of a group from another or within a particular set of, defined conditions. Suppose we are interested in religious attitude among students in a college. Of prime concern is the variable of religious attitude. However other variables such as sex, age may have to be taken into account in interpreting our findings. The membership in a specific college and the time when data were collected are constants.

(e) *Sample : A sample is any subgroup or sub aggregate drawn by some appropriate method from a population.* It is some segment of the population that is examined and measured as an aid to investigating properties of the parent population. As a sample of college students, one might take all the students of first year zoology class in a college. Similarly 2 ml milk from a bottle of milk taken for bacteriological studies is a sample.

(f) *Representativeness and Randomness* : Though any part of a population can be regarded as a sample, for the sample to be useful in making inferences about the population it must in some sense be representative. In some respects, the first year zoology students might prove representative at least of the students body of their college in height and weight for example, but not in age or in knowledge of modern history. 2 ml milk will not be representative bacterial population unless one ensures that the bottle is well mixed before pouring.

The only way of ensuring that a sample is representative is to introduce an element of *randomness* into its selection. After the population has been exactly defined the sample is drawn by tots. Sampling methods are discussed in chapter 2 of this book.

(g) *Parameters and Statistics : A parameter is any numerical property descriptive of a population.* Population mean, median and mode are the *paramters of location.* They relate to the general location of numerical values of a variable.

A statistic is any numerical propert, calculated for a sample. The statistics one calculates varies from sample to sample. Often a correspondence exists between sample statistics and population parameters and inferences about an unknown parameter may be based on calcul.ations with the known sample statistic. Mean height of students of a college is parameter, and mean height of students of first year zoology students of the college is a statistic.

(h) *Accuracy and Precision :* "Accuracy" and "precision" are used synonymously in everyday speech, but in statistics we define them more rigorously. *Accuracy is the closeness of a measured or computed value to its true value. Precision is the closeness of repeated measurement of the same quantity.* A biased but sensitive scale might yield inaccurate but precise weight. By chance an insensitive scale might result in an accurate reading, which would however be imprecise, since a repeated weighting would be unlikely to yield an equally accurate weight. Unless there is bias in a measuring instrument, precision will lead to accuracy. We need therefore mainly be concerned with the former.

Precise variates are usuallly but not necessarily whole numbers. Thus, when we count four eggs in a nest, there is no doubt about the exact number of eggs in the nest if we have counted correctly; it is four, not three nor five and clearly it could not be four plus or minus a fractional pan. Meristic variables are generally measured as exact numbers. Seemingly, continuous variables derived from meristic ones can under certain conditions also be exact numbers. For instance, ratios between exact numbers are themselves also exact. If in a colony of rats there are 18 females and 12 males, the ratio of females to males is 1.5, continuous variate but also an exact number.

For example, if we report that the winglength of a bird is 8 cm long, we are stating the number 8 (a value of a continuous variable) as an estimate of the the true wing length. This estimate was made using some sort of a measuring device. Had the device been capable of more accuracy, we might have concluded that the wing was 8.32 cm long." When recording values of continuous variables, it is important to designate the accuracy with; which the measurement have been made. By convention the value 8 denotes a measurement in the range of 7.50000... to 8.49999..., the value 8.3 designates a range of 8.25000... to 8.34999..., and the value 8.32 implies that the true value lies within the range of 8.31500... to 8.32499... . That is, the reported value is the midpoint of the implied range, and the size of this range is designated by the last decimal place in the measurement. The value of 8 cm implies a range of accuracy of 1 cm, 8.3 cm implies a range of 0.1 cm and 8.32 cm implies a range of 0.01 cm. Thus, to record a value of 8.0 implies greater accuracy of measurement than does the recording of a value of 8, for in the first instance the true value is said to be between 7.95000... and 8.049999... (e.g. within a range of 0.1 cm), whereas 8 implies a value between 7.5000... and 8.49999... (i.e., within a range of 1 cm). To state 8.00 cm implies an accuracy in measurement which ascertains the birds wing length to between 7.99500... and 8.00499... cm (i.e., within a range of 0.01 cm). Those digits in a number that denote the accuracy of the measurement are referred to as *significant figures*. Thus, 8 has one significant figure, 8.0 and 8.3 each have two significant figures, and 8.00 and 8.32 each have three.

In working with exact values of discrete variables, the preceding considerations do not apply. That is, it is sufficient to state that a frog has 4 limbs or that its left lung contains 13 flukes. The use of 4.0 or 13.00 would be inappropriate, for since the numbers involved are exactly 4 and 13, there is no question of accuracy or significant figures.

But there are instances where significant figures and implied accuracy come into play with discrete data. An entomologist may report that there are 72,000 moths in a particular forest area. In doing so, it is probably not being claimed that this is the exact number but an estimate of the exact number, perhaps accurate to 2 significant figures. In such a case 72,000 would imply a range of accuracy of 1000, so that the true value might lie anywhere from 71,500 to 72,500. If the entromologist wished to convey the fact that this estimate is believed to be accurate to the nearest 100 (i.e., to 3 significant figures), rather than to the nearest 1000, he had better present his data in the form of *scientific notation*, as follows. If the number $7.2 \times 10^4 = 72{,}000$) is written, a range of accuracy of 0.1×10^4 (= 1000) is implied, and the true value is assumed to lie between 71,500 land 72,500. But if 7.20×104 were written a range of accuracy of 0.01×10^4 (= 100) would be implied, and the true value would be assumed to be in the range of 71,950 to 72,050. Thus, the accuracy of large values (and this applies to continuous as well as discrete variables) cart be expressed succinctly using scientific notation.

APPLICATIONS OF BIOSTATISTICS

Biological experiments deal with living things and the variability among the individuals is quite common. Variations are the basis of evolution of new forms of life. It is quite possible that these variations may mislead the experimenter. He might consider the natural variations as the differences resulting from his treatment or vice-versa. With the application of statistical methods such confusions are unlikely to occur. Statistics is used extensively in different branches of biology.

In *genetics* we often study whether or not the new generations follow the Mendelian ratios with the use of chi-square test which deals with observed and expected frequencies we can test such hypothesis. The *frequencies* of different genes and their changes due to the effects of forces like mutations, migration, isolation and selection can be estimated with the application of different statistical methods.

In ecology the statistical approaches deal with different aspects of populations and their environment. Ecologists often measure variations in envirionment and their effects on the different characteristics of organisms. Also association or lack of it among pairs of groups are of ecological interest. Various procedures like correlation analysis, chi square, frequency along with binomial and geometric distributions are frequently used by ecologists.

In taxonomy the statistics is of great help where one has to classify the taxonomic units on the basis of different characteristics and also find out association or lack of it among various characters of groups. With the help of statistics one can draw inferences about the populations. Similarly statistics is used in almost alll the branches of biology, like ethology, environmental biology, cell biology physiology etc.

STATISTICAL SYMBOLS

There are a number of symbols used in statistics. However there is no absolute conformity in the use of these symbols and notatiunal usage will vary somewhat from author to author. A common and simple statistical equation is

$$\Sigma f = N$$

This equation is read as, capital sigma (Σ), summation of, or the sum of the frequencies (f) is equal to N, the number of individuals. Summation sign, Σ is one of the most commonly used statistical symbols.

If we are dealing with one variable (a variable is a characteristic that manifests differences in magnitude or quantity), it is designated as "x". One measurement or observation of a variable is usually symbolized by the corresponding capital letter; hence X is one measure for variable x and Y is one measure for variable y. Different individual measures on the same variable are denoted by numeric subscripts : X_1, X_2, X_3...... X_N.

Different symbols are used for parameters and statistics.

Characteristic	*Parameter*	*Statistic*
Number of individuals	N	n
Mean	μ	$\overline{X}$
Variance	σ^2	S^2
Standerd Deviation	σ	s
Proportion	P	P
Pearson correlation coefficient	R	r
Number of cases	n	N

SCALES OF MEASUREMENT

We conclude this introductory chapter with a note on the types of data which may be subjected to statistical analysis. Data may be classified according to their scales of measurement as follows :

(a) Nominal Scale

On a *nominal scale* numbers are used as labels and no ranking is possible; for example groups of patients could be labelled 1, 2 and 3 corresponding to the classification, normal, diabetic and hypertensive.

(b) Ordinal Scale

On an ordinal scale scores can be assigned to the observations in such a way that the rankings obtained are meaningful, but the intervals between the rankings need not be defined; for example in a survey people's opinion of a product may be ranked according to the scale (1) poor, (2) fairly good (3) good(4) very good, (5) excellent.

(c) Interval Scale

On an interval scale both the ranking and intervals are meaningful, but there is no meaningful zero; an example is the measurement of temperature in degrees Celsius.

(d) Ratio Scale

On a ratio scale the measurements may be ranked, the intervals are meaningful and there is also a meaningful zero. Thus the ratios of measurements are also meaningful if they are made on a ratio scale. Measurements of quantities such as length, area and time are made on ratio scales, as are counts, such as the number of arrivals of cars at a service station in a given period of time.

EXERCISES

1. What is biostatistics ?
2. How is biostatistics helpful to a biologist ?
3. Define parameter, variable, population and sample ?
4. What is difference between accuracy and precision ?
5. What are the applications of biostatistics ?
6. Define, explain and mention uses of biometry.
7. What do you mean by data, population, "sample, variable, parameter, class interval, frequency distribution, cumulative frequency distribution, primary data, secondary data.
8. What do you mean by grouping of data ? RBCs number in lac/mm^2 of 30 fishes of a species were measured as follows :

32, 30, 33, 31, 28, 33, 33, 32, 30, 32, 34, 31, 35, 35, 35, 35, 36, 36, 37, 34, 32, 34, 34, 36, 37, 38, 37, 36.

(i) Prepare simple frequency distribution table.

(ii) Prepare overlapping and non-overlapping frequency distribution table.

(iii) Prepare overlapping and non-overlapping cumulative frequency distribution table.

9. The lowest and highest levels of few class intervals are given below. Mention length and mid-points of each class interval :

   40–50,32–44, 20–32, 10–22, 20–30, 30–39, 40–49, 40–52, 53–64.
10. What is the difference between parameter and statistics? Give examples.
11. Explain clearly what do you mean by descriptive and inferential statistics ?

2

Populations and Samples

INTRODUCTION

Consider a research worker who is interested to study the weights of fishes in a new pond after one month of growth. If there are about 2500 fishes in the pond, he may decide to select a smaller group of fishes and measure their weights. From these weights he may try to make inference about the weights of all fishes. Suppose that he decides to take the weights of 150 fishes. These 150 weights constitute Bis *sample* and weights of 2500 fishes are the *population* of his interest.

The population refers to a group of all items (individuals, objects, household etc), about which the. investigator wishes to draw conclusions.

A sample is a portion of the population selected to represent the population.

Each individual numerical value (weight of each fish) in the above example is called an *observation.*

In statistics population always means totality of individual observations about which inferences are to be made, existing anywhere in the world or atleast within a definitely specified sampling area limited in space and time.

Often the group at hand is the group of interest. If an experimenter wants to study the tail length of mature rats in a rat colony all the mature rats in the colony constitute the population and not a sample.

As another example of population and sample consider a research worker who wishes to evaluate a new treatment for a certain illness. If the new treatment is given to thirty patients who visit the hospital. These thirty patients constitute the sample. The researcher is not primarily interested in these particular patients, but rather how good the treatment might be for any patient having the disease. Here the population of interest consists of all patients who might have this particular disease and given the new treatment. Here the population of interest is *hypothetical* and does not exist as such. The researcher may not come across any patient other than these 30 patients, but, nevertheless, there may be persons with this disease in some other parts of the world.

ADVANTAGES OF USING A SAMPLE

There are many reasons for using a sample rather than the entire population. Generaly the population is very large and it is not possible to study the entire population. Also taking observations on the whole population may involve a large amount of money and time. If the population is hypothetical, one has no option but to study the sample only. In our example on the effects of a new treatment on the patients suffering from a particular disease; even if a large number of patients are studied, they still will be considered to be a sample from a very large population *existing* or *likely to exist* in future anywhere in the world.

It is possible to study a sample more carefully and accurately than studying a large number of individuals of the whole population. However the studies based on samples sometimes can lead to wrong conclusion if:

(a) Sample size is not appropriate.

(b) Sample selection is not correct.

Sample Size

Sample size is the number of items to be included in the study in order to secure meaningful results. It is very important to take proper sample size. There is a statistical method to determine the sample size based on certain characteristics. We shall discuss it at a later stage, but for the present one must remember that sample size cannot be fixed arbitrarily.

Secondly the way the individual items are selected also has a significant effect on the conclusion drawn. *The sample must always be a true representative of the whole population.* This important aspect of the sample is sometimes ignored by the researchers. There are many types of samples, however we shall discuss the simplest kind of a sample which is called a *simple random sample.*

Selection of Samples

Samples may be of two types in the broad sense :

1. Qualitative sample
2. Quantitative sample.

1. *Qualitative sample :* Qualitative samples are representing the quality of the populations or when we compare the two samples or objects. For example, plants are taller in one area than in another, colour of fruit; smaller or bigger size of cells, etc. These are qualitative samples.
2. Quantitative sample : When a sample is representing the quantity, the number is involved such as height of plant, total number of fruits per plant, size of fruit, etc. In this sample, numerical measurements are playing their important role in drawing useful conclusions.

Selection of Appropriate Method of Sampling

After discussing the different methods of sampling, the objective is which method is to be adopted in a particular situation? It is observed that the selected single method will prove to be the best under all situations.

Sample size : Sample size is an important matter to be taken while adopting a sample technique. Some experts are of the opinion that sample size should be 5% of the size of the population while others believe that sample size should be at least 10%. The following two considerations may be kept in mind in determining the appropriate size of the sample.

1. The size of the sample should increase as variation in the individual items increases.
2. The greater the degree of accuracy desired, the larger should be the sample size.

Merits of Sampling

The sampling technique has the following merits :

(a) Less time

(b) Less cost

(c) More reliable results

(d) More detailed information.

Limitations of Sampling

Despite the various advantages of sampling, it does have certain limitations. Some of the difficulties involved are given below :

1. Sample survey must be carefully planned and executed, otherwise the results obtained may be inaccurate.
2. Sampling generally requires the services of an expert.

Random Sample and Sampling Methods

A random sample is a sample, chosen in a very specific way. In fact, if a sample is selected at random, the known principles of inference apply, and if it is not so chosen, they do not apply. *For a sample to be random, it must have been selected in such a way that every element in the population had an equal opportunity of being included in the sample.*

For example suppose a deck of 52 cards is thoroughly shuffled and four cards are drawn.

The hand thus obtained may be considered to be a random sample of the population, because every card in the deck had an equal opportunity of inclusion in the hand, and every possible set of four cards had an equal opportunity of selection.

There are two properties of random samples that we need to know now. ***First**, if several random samples are drawn from the same population the* *elements of the sample will differ, and therefore the statistical characteristics will change from sample to sample.* Thus if we take two samples from a population of seeds and note whether their coat is wrinkled or smooth, the elements of the two samples may differ, that is, number of seeds with wrinkled seed coat may be different in the two samples. If we take a random samples of fishes from a pond and measure their length, we will get different values for different samples. *Second property of a random sample is, larger the smaple, lesser will be the variation of characteristics of the sample from one random sample to another.*

Advantages of Random Sample and its Selection

The main advantage of using a random sample is that there are mathematical procedures for random samples which enable the researcher to draw inferences about the whole population. These methods do not apply to samples other than random samples. A random sample eliminates personal bias- The researcher is not given the opportunity to reject those observations which may not support his theory or alternately select only those observations which may support his theory.

Consider a research worker who is interested to study the effect of a new fertiliser on the growth of some variety of plants. While selecting the sample he may include only those plants which show considerable growth and may reject the plants with less growth, such a smaple cannot be called as a random sample. To take a random sample the worker has to take plant from different area and include all sizes of plants.

Suppose one is interested to study the seed coat colour. If there are 1000 seeds with the researcher, they constitute the population of interest. If he has to select a random sample of size 30, he can do so in two ways. He can pick up 30 seeds from the group of 1000 one by one without looking at them. Such sampling is called *sampling without replacement; in which an observation is included in the sample only once.*

In the second method he can pick up a seed from the group, note its coat colour and put it back (replace) into the population. Such *sampling is called sampling with replacement, in which all members of the population have a chance to be selected at each draw.* Normally sampling without replacement is used for experimental purposes. As another example of random sampling suppose a doctor wisies to select a sample of size 20 from the 1500 records on heart diameters of asthmatic patients. The 1500 heart diameters constitute the population of interest. To take a random sample of size 20, one method is to write each diameter from 1-1500 on a standard piece of paper and then place 1500 pieces of papers in a container, mix thoroughly and pick up 20 pieces one by one. If the each piece of paper taken, randomly is put back into the container after noting its value the sampling will be called *sampling with replacement.* On the other hand if the papers are not put back into the container the sampling will be *without replacement.*

Disadvantages of Random Sample and its Selection

Despite great advantages of probability sampling technique, it has certain limitations, too. These are as follows :

1. It requires a high level of skill and experience for its use.
2. The costs involved in sampling are large.
3. It requires a lot of time to plan and execute the probability sampling.

Non-Random Sampling

Non-random sampling is the process of sample slection without the use of randomization. The selection is done on the basis of consequences or expert judgement.

The most important difference between random and non-random sampling is that in case of random sampling, the pattern of sampling can be ascertained while in case of non-random sampling, there is no way of knowing the pattern of variability of the process.

Non-random sampling includes the following three types :

1. Judgement sampling
2. Convenience sampling
3. Quota sampling.

Let us discuss these one by one.

1. *Judgement sampling :* The choice of sample items basically depends on the discretion of the Investigator. For example, if a sample of ten students is to be selected from the class of sixty for analysis of the spending habit of students, the investigator would select 10 students who, in his own opinion, are representative of the class.

   This method is often utilized in solving business and economic problems. This method has several limitations.

1. Though this method is simple, but it is not a scientific method.
2. It involves the risk of wrong conclusions.
3. Since an element of subjectivness is possible, this method cannot be recommended for general use.

2. *Convenience sampling :* In this method, a fraction of the population being investigated which is selected neither by probability nor by judgement, but by convenience. Convenience sample is done in case of interview. The results obtained by following convenience method can hardly be representative of the populations. They are generally biased and unsatisfactory.

3. *Quota sampling :* Quota sampling is a type of judgement sampling. In quota sample, quota are set up according to some specified characteristics, such as so many in each age, so many in each of several income groups, etc.

Quota sample is often used in public opinion studies. It provides fruitful and satisfactory results, if the interviewers are fully-trained and follow the instructions carefully.

USE OF TABLES OF RANDOM NUMBERS FOR SAMPLING

In larger populations it is impractical to make a tag or write the values on pieces of paper, put them in a container and take the sample. Tables of random numbers can be used to draw a random sample if we have a list of the population. Tables of random numbers are available. One such random number table has been given in the Appendix (Table I).

The single digits in the random numbers table have been obtained by some process which is equivalent to drawing repeatedly a piece of paper from a container.

To draw a sample, the values of the population are first assigned number 1, 2, 3, ... N. Then depending upon the total number of observations in the population, we select the digits from the table, e.g., if we have 1500 observations in our population of interest we will select four digits starting from any where in the table. Suppose we start from the first row of the first page of random numbers and note down the numbers as they appear in the row,

1009 7325 3376 5201 3586 3467 3548

Now since the largest number assigned to the observations of population is four digit number 1500. We make groups of these noted random numbers in four digits and we get

1009 7325 3376 5201 3586 3548

Since the four digits number 1009 is less than 1500, we select this number and include the observation number 1009 in our sample. The other numbers are more than 1500 we ignore them. Similarly we keep on noting random numbers making the groups and selecting the values till the desired sample size has been selected with the random numbers for simple random sampling a random number previously used to identify an item for the sample may reappear in the random number table. We can ignore these numbers if we have to select a sample without replacement, because the corresponding element is already included in the sample.

Example:

Select a sample of size 20 from the following data on the heights (in centimetres) of 500 persons, using random numbers table, by without replacement sampling mehtod

175	180	170	160	135	175	143
110	160	178	165	156	110	147
120	130	171	165	170	147	151
135	150	169	165	169	153	154
170	120	168	170	171	155	156
172	180	172	170	172	156	112
165	170	171	172	173	150	116
160	166	155	179	174	152	176
155	159	172	184	172	159	129
176	160	169	170	172	160	152
179	170	165	169	170	179	164
130	169	156	182	165	175	174
150	165	157	176	166	172	174
170	158	150	177	169	179	176
168	157	176	170	175	168	175
162	167	178	160	176	165	173
161	173	176	170	160	162	173
155	179	177	165	160	155	173
120	181	176	155	168	156	173
171	180	120	140	168	175	173
155	185	172	172	115	115	112
120	182	179	155	179	172	150
129	189	179	121	190	176	115
149	172	160	110	180	179	170
149	169	175	155	172	177	160
162	180	172	180	175	170	180
168	190	175	182	177	172	160
170	163	170	172	177	171	175
174	162	179	172	176	179	155
172	165	169	169	177	176	185
179	170	189	175	176	175	185
180	171	185	176	172	180	155
170	176	180	173	170	180	190
175	176	169	173	172	189	179
182	174	179	174	169	178	175
170	182	162	176	167	177	175

160	189	175	175	167	176	171
152	179	171	169	179	176	172
139	139	130	172	152	176	172
145	158	175	168	140	175	171
190	170	155	150	160	176	170
185	176	156	172	168	172	179
189	169	156	176	178	170	171
179	166	176	171	172	179	179
171	185	176	171	172	176	180
169	115	149	179	182	149	169
172	125	152	170	181	150	172
172	120	162	180	179	159	175
176	130	172	182	175	169	176
175	169	172	181	169	171	177
180	172	175	189	179	181	176
182	175	175	185	182	185	176
115	178	175	173	183	177	175
160	178	171	174	189	176	176
162	177	172	179	183	174	175
162	176	180	172	172	175	175
159	171	189	174	171	176	175
152	169	170	176	174	172	175
155	171	140	175	174	171	179
170	131	170	176	174	173	180
149	161	139	182	182	176	180
172	172	152	185	189	186	186
165	171	190	180	173	171	183
162	169	170	149	174	169	183
165	179	185	159	175	170	173
172	180	183	159	169	179	171
172	172	183	158	166	182	172
179	172	171	169	176	180	177
179	176	179	169	178	180	175
175	176	170	172	177	180	175
160	165	169	179	188	182	172
175	177	180				

Solution:

First we will number the heights starting from the first observation, so the number 1 is 175, number 2 is 180, number 3 is 170, ... number 100 is 157, number 140 is 173 and, number 500 is 180. For a random sample of size 20, we need 20 three digit numbers to designate the twenty measurements being chosen. To obtain the numbers, we see the random tables from any where with the drop of a pencil. We start, say from the first column 31st row (of the random table) the numbers after grouping are :

044 935 249 475 246 338 244 458
625 102 561 962 793 356 533 712 472
005 499 765 464 051 881 599 611 963 596
546 928 239 123 287 295 293 596 315
307 268

Now the values to be included in the sample are the values corresponding to the following numbers of our data :

044	464
249	051
475	239
246	123
338	287
244	295
458	293
102	315
356	005
472	499

Note : that we have ignored the numbers greater than 500.

The values corresponding to these numbers are :

170, 176, 180, 170, 130, 178, 183, 170, 181, 171, 135, 177, 172, 166, 182, 165, 170, 189,172, 180.

OTHER SAMPLING METHODS

We have described the simple random samples and random sampling procedures, however, it should be noted that simple random sampling is not the only sampling method available. Some of the alternative sampling methods are : stratified simple random sampling; cluster sampling, and systematic sampling.

Stratified Simple Random Sampling

In this sampling procedure the population is divided into groups of elements called strata such that each item in the population belongs to only one stratum. The strata are formed on the basis of age, height, type and so on. The results depend upon the similarities among the elements within each strata. A simple random sample is taken from each stratum. There are statistical procedures, quite different from those for simple random samples;

for combining the results of individual samples. The single estimate of the population parameter of interest is obatined by the application of these formulas. The value of stratified simple random sampling depends upon the homogeneity within the strata. If units within strata are alike, the strata will have low variances. The homogeneous strata will provide results similar to simple random sampling procedure, the sample sizes however should be smaller.

Cluster Sampling

In the cluster sampling the population is divided into separate groups of elements called clusters. Each element of the population belongs to one and only one cluster. A simple random sample is taken from each cluster. Cluster sampling tends to provide best results whenever the elements within the cluster are heterogenous. In the ideal case, each cluster is a representative small-scale version of the entire population. The value of cluster sampling depends upon how representative each cluster is of the entire population. If each cluster is true representative of the population, sampling a small number of clusters will provide good estimates of the population parameters. If the clusters are geographic subdivisions, the sampling is called area sampling. For example, we may be interested in the behaviour of stray dogs in a city. Since the dogs are scattered over a wide area, it is difficult to take a simple random sample. We can divide the total area into a number of smaller non-overlapping areas, say, city blocks. These city blocks are the clusters of still smaller units. Cluster sampling requires a larger total sample size than either simple random sampling or stratified simple random sampling.

Systematic Sampling

In some instances, the most practical way of sampling is to select every 20th, 50th or 100th item on the list. Such sampling is called systematic sampling. For example if a sample of size 25 is required from a population of 5000 elements a systematic sample would involve selecting randomly 1 of the first 100 elements from the population list. Other sample elements are identified by starting with the first sampled element and then selecting every 100th element that follows the list. The systematic sampling may provide misleading results if there is hidden periodcity in the population. For example, if we are interviewing the residents of every 10th house it may so happen that each house thus chosen is a corner house. Such hidden periodicities are present in many populations. So care should be taken while selecting sampling procedure.

SAMPLING TECHNIQUES

It is very difficult to deal about sampling methods in a condense form. But it is equally important aspect in some of the disciplines such as genetics, ecology, physiology, plant-breeding etc.

If a character of living population is to be studied, it is not possible to measure the variations found in all the individuals, however it can be done by simply studying the representative samples drawn from each population. If sampling technique is correct, the conclusion drawn from the data will be truly representative. The most important thing which one should bear in mind, at the time of collecting data, is that the sample must be of proper size in relation to the population.

The samples may be drawn at random or at a regular intervals. In random sampling every individual of population has equal chances for its representation. In case of sampling by design, samples at regular intervals are taken, i.e., in a row of plants, every plant may be taken as a sample. If a population is heterogeneous in nature, different kinds of plants or every type must be represented in the sample.

The sample must be such that if conclusion is drawn from different sets of sample, it must lead to close relation to all aspects of any character. If they differ .widely, it may be due to wrong techniques in sampling. It is important that sampling techniques must be correct and samples must be large enough to give reliable data about a population.

SAMPLING AND NON-SAMPLING ERRORS

The error arising due to drawing inference(s) about the populations on the basis of few observations (sampling) is termed as sampling error. However, the error chiefly arising due to ascertainment is termed as non-sampling error.

Sometimes, care has not been taken in selecting sample. The result of sample study may not be accurate to true value. Hence; sampling gives rise to certain errors known as sampling errors (for sampling fluctuations).

Sampling errors are of two types :

(i) Biased errors

(ii) Unbiased errors.

Biased errors arise from any bias in selection elimination, etc. while unbiased errors arise due to chance of differences between the members of a population included in the sample and those; not included in the population.

Biased errors may arise due to :

(a) Faulty process of selection

(b) Faulty work during the collection of information and

(c) Faulty methods of analysis.

Non-sampling errors : When a complete enumeration of units in an area is prepared as data from errors, errors arising in this way are termed as non-sampling errors. They are dut to factors other than the inductive process of inferring about the population from sample.

Non-sampling errors can occur at every stage of planning. Such errors can arise due to a number of causes such as defective method of collecting the data and its tabulations, incomplete coverage of sample, etc. The non-sampling errors may arise from one or more of the following factors :

1. Inaccurate methods of interview or observation of measurement.
2. Lack of trained and experienced investigators.
3. Lack of adequate supervision of primary staff.
4. Errors due to non-response, i.e., imcomplete coverage with respect to units.
5. Errors in data processing operations such as codıng, branching, verification, tabulation, etc.
6. Errors committed during presentation and printing of tabulated results.

These sources are not exhaustive but are able to point out the possible source(s) of error.

Control of non-sampling errors : In some situations, the non-sampling errors may be large and deserve greater attention than the sampling errors. Sampling errors decrease with increase in sample size, non-sampling errors tend to increase with the sample size. Non-sampling errors require to be controlled and reduced to a level at which their presence does not spoil the use of final results.

EXERCISES

1. Define a sample, random sample and population?
2. What are the advantages of using a random sample?
3. "A random, sample is a better representative of the population than a simple sample" comment?
4. A researcher is interested to study weights of new born rats in a rat colony having 600 rat cages kept in 20 rooms :
   (a) What is the population for this study?
   (b) How can the researcher take a simple random sample?
   (c) Can he take a sample of size 30 from one room only? If no why?
5. What is meant by sampling with replacement and without replacement ? Give one example each.
6. Give four attributes which a good sample must possess.
7. Normally the experimenters study a sample rather than the whole population to make inferences. Why?

8. Define simple random sampling, stratified simple random sampling and cluster sampling?
9. What are the differences between stratified simple random sampling and cluster sampling?
10. A researcher is interested to study weights of new born rats in a colony having 600 cages kept in 20 rooms.
    (a) What is the population of this study?
    (b) How can the researcher take a simple random sample?
    (c) Can he take a sample of size 30 from one room only? If not why?

3

Describing the Data
(Tabular and Graphical Approaches)

INTRODUCTION

There are several simplified procedures like rearrangement, tabulation, graphs and summarization which are commonly used to make the data readable and easily comprehensible. This chapter discusses the arrangement of data in the form of frequency distributions, the graphic representation of frequency distributions, and the ways in which frequency distributions may differ. In the second section of this chapter we will discuss the statistics used to describe the properties of frequency distributions, or the properties of the collections of numbers which these distributions comprise.

TABULAR REPRESENTATION

Statistical data are frequently arranged and presented in the form of tables. Tables should be designed to enable the reader grasp with minimal effort the information that they intend to convey. In constructing a table certain points should be kept in mind. Some of these points are:

(1) Every table should be self-explanatory.

(2) Columns of numbers should be appropriately labelled and arranged.

(3) Necessary explanatory foot-notes should be given at the bottom of the table.

(4) The information contained in the table may be separated by the insertion of horizontal or vertical lines.

(5) Tables should be appropriately numbered and should be inserted in the text, close to where they are first mentioned.

Array

Suppose a researcher collects 100 observations on the number of flowers on 100 antirrhinum plants and records his observations as given below in Table 3.1

The data given in this form is called *ungrouped data* and does not give much information. This data set needs rearrangement. The most elementary rearrangement of data is an array.

Table 3.1 : Number of flowers on 100 antirrhinum plants.

109	111	82	105	134
113	90	79	100	117
80	90	121	75	93
99	90	92	96	82
101	104	80	81	83
104	93	109	72	110
111	91	109	111	81
122	83	92	101	77
99	103	93	91	67
108	93	84	84	100
102	84	96	89	81
107	95	91	107	102
109	93	82	103	116
86	78	73	104	104
103	108	76	94	108
72	87	121	80	127
105	103	106	119	90
93	89	110	103	100
99	79	117	114	117
93	82	98	89	119

An array is simply an arrangement of the observations according to size from the smallest to the largest.

In an array the data is put in some kind of order. It becomes easier to give some information about the observation. The rearranged data is called *arrayed data.* Table 3.2 gives arrayed data on the number of flowers on 100 plants.

Table 3.2 : An array of number of flowers on 100 plants.

67	72	72	73	75	76	77	78	79
79	80	80	80	81	81	82	82	82
82	83	83	84	84	84	86	87	89
89	89	90	90	90	90	91	91	91
92	92	93	93	93	93	93	93	93
94	95	96	96	98	99	99	99	100
100	100	101	101	102	102	103	103	103
103	103	104	104	104	104	105	105	106
107	107	108	108	108	109	109	109	109
110	110	111	111	111	113	114	116	117
117	117	119	119	121	121	122	127	134

Frequency Distributions

Frequency distribution is one of the most useful methods of data summarization. It provides some insight of the data that cannot be obtained easily in an arrayed data. Suppose one takes observations on the numbers of peas in 24 pods and records following observations and presents it in arrayed from.

Number of peas : 1, 1, 1, 2, 2, 2, 3, 3, 3, 3, 4, 4, 4, 4, 5, 5, 5, 5, 5, 5, 5, 6, 6, 6,

To make it more precise, this data can be presented in a tabular form.

Number of Peas	Frequency
1	3
2	3
3	4
4	4
5	7
6	3
Total	24

This arrangement of data is called *frequency distribution.*

Consider the data of Table 3.2. These are number of flowers on 100 antirrhinum plants.

It is clear from this arrayed data that the number of flowers range from 67 to 134. By counting the number of times each value occurs, an arrangement of the data as shown in Table 3.3 is obtained. This arrangement is also a frequency distribution.

Note that the number of groupings of flower numbers is large. Therefore, *to reduce the number of classes, the data is arranged into arbitrarily defined groupings of the variables.*

The arbitrarily defined groupings of the variable are called class-intervals. There are no hard and fast rules in constructing a frequency table. The obejctive is to present the data in a tabular format so that it contains as much information as possible. The choice of class width and number of classes cannot be made independently.

Normally two guide lines are followed in this regard :

(1) Use between 5 and 20 classes.

(2) Make the classes with equal width.

To determine the approximate number of classes, first determine the range of the data. *Range is defined as the difference between the largest and the smallest observation.*

Table 3.3 : Frequency distribution of number of flowers of Table 3.2 with as many classes as the number of plants.

Number of Flowers	*f*	*Number of Flowers*	*f*	*Number of Flowers*	*f*	*Number of Flowers*	*f*
134	1	117	3	100	3	83	2
133	0	116	1	99	3	82	4
132	0	115	0	98	1	81	3
131	0	114	1	97	0	80	3
130	0	113	1	96	2	79	2
129	0	112	0	95	1	78	1
128	0	111	3	94	1	77	1
127	1	110	2	93	7	76	1
126	0	109	4	92	2	75	1
125	0	108	3	91	3	74	0
124	0	107	2	90	4	73	1
123	0	106	1	89	3	72	2
122	1	105	2	88	0	71	0
121	2	104	4	87	1	70	0
120	0	103	5	86	1	69	0
119	2	102	2	85	0	68	0
118	0	101	2	84	3	67	1

By fixing the trial width of each class (or class-interval) divide the range by classes to be made. Round off the figures if required.

$$\text{Approximate number of class} = \frac{\text{Highest value of the data - Lowest value}}{\text{Width of each class}}$$

Class Mid Point

Mid point of class is the value that falls in the middle of the class interval. It is the average of the class limit. The class mid point for the class

$$100\text{-}105 \text{ is } \frac{100+105}{2} = 102.5.$$

Working Rules to Prepare a Frequency Table

The steps listed below are widely used in the selection of class intervals and lead in most cases to a convenient handling of the data.

Step 1 : Arrange the data in an ordered array and note down the lower and upper limits of the data.

Step 2 : Select a class interval of such a size that between 5 and 20 intervals will cover the total range of observations. For example, if the smallest observation in a set were 7 and the largest 156 a class interval of 10 would be appropirate. If the smallest observation were 2 and the larges 38, a class interval of 3 would result in arrangement of 12 intervals. If the observations ranged from 9 to 20, a class interval of 1 would be convenient.

Step 3 : Select the number of classes (class-intervals) using the formula : Number of Classes (or class-intervals)

$$= \frac{\text{Highest value - Lowest value}}{\text{Class width}}$$

Step 4 : Start the class interval at a value which is a multiple of the size of that interval. For example, with a class interval of 5, the intervals would start with the values 5, 10, 15, 20 etc.

Step 5 : Arrange the class intervals according to the order of magnitude of the observations they include, the class interval containing the largest observation being placed at the top.

Step 6 : Against each interval wirte down as many vertical lines in the tally column as the number of items it contains.

Step 7 : In the next column titled frequency, write the frequency that is, number of items present in a particular class interval. For example, in a class interval 130-134, there may be one item. Write 1 in front of this class in the second column. The frequency is counted from the number of vertical lines in the tally column against each class interval.

Example 1:

Let us now construct a frequency table for the data given in Table 3.1 and arranged in an array in Table 3.2.

Solution:

From this arrayed data we note that the highest limit is 134 and the lowest limit is 67. As we want to make classes between 5 and 20 it will be appropriate to keep the class interval of size 5.

Next we select the number of classes by using the formula given for the purpose and select the number as 14. Then we arrange the classes according to their magnitude and begin with the first class as 130-135 as given below in Table 3.4.

Table 3.4 : Frequency distribution of the number of flowers of Table 3.2

Class interval	*Tally*	*Frequency*
130-134	|	1
125-129	|	1
120-124	|||	3
115-119	~~||||~~ |	6
110-114	~~||||~~ ||	7
105-109	~~||||~~ ~~||||~~ ||	12
100-104	~~||||~~ ~~||||~~ |||| |	16
95-99	~~||||~~ ||	7
90-94	~~||||~~ ~~||||~~ ~~||||~~ ||	17
85 -89	~~||||~~	5
80-84	~~||||~~ ~~||||~~ ~~||||~~	15
75-79	~~||||~~ |	6
70-74	|||	3
65 -69	|	1
Total		100

Exact Limits of The Class Interval

Where the variable under consideration is continuous and not discrete, we select a unit of measurement and record our observations as discrete values. When we record an observation in discrete form and the variable is a continuous one, we imply that the value recorded represents a vlaue falling within certain limits. These limits are usually taken as one-half unit above and below the value reported. Thus when we report a measurement of height to the nearest inch, say, 16 inches. We mean that if a more accurate form of measurement had been used, the value obtained would fall within the limits 15.5 and 16.5 inches.

Strictly speaking the limits are 15.5 to 16.499, where the latter figure is a recurring decimal, but for convenience we write the limits as 15.5 to 16.5. Similarly, a measurement made to the nearest tenth part of an inch, say, 31.7 inches, is understood to fall within the limits 31.65 and 31.75 inches. In a reaction-time experiment a particular observation measured to the nearest thousandth of a second might be, say 0.196 second. This assumes that had a more accurate timing device been used the measurement would have been found to fall somewhere within the limits .1955 and .1965 second.

Class intervals are usually recorded to the nearest unit and thereby reflect the accuracy of measurement. For various reasons it is frequently necessary to think in terms of so-called exact limits of the class interval. These are sometimes referred to as class boundaries, or end values, and sometimes as real limits. Consider the class interval 95 to 99 in Table 3.4. We grouped within this interval all measurements taking the values, 95, 96, 97, 98 and 99. The limits of the lower values are 94.5 and 95.5, while those of the upper values are 98.5 and 99.5. The total range, or exact limits, which the interval is presumed to cover is then clearly 94.5 and 99.5 which means all values greater than or equal to 94.5 and less than 99.5.

The above discussion is applicable to continuous variables only. With discrete variables no distinction need be made between the class interval and the exact limits of the interval, the two being identical.

Example 2:

Given below is the data on weights fishes reared in a pond. Construct a frequency distribution table showing true class limits.

68 42 30 28 79 22' 24 44 43 74 36 28 31 25 12
51 32 38 42 31 38 16 69 23 22 27 28 19 30 49
63 27 36 32 27 23 25 65 25 51 42 28 45 57 12
49 27 50 21 24 47 43 49 23 46 43 12

Solution:

First of all arrange the data in an array :

12	27	36	49
12	27	38	50
12	27	38	51
16	27	42	51
19	28	42	57
21	28	42	63
22	28	43	65
22	28	43	68
23	30	43	69
23	30	44	74
23	31	45	79
24	32	46	
24	32	47	
25	32	49	
25	36	49	
25			

The uper limit is 79 and the lower limit is 12. Let the number of items in a class interval be 10. The number of class will be $\frac{79-12}{10} = 6.7 = 7$

Since the weight is a continuous variable we use exact class limits and chose the classes as 69.5-79.5, or 9.5-19.5 etc and construct the table as given below in Table 3.5.

Table 3.5 : Frequency distribution table of weights of 57 fishes showing true class limits.

Exact Class Limits	*Tally*	*Frequency*
9.5-19.5	𝍸	5
19.5-29.5	𝍸 𝍸 𝍸 ||||	19
29.5-39.5	𝍸 𝍸	10
39.5-49.5	𝍸 𝍸 |||	13
49.5-59.5	||||	4
59.5-69.5	||||	4
69.5-79.5	|||	2
Total		57

Relative Frequency Distribution

We have been expressing the frequency of each class in terms of the total number of data items in the data set that fall within that class. Sometimes we may find it of interset to consider the relative frequency of items in each class. The *relative frequency is simply the fraction or proportion of the total number of items belonging to the class*. For a data set having a total of n observations or items, the relative frequency of each class is given by

Frequency of the Class

$$\text{Relative Frequency of a class} = \frac{\text{Frequency of the class}}{n}$$

For example the relative frequency of the first class in the Table 3.5 is

$$= \frac{\text{Frequency of class}}{\text{Total observation}} = \frac{5}{57} = 0.087$$

Using Table 3.5 we can find *relative frequency* distribution for the weights of fishes. It is shown in Table 3.6.

Table 3.6 : The data of showing relative frequency distribution.

Class-interval	*Tally*	*Frequency*	*Relati ve Frequency*
9.5-19.5	𝍸	5	0.087
19.5-29.5	𝍸 𝍸 𝍸 IIII	19	0.333
29.5-39.5	𝍸 𝍸	10	0.175
39.5-49.5	𝍸 𝍸 III	13	0.228
49.5-59.5	IIII	4	0.070
59.5-69.5	IIII	4	0.070
69.5-79.5	II	2	0.035
Total		57	1

Cumulative Frequency Distribution

Situations occassionally arise where our concern is not with the frequencies within the class intervals themselves or the relative frequency but rather with the number or percentage of values "greater than" or "less than" a specified value. Such information is made available by the preparation of a cumulative frequency distribution.

The cumulative frequencies are obtained by adding successively, starting from the top, the individual frequencies.

From a cumulative frequency distribution we can obtain the number of cases falling below or above a particular score values. For example in the Table 3.7 we observe that 34 fishes have weight less than 39.5 gm and 55 fishes less than 69.5 gms.

Table 3.7 : Frequency, cumulative and Relative frequency distributions of weights of 57 fishes.

Class-interval	*Frequency*	*Cumulative Frequency*	*Relative Frequency*	*Cumulative Relative Frequency*
9.5-19.5	5	5	0.087	0.087
19.5-29.5	19	24	0.333	0.421
29.5-39.5	10	34	0.175	0.596
39.5-49.5	13	47	0.228	0.824
49.5-59,5	4	51	0.070	0.894
59.5-69.5	4	55	0.070	0.964
69.5-79.5	2	57	0.035	1.00
Total	57		1.00	

As a final point we note that a *cumulative relative frequency distribution is the one that shows the fraction of items with value less than the upper class limit.* It can be developed from the relative frequency distribution in the same way that a cumulative frequency distribution is developed from the frequency distribution. It can also be developed by dividing the cumulative frequencies by the number of items in the data set.

The class intervals can be written in either ascending or decending order of magnitude. In whatever way it is wirtten the cumulative frequency of the class highest magnitude is always equal to the total number of observations. For example, in Table 3.7 the cumulative frequency of the last class (69.5-79.5) is 57. In Table 3.4 the cumulative frequency of the first class (130-134) which has highest magnitude is 100.

Example 3:

The allowing data give the number of florets on 40 sunflowers.

42	*88*	*37*	*75*	*98*	*93*	*73*	*62*
96	*80*	*52*	*76*	*66*	*54*	*73*	*69*
83	*62*	*53*	*79*	*69*	*56*	*81*	*75*
52	*65*	*49*	*80*	*67*	*59*	*88*	*80*
44	*71*	*72*	*87*	*91*	*82*	*89*	*79*

Prepare a frequency distribution and a cumulative frequency distribution for these data.

Solution :

On arranging the data in an array we find that the lowest limit is 37 and the highest limits is 98. Let the class interval be 5. The number of class intervals will be approximately equal to $\frac{98-37}{5} = \frac{61}{5} = 12.2 \approx 13$

Class-intervals	*Frequency*	*Cumulative Frequency*
35-39	1	1
40-44	2	3
45-49	1	4
50-54	4	8
55-59	2	10
60-64	2	12
65-69	5	17
70-74	4	21
75-79	5	26
80-84	6	32
85-89	4	36
90-94	2	38
95-99	2	40
Total	40	

Example 4:

Consider the above data as that of continuous variable, write down the exact limits and the mid-points of the class intervals for the frequency distribution.

Solution :

The exact limits and the mid-points are as given in the table below.

$$\text{Mid-point of a class} = \frac{\text{Upper limits of the class} + \text{Lower limit of the class}}{2}$$

Class-intervals	*Exact limits*	*Mid-points*
35-39	34.5-39.5	37
40-44	39.5-44.5	42
45-49	44.5-49.5	47
50-54	49.5-54.5	52
55-59	54.5-59.5	57
60-64	59.5-64.5	62
65-69	64.5-69.5	67
70-74	69.5-74.5	72
75-79	74.5-79.5	77
80-84	79.5-84.5	82
85-89	84.5-89.5	87
90-94	89.5-94.5	92
95-99	94.5-99.5	97

Mid point of the class 35 is calculated as

$$\text{Mid point} == \frac{34.5 + 39.5}{2} = 37$$

GRAPHICAL REPRESENTATION

Large sets of data are condensed by constructing frequency distribution. These tabular approaches can be used to increase our understanding of the information contained in the data. Graphs often provide additional insight about the nature of data set. A graph is a visual portrayal of a collection of numerical data or some aspect of the collection. Graphs enable us to think coherently about data problems in visual terms. Many sets of data are complex. A graph is meant for simplification.

The representation of quantitative data suitably through charts and diagrams is known us graphical representation of statistical information.

Although, graphs can be easily understood and often show relationship between two or more sets of figures it takes a lot of time to prepare a graph.

Some of the finer details are lost while making a graph. If the graph is well done, it is usually easier to read and interpret than the table.

Working Rules for the Construction of Graphs

Step 1 : In the construction graphing of frequency distributions it is customary to let the horizontal axis represent values of a variable and the vertical axis frequencies.

Step 2 : The arrangement of the graph should proceed from left to right. The low numbers on the horizontal scale should be on the left, and the low numbers on the vertical scale should be toward the bottom.

Stpe 3 : The distance along either axis selected to servel as a unit is arbitrary and affects the appearance of the graph. Some writers suggest that the units should be selected such that the ratio of height to length is roughly 3:5. This procedure seems to have some g aesthetic advantages.

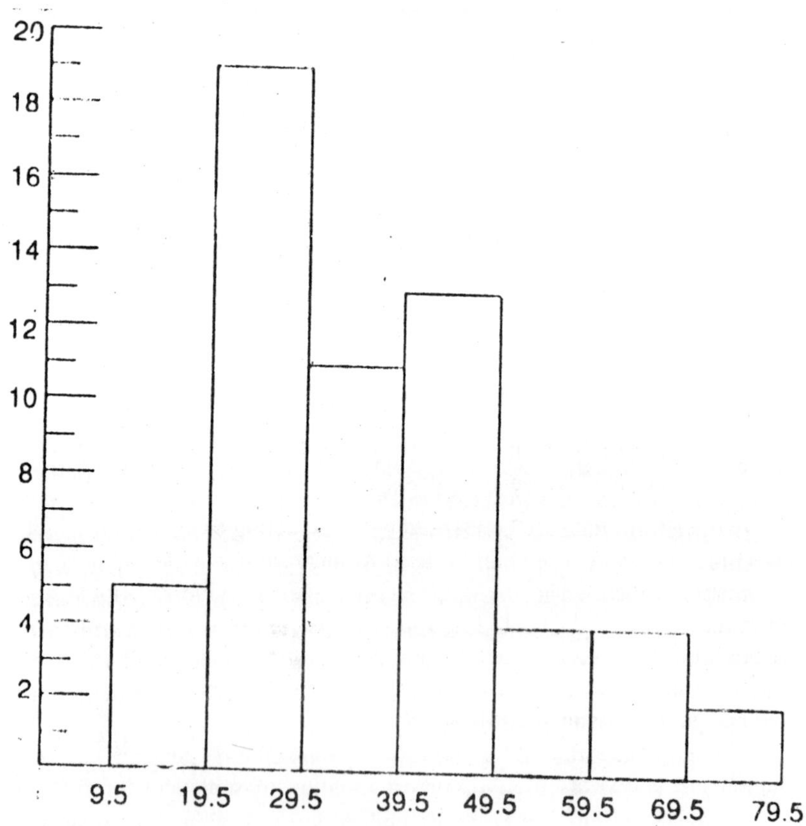

Fig 3.1 : Histogram of weights of fishes as given in Table 3.7.

Step 4 : Whenever possible the vertical scale should be so selected that a zero point falls at the point of intersection of the axis. With some data this procedure may give rise to a most unusual-looking graph. In such cases it is customary to designate the point of intersection as the zero point and make a small break in the vertical axis.

Step 5 : Both the horizontal and vertical axes should be appropriately labelled.

Stpe 6 : Every graph should be assigned a descriptive title which states precisely what it is about.

Histogram

A histogram is a graph in which the frequencies are represented by areas in the form of bars. Steps to prepare a histogram are as follows :

Working Rules to Prepare a Histogram

Step 1 : Take a piece of suitably cross-sectioned graph paper. Paper subdivided into tenths of an inch with heavy lines 1 inch apart are convenient.

Step 2 : Draw a horizontal line to represent the variable and vertical line (Y-axis) to represent frequencies.

Step 3 : Select an appropriate scale for both the variable and the frequency.

Step 4 : Write the mid points of intervals along .Y-axis and the frequency scale on the V-axis. Plot the frequency corresponding to each class interval and draw the horizontal line the full length of the interval.

Step 5 : Join the ends of these lines to the corresponding ends of intervals on the horizontal axis. So above each class interval on the horizontal axis a rectangular bar or cell is erected. The height of each bar corresponds to the respective frequency. The rectangles of a histogram must be joined. To accomplish this we must take into account the true or exact limits of the class intervals. Histogram for the data on weights of 57 fishes as given in the Table 3.7 is given below in the Fig. 3.1.

Frequency Polygon

Frequency polygon provides an alternative to a histogram as a way of presenting a frequency distribution graphically. Again, the data values are placed on the horizontal axis and the frequencies on the vertical axis. However, instead of using rectangles, as with the histogram, we find class mid points on the horizontal axis and then plot the points directly above the class mid points at a height corresponding to the frequency of the class. Classes of zero frequency are added at each end of the frequency distribution so that the frequency polygon touches the horizontal axis at both ends of the graph. The frequency polygon is then formed by connecting the points with straight lines. The frequency polygon for the weights of fishes is shown in Fig. 3.2.

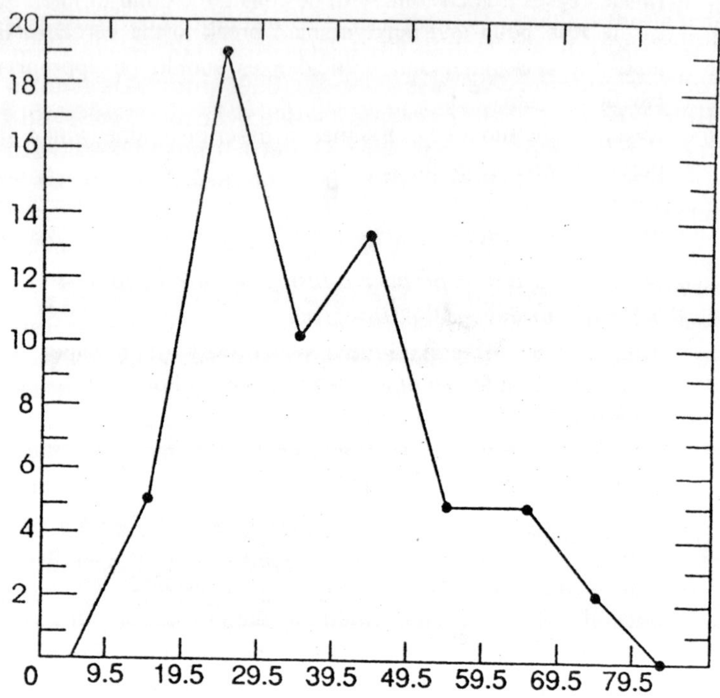

Fig 3.2 : A frequency polygon of weights 57 of fish as given in Table 3.7

Cumulative Frequency Polygon

The cumulative frequency polygon is a broken line graph like the frequency polygon. The vertical scale in this device is a scale of cumulative frequency, and the horizontal axis is marked off at class boundaries. Above each boundary point the cumulative frequency, that is the total number of observations with values less than that boundary value is plotted.

The cumulative frequency polygon as we pass from left to the right along the horizontal axis is never decreasing, that is, the polygon itself either rises away from the horizontal axis or stays level but never dips down. Unlike histogram and frequency polygon the basic unit in the cumulative frequency polygon is a unit length rather than area. The height of the polygon above the horizontal axis is the critical feature.

PROPERTIES OF GRAPHICALLY REPRESENTED FREQUENCY DISTRIBUTIONS

The differing characteristics of frequency distributions can be readily represented in graphic form. Consider the three distributions in Fig. 3.3. These distributions appear identical in shape. They are markedly different, however, in terms of the central values about which the observations in each distribution appear to concentrate; that is, they have different averages although they may be identical in all other respects. Distribution A has a lower average than B and B than C.

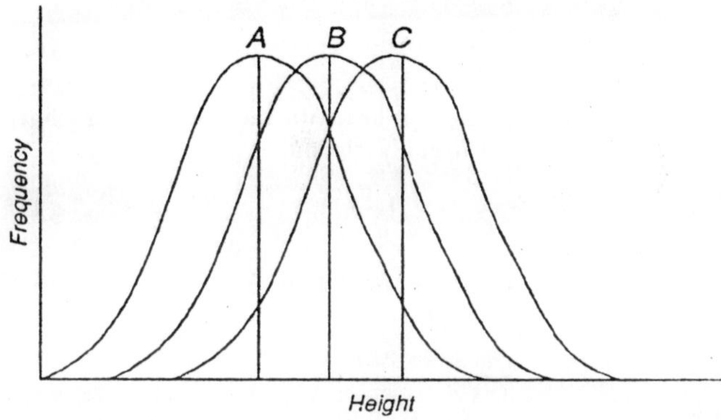

Fig. 3.3 : Three frequency distributions Identical in shape but with different averages.

Now consider the distributions in Fig. 3.4. Inspection of these three distributions suggests that while the observations in each case appear to concentrate about the same average, they are nonetheless markedly different one from another. In the case of distribution A the observations appear to be more closely concentrated about the average than in the case ofB, and the same applies to B in relation to C. Thus these distributions differ in variation. The observations in A are less variable than the observations in B, and those in B are less variable than those in C.

Examine now the distributions in Fig. 3.5. These three distributions hâve different averages and possibly different measures of variation. They differ also in skewness. Distribution B is symmetrical about the average; that is, if we were to fold it over about the average, we should find that it had the same shape on both sides. Distributions A and C are asymmetrical, the shape to the left of the average being different from the shape to the right. Distribution A is positively skewed, the longer tail is extending towards the low end of the scale.

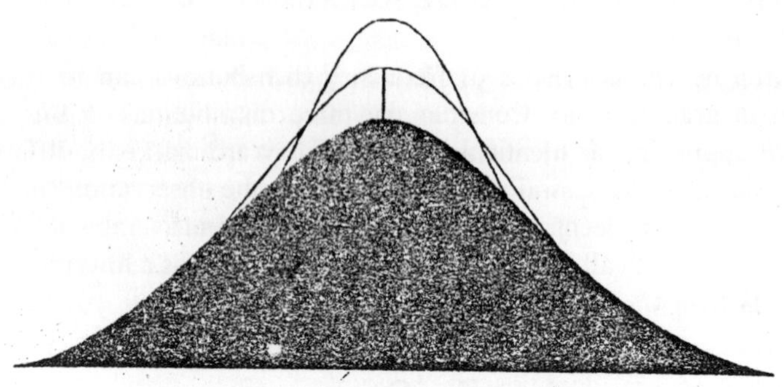

Fig. 3.4 : Three frequency distributions with the same average but with diffrent variation.

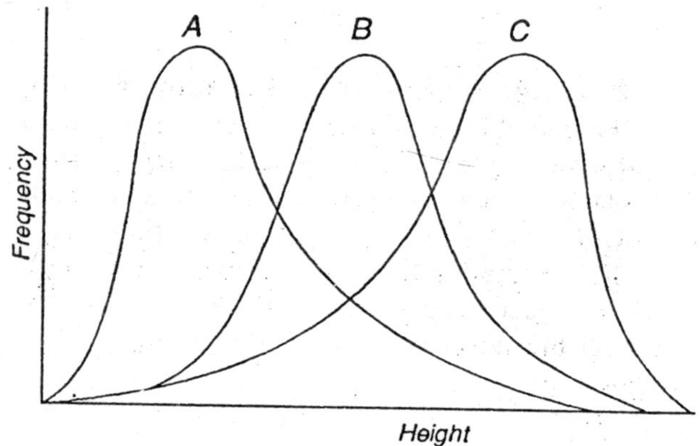

Fig. 3.5 : Three frequeny distributions differing in skewness.

Consider now the graphic representation of kurtosis as shown in Fig. 3.6. Distribution A is a symmetrical bell-shaped distribution known as the normal distribution. Distribution B is observed to be flatter on top than the normal distribution and is referred to as *platykurtic*, while distribution C is more peaked than the normal and is spoken of as *leptokurtic*.

The meaning in which attached to the descriptive properties of collections of measurements arranged in frequency distributions is largely intuitive and is derived from the inspection of distributions in tabular or graphic from. To proceed with the study of data interpretation we require precisely defined numerical measures of central location, variation, skewness and kurtosis.

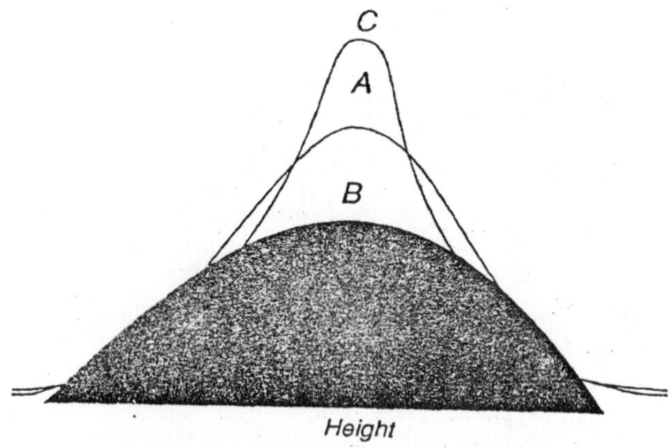

Fig. 3.5 : Three frequeny distributions differing in kurtosis.

EXERCISES

1. The following table shows the results of B. Com. students of a college for the last three years. Present the data by a suitable diagram.

Years	Ist Class	2nd Class	Pass	Failed
1996	20	30	40	10
1997	30	50	20	20
1998	25	60	40	25

2. Draw histogram for the following data and superimpose on it the frequency curve:

Variable:	5-9	10-14	15-19	20-24	25-29	30-34	35-39
Frequency:	8	15	18	30	16	12	6

3. The following data relate to the expenditure of a family. Represent this data by angular (pie) diagram:

Items of expenditure	Rupees (per month)
Food	160
Clothing	110
Rent	80
Recreation	60
Miscellaneous	40

4. Draw a (rough) Pie chart to represent the following data relating to the production cost of manufacture:

Cost of Materials Rs. 38,400

Cost of Labour	:	Rs. 30.720
Direct Expenses	:	Rs. 11.520
Overhead Expenses	:	Rs. 15.360

5. (a) What are different methods of graphical representation of data? Explain them.
   (b) Distinguish Histogram and Histogram clearly with illustrations.
   (c) What is histogram? How could you represent a grouped frequency distribution by means of a histogram when the class widths are (i) equal, and (ii) not equal.
6. (a) Explain the various methods that are used for graphical representation of frequency distribution.
   (b) What are the advantages of presenting data through diagrams and graphs?
   (c) Distinguish between diagrammatic and graphic representation of data.
7. (a) Illustrate graphically the distinction between a frequency polygon, a histogram and an ogive curve. Comment on their uses.
   (b) Prepare a 'more than ogive' curve with imaginary figures.
   (c) Explain the various diagrams which are used in statistics to show the salient features of data.
8. Represent the following data by a suitable diagram:

Item of expenditure	Family A	Family B	Family C
Food	20	60	120
Clothing	4	15	70
House Rent	3	20	80
Education	2	6	25
Books	2	10	70
Misc	2	8	50

9. Represent the following data by a suitable diagram:
   GNP–Industrial origin–percentages

Countries	Agriculture	Industry	Services	Others
England	3	40	44	13
America	3	35	61	1
Japan	6	48	43	3
India	45	19	28	8

10. Draw a histogram of the frequency distribution given below:

Variable :	10-24	15-19	20-29	30-49
Frequency :	5	10	30	20

11. The following table gives the country of origin of feature films exhibited in India:

Country :	India	U.S.A.	U.K.	Other Countries
No. of Films:	144	81	64	16

Represent them by Square or Circle diagram.

12. (a) What are 'ogive' curves? How are they used in reclassification of data?

(b) "Graphs and diagrams are more effective than any other method of presenting data". Why?

13. (a) Discuss the merits and Imitations of representing statistical data through graphs and diagrams. How do you represent data by means of a pie diagram?

(b) Point out the role of diagrammatic presentation of data.

Explain briefly the different types of bar diagrams known to you.

14. (a) "Charts and graphs are more effective in attracting attention than any of the other methods of presenting data." Do you agree? Give reasons in support of your answer.

(b) What are the different methods of a graphical presentation of data? Explain them.

15. Represent the following data by a suitable diagram:

Item of expenditure	Family A	Family B	Family C
Food	50	45	60
Clothing	20	25	20
Rent	10	10	10
Education	5	10	5
Miscellaneous	15	10	5
Total	100	100	100

16. Represent the following data by a pie diagram:

Items	Expenditure in Rupees
Food	84
Clothing	27
Recreation	10
Education	15
Rent	23
Miscellaneous	21

17. Draw a histogram and a frequency polygon for the following data:

Class :	0-10	10-20	20-30	30-40	40-50	50-60	60-70
Frequency :	6	8	10	15	13	8	5

18. Represent the data given below with a suitable diagram.

Year	Value in Rs. in 1997-98	Value in Rs. in 1998-99
Raw material	150	200
Labour	100	200
Power	75	150
Advertisement	25	100
Other charges	50	200
Total cost of manufacturing	400	900

19. The annual profits in lakhs of rupees of 100 companies are distributed as follows: Profits per Co.

(Rs. lakhs) :	0-50	50-100	100-150	150-200	200-250	250-300
No. of Cos :	12	18	27	20	17	6

Draw a histogram and frequency polygon.

20. The following table shows the Sixth Five-Year Plan public sector outlays by heads of development:

Heads of Development	(Rs. Crores) Centre	States
Agriculture	4,765	7,039
Irrigation and Flood Control	6,635	11,395
Energy	9,995	8,293
Industry and Minerals	12,770	2,985
Transport and Communication	12,200	5,120
Social Services	8,216	1,420
Total	54,581	36,252

Represent the data by some suitable diagram and write a report on the data bringing out the salient features.

21. Which of the following statements are True of False?
    (i) The area of a rectangle is equal to the product of its length and width. T/F
    (ii) Cubes are two-dimensional diagrams. T/F
    (iii) A frequency polygon has more than four sides. T/F
    (iv) A pie diagram is a circle broken down into component sectors. T/F

(v) Data classified geographically or qualitatively cannot be presented on a line graph. T/F

(vi) A cumulative frequency distribution enables us to see how many observations lie above or below certain values. T/F

(vii) Squares are one dimensional diagrams. T/F

(viii) Bar diagram represents a frequency distribution. T/F

22. Tick the correct answer:

(a) Diagram and graphs are tools of (i) collection of data. (ii) analysis. (iii) presentation. (iv) summarisation. (v) None of these.

(b) In a two-dimensional diagram (i) only height is considered. (ii) only width is considered, (iii) height, width and thickness are considered, (iv) both height and width are considered.

(c) Which of the following should be avoided as methods of presenting data: (i) spheres and cubs. (ii) bars, (iii) pie diagrams, (iv) petrographers. (v) rectangles?

23. (a) Discuss various types of two dimensional diagrams. Explain with example, how these are prepared?

(b) Distinguish between 'Natural scale and 'Ratio Scale'. What are the methods of constructing graphs on ratio scale? How do you interpret graphs on ration scale? In which cases should ration scale the used?

24. (a) Discuss the methods of presentation of data through graphs and diagrams.

(b) Discuss the various types of graphical presentation of data.

25. Proportions of males and females in India are given below according to occupation. Draw a suitable diagram:

Serial No.	Occupation	Males	Females
1	Manufacturing	47%	53%
2	Agricultural labour	55%	45%
3	Household industries	61%	39%
4	Miscellaneous	88%	12%

26. Present the following data of results of the II Yr. B. Com. Statistics examination of Bangalore University held in November, 1999 and November, 2000 by means of multiple bars:

Year	I Class	II Class	III Class	Final
November, 1998	100	300	500	300
November, 1999	120	400	600	280
November, 2000	100	500	700	300

27. Represent the following data with a suitable diagram:

Index Numbers of wholesale prices

Year	Cereals	Pulses	Fibres	Oilseeds
1992	443	424	432	499
1993	465	438	446	593
1994	471	449	476	665
1995	483	506	622	679
1996	450	483	454	483
1997	451	494	420	573

28. Draw a histogram of the following distribution:

Life of electric Imps (in hrs.)	1,010	1,030	1,050	1,070	1,090
Firm A:	13	130	482	360	18
Firm B:	287	106	26	230	352

29. Fill in the blanks:
   (i) "A picture is worth ...... words".
   (ii) For constructing a graphs, we generally make use of..... whereas a diagram is generally constructed on a......
   (iii) Bar diagrams are ...... dimensional diagrams.
   (iv) Cartograms are used to give quantitative information on a....
   (v) Graphs of time series are called...... whereas graphs of frequency distribution are called.......
   (vi) Natural scale indicates......changes whereas ratio scale indicates.......changes.
   (vii) Point out which is correct:
   Diagrams are for.......
   (a) the use of experts.
   (b) better mental appeal.
   (c) None of these.
   Ans. (i) 10,000,
   (ii) graph paper, plain paper,
   (iii) one.

(iv) geographical basis,
(v) historigrams histogr·ms,
(vi) absolute, relative,
(vii) better mental appeal.

30. (a) State the difference between Natural sale and Ration scale.
    (b) What purpose is served by a semi-logarithmic graph paper?
    (c) Explain a pictogram.
31. (a) Define an Ogive and how to obtain the value of median graphically.
    (b) Illustrate graphically the distinction between a frequency polygon. A histogram and ogive curve. Comment on their uses.
    (c) Define angular diagram. Discuss the usefulness of diagrammatic representation of facts.
    (d) Explain and illustrate a Histogram.
32. Draw a rectangular diagram to represent the following information:

	Factory A	Factory B
Price per unit	Rs. 15.00	Rs. 12.00
Units produced	1,000 Nos.	1,200 Nos.
Raw material/unit	Rs. 5.00	Rs. 5.00
Other expenses/unit	Rs. 4.00	Rs. 3.00
Profit/unit	Rs. 6.00	Rs. 4.00

33. (a) Describe the utility of diagrammatic representation of statistical data.
    (b) Following data relate to the expenditure of two families per month:

Items	Family A	Family B
Food	160	60
Rent	150	40
Clothing	100	30
Education	80	40
Lighting	30	10
Others	80	20

Represent the data by percentage bar diagram or pie chart.

34. Represent the following data relating to the expenditure of two families by means of a percentage bar diagram:

Expenditure items	Family A Income Rs. 400	Family B Income Rs. 600
Food	120	150
Clothing	80	100
Housing	60	100
Health & Education	40	80
Fuel & Lighting	40	40
Miscellaneous	40	60

35. Draw histogram, frequency polygon, cumulative frequency curve with the help of following two tables :

Table I

Class interval	Frequency	Class interval	Frequency
1–10	3	51–60	40
11–20	14	61–70	47
21–30	21	71–80	50
31–40	25	81–90	50
41–50	40		

Table II

Class interval	Frequency	Class interval	Frequency
24.5–29.53	48.5–53.5	40	
30.5–35.56	54.5–60.5	47	
36.5–41.514	61.5–66.5	50	
42.5–47.520	67.5–72.5	50	

36. The percentage of water, lipid, protein and other materials are 66.35%, 6.66%, 5.2%, 21.79% respectively in the body of a species of fish. Draw a pie chart with the help of the given data.

37. Milk production of four states per day (litres) are as follows. Prepare a pie chart on the basis of following data:

    States Bihar Bengal Delhi U.P.

    Production (in 100 litres) 700 620 328 640

38. Rainfall in seven towns of Bihar were recorded as follows in the year 1997. Draw a pie chart on the basis of data :

Gaya	Patna	Nawada	Ranchi	Arrah	Jelwnabad	Sasuram
150cm	230cm	125cm	360cm	240cm	175 cm	210cm

39. Following data were obtained in a hospital of Bombay in respect of age and frequency of cancer. Make a frequency polygon.

Age	39–49	50–59	60–69	70–79	80–89
No. of cancer patients	2	3	15	21	18

40. Draw histogram with the help of following data:

No. of pods	No. of plants	No. of pods	No. of plants
15–17	5	30–32	15
18–20	6	33–351	5
21–23	8	36–38	9
24–26	12	39–41	5
27–29	22		

41. What is abscissa and ordinate in a graph. Mention 4 quadrants giving a figure.

42. Present the following data by a suitable diagram:

Items of Expenditure	Family A	Family B
Food	2000	2500
Clothing	1000	2000
House Rent	800	1000
Fuel and Lighting400	500	
Miscellaneous	800	2000
	5000	8000

43. The weight of twenty new born child (in kilogram) of Banaras Medical College in a particular day were as follows :

3.0, 3.5, 3.0, 2.5, 2.75, 2.0, 2.25, 2.25, 2.0, 2.2, 2.75, 2.0, 3.2, 3.5, 3.5, 4.0, 3.5, 3.2, 2.0, 2.5 and 2.0

Make cumulative frequency table and draw ogive.

44. Represent the following data by means of percentage subdivided bar diagram:

Cost per equipment	1994 Rs.	1995 Rs.	1996 Rs.
Raw materials	2160	2600	2700
Labour	540	700	810
Direct expenses	600	300	350
Factory expenses	360	200	300
Office expenses	180	200	270
Total	3840	4000	4490

45. (a) Discuss the methods of presentation of statistical data through graphs and diagrams.

(b) When do we use bar diagrams? How many different types of them are in use? Give some other forms of diagrammatic representations.

46. (a) Explain how tables, graphs and charts help in the effective presentation of data.

(b) What are the common methods for grouping and presentation of data?

(c) Explain the need and usefulness of diagrammatic representation of statistical data. What are the different types of diagram you know?

47. (a) Mention the advantages of graphic presentation of statistical data.

(b) What are the general rules for graphing the data?

48. As per Railway Budget how rai!ways spend each rupee is given below. Draw a pie chart to represent the data. "Every one rupee the Railways spend, a hefty 3 paise is accounted for by staff wages and other allowances. 16 praise for depreciation reserve fund, 13 praise each for fuel and miscellaneous, 7 paise for dividend, 8 paise for stores and 23 paise for development works".

49. The following table gives the average monthly earnings of the mill workers in a certain city:

Monthly earnings (Rs.)	No. of workers	Monthly earnings (Rs.)	No. of workers
800-850	21	1200-1250	36
850-900	29	1250-1300	45
900-950	19	1300-1350	27
950-1000	39	1350-1400	48
1000-1050	43	1400-1550	21
1050-1100	94	1450-1500	12
1100-1150	73	1500-1550	9
1150-12C0			

(i) Draw ogives by less than and more than methods for the data given below.

(ii) Find the number of workers whose wages lie between Rs. 1.180 and Rs. 1,480.

50. (a) What are the advantages of diagrammatic and graphic presentation of data?

(b) The merits of diagrammatic representation of data are classified under three main headings: attraction, effective impression and comparison. Explain and illustrate these points.

51. (a) Indicate the method of constructing histogram, frequency polygon and ogive.

52. Prepare a frequency distribution for the following data on number of seeds in a beans pod.

2	11	6	4	18	1	9	2	2	15
8	16	12	11	17	3	3	5	3	7
11	9	5	16	16	16	4	9	5	7
4	10	4	4	15	15	5	5	11	18
5	10	9	8	7	7	2	6	13	1.

4

Measures of Central Location

VARIOUS MEASURES OF CENTRAL TENDENCY

In the previous chapters we discussed the tabular and graphical approaches to describe a sample. Although these are effective means of data summarization, nonetheless numerical values are often preferred. Such numbers which describe a population are called *parameters*; and if they describe a sample they are called *statistics*.

The first number that might be desirable is the one which conveys a fairly adequate idea about the whole group. This single expression in statistics is known as the *average*. Averages are generally the central part of the distribution therefore, they are called *measures of central tendency*.

The most common measures of central tendency are :

(1) Arithmetic Mean

(2) Median

(3) Mode.

Each of them, in its own way can be called a representative of the characteristic of the whole group and thus the performance of the group as a whole can be described by the single value which each of these measures gives. The values of mean, median and mode also help us in comparing two or more groups or frequency distribution in terms of typical or characteristic performance.

Professor G.V. Yule has given certain characteristic of a measure of central tendency. These are :

(1) The measure of central tendency should be based on all observations of data.

(2) It should be rigidity defined.

(3) It the should be easy to calculate and understand.

(4) The sampling variations should not affect the measure.

(5) It should be easy to interpret and subject it to further calculations.

ARITHMETIC MEAN

The most popular and widely used measure of representing the entire data by one value is what most laymen call an 'average' and what the statisticians call the arithmetic mean. Its value is obtained by adding together all the items and by dividing this total by the number of items. Arithmetic mean may either be

(i) simple arithmetic mean, or

(ii) weighted arithmetic mean.

Calculation of Simple Arithmetic Mean—Individual Observations

The process of computing mean in case of individual observations (i.e., where frequencies are not given) is very simple. Add together the various values of the variable and divide the total by the number of items. Symbolically:

$$\overline{X}* = \frac{X_1 + X_2 + X_3 + \ldots + X_n}{N} \text{ or } \overline{X} = \frac{\sum X}{N}$$

Here $\overline{X}$ = Arithmetic Means, $\sum X$ = Sum of all the values of the variable X, i.e., X_1, X_2, X_3, ... X_n; N = Number of observations.

Steps. The formula involves two steps in calculating mean:

(i) Add together all the values of the variable X and obtain the total, i.e., $\sum X$.

(ii) Divide this total by the number of the observations, i.e. N.

Objectives of Average

There are two main objectives of the study of averages:

(i) *To get single value that describes the characteristic of the entire group.* Measures of Central Tendency, by condensing the mass of data in one single value, enable us to get a bird's-eye view of the entire data. Thus one value can represent thousands, lakhs and even millions of values. For example, it is impossible to remember the individual incomes of millions of earning people of India and even if one could do it there is hardly any use. But if the average income is obtained by dividing the total national income by total population we get one single value that represents the entire population. Such a figure would throw light on the standard of living of an average Indian.

(ii) *To Facilitate Comparison Measures of Central Tendency,* by reducing the mass of data to one single figure, enable comparison to be made. Comparison can be made either at a point of time or over a period of time. For example, we can compare the percentage results of the students of different colleges in a certain examination, say. B. Com. for 1999, and thereby conclude which college is the best or we can compare the pass percentage of the some college for different time periods and thereby conclude as to whether the results are improving or deteriorating. Such comparisons are of immense help in framing suitable and timely policies. For example, if the pass percentage of students in College A in B. Com. was 80 in 1998 and 75 in 1999, the authorities have sufficient reason for investigating the possible cause of the deterioration in results.

However, while making comparison one should also take into consideration the multiplicity of forces that might be affecting the data. For example, if per capita income is rising in absolute terms from one period to another, it should not lead one to think that the standard of living is necessarily improving because the prices might be rising faster than the rise in per capita income and so in real terms people might be worse off. Moreover, the same measure should be used for making comparison between two or more groups. For example, we should not compare the mean wage of one factory with the median wage of another factory for drawing any inference about wage levels.

Requisites of a Good Average

Since an average is a single value representing a group of values, it is desired that such a value satisfies the following properties:

(i) *Easy to Understand.* Since statistical methods are designed to simplify complexity. It is desirable that an average be such that can be readily understood; otherwise, its use is bound to be very limited.

(ii) *Simple to Compute.* An average should not only be easy to understand but also simple to compute so that it can be used widely. However, though ease of computation is desirable, it should not be sought at the expense of other advantages, i.e. if in the interest of greater accuracy, use of a more difficult average is desirable, one should prefer that.

(iii) *Based on all the Items.* The average should depend upon each and every item of the series so that if any of the items is dropped the average itself is altered. For example, the arithmetic mean of 10, 20, 30, 40, 50, is, $\frac{10 + 20 + 30 + 40 + 50}{5} = \frac{150}{5} = 30$. If we drop one item, say, 50, the arithmetic mean would be = 25.

(iv) *Not be unduly Affected by Extreme Observations.* Although each and every item should influence the value of the average, none of the items should influence it unduly. If one or two very small or very large items unduly affect the average, i.e., either increase its value or reduce its value, the average cannot be really typical of the entire series. In other words, extremes may distort the average and reduce its usefulness.

(v) *Rigidly Defined.* An average should be properly defined so that it has one and only one interpretation. It should preferably be defined by algebraic formula so that if different people compute the average from the same figures they all get the same answer (barring arithmetical mistakes). The average should not depend upon the personal prejudice and bias of the investigator, otherwise the results can be misleading.

(vi) *Capable of Further Algebraic Treatment.* We should prefer to have an average that could be used for further statistical computations so that its utility is enhanced. For example, if we are given the data about the average income and number of employees of two or more factories, we should be able to compute the combined average.

(vii) *Sampling Stability.* Last, but not the least, we should prefer to get a value which has what the statisticians call 'sampling stability'. This means that if we pick 10 different groups of college students, and compute the average of each group we should expect to get approximately the same value. It does not mean, however, that there to get approximately the same value. It does not mean, however, that there can be no difference in the values of different samples. There may be some difference but those samples in which this difference (technically called sampling fluctuation) is less are considered better than those in which this difference is more.

Types of Averages

The following are the important types of averages:

(a) Harmonic mean

(b) Geometric mean

(c) Mode

(d) Median

(e) Arithmetic mean :

 (i) simple, and

 (ii) weighted.

Besides these, there are less important averages like moving average, progressive average, etc. These average have a very limited field of application and are, therefore, not so popular.

Example 1:

From the following data of the marks obtained by 60 students of a class, calculate the arithmetic mean:

Marks	*No. of Students*	*Marks*	*No. of Students*
20	8	50	10
30	12	60	6
40	20	70	4

Let the marks be denoted by X and the number of students by f.

Solution:

Calculation of Arithmetic Mean

Marks X	*No. of students* f	fX
20	8	160
30	12	360
40	20	800
50	10	500
60	6	360
70	4	280
	N = 60	ΣfX = 2,460

$$\overline{X} = \frac{\Sigma fX}{N} = \frac{2,460}{60} = 41$$

Hence the average marks = 41.

Short-cut Method According to this method,

$$\overline{X} = A + \frac{\Sigma fX}{N}$$

where A = Assumed mean; d = (X – A); N = Total of observations, i.e., Σf.

Steps:

(i) Take an assumed mean.

(ii) Take the deviations of the variable X from the assumed mean and denote the deviations by d.

(iii) Multiply these deviations with the respective frequency and take the total Σfd.

(iv) Divide the total obtained in third step by the total frequency.

Example 2:

The following table gives the monthly income of 10 employees in an office:

Income (Rs.) 1,780 1,760 1,690 1,750 1,840
1,920 1,100 1,810 1,050 1,950

Calculate the arithmetic mean of incomes.

Let income be denoted by the symbol X.

Solution:

Calculation of Arithmetic Mean

Employee	*Monthly Income (Rs.)*	*Employee*	*Monthly Income (Rs.)*
1	1,780	6	1,920
2	1,760	7	1,100
3	1,690	8	1,810
4	1,750	9	1,050
5	1,840	10	1,950
		N = 10	ΣX = 16,650

$$\overline{X} = \frac{\Sigma X}{N} \neq \Sigma X = 16{,}650,\ N = 10$$

$$\overline{X} = \frac{16{,}650}{10} = 1{,}665.$$

Hence the average income is Rs. 1,665.

Short-cut Method

The arithmetic mean can be calculated by using what is known as an arbitrary origin. When derivations are taken from an arbitrary origin, the formula for calculating arithmetic mean is

$$\overline{X} = A + \frac{\Sigma d}{N}$$

where A is the assumed mean and d is the deviation of items from assumed mean, i.e., d = (X – A).

Steps.

(1) Take an assumed mean.

(2) Take the deviations of items from the assumed mean and denote these deviations by d.

(3) Obtain the sum of these deviations, i.e. Σd.

(4) Apply the formula : $\overline{X} = A + \frac{\Sigma d}{N}$

Calculate arithmetic mean by taking 1,800 as the assumed mean.

Solution:

Calculation of Arithmetic Mean

Employee	*Income*	*(X – 1800)*
1	1,780	–20
2	1,760	–40
3	1,690	–110
4	1,750	–50
5	1,840	+40
6	1,920	+120
7	1,100	–700
8	1,810	+10
9	1,050	–750
10	1,950	+150
N = 60		Σd = –1350

$$\overline{X} = A + \frac{\sum d}{N}$$

$A = 1{,}800,\ \sum d = -1{,}350,\ N = 10$

$$\overline{X} = 1{,}800 - \frac{1{,}350}{10} = 1{,}800 - 135 = 1{,}665.$$

Hence the average income is Rs. 1,665.

Note. The reader will find that the calculations here are more than what we had when we used the formula

$$\overline{X} = \frac{\sum X}{N}.$$

This is true for ungrouped data. But for grouped data considerable saving in time is possible by adopting the short-cut method.

Calculation of Arithmetic Mean—Discrete Series

In discrete series arithmetic mean may be computed by applying

(i) Direct method, or (ii) Short-cut method.

Direct Method

The formula for computing mean is

$$\overline{X} = \frac{\sum fX}{N}$$

where, f = Frequency; X = The variable in question;

N* = Total number of observation, i.e., Σf.

Steps:

(i) Multiply the frequency of each row with the variable and obtain the total ΣfX.

(ii) Divide the total obtained by step (i) by the number of observations i.e., total frequency.

Example 3:

From the following data compute arithmetic mean by direct method.

Marks	*0–10*	*10–20*	*20–30*	*30–40*	*40–50*	*50–60*
No. of students	*5*	*10*	*25*	*30*	*20*	*10*

Solution:

Calculation of Arithmetic Mean by Direct Method

Marks	***Mid-point*** *m*	***No. of Students*** *f*	*fm*
0–10	5	5	25
10–20	15	10	150
20–30	25	25	625
30–40	35	30	1,050
40–50	45	20	900
50–60	55	10	550
		N = 100	Σfm = 3,300

$$\overline{X} = \frac{\Sigma fm}{N} = \frac{3{,}300}{100} = 33.$$

Short-cut Method. When short-cut method is used, arithmetic mean is computed by applying the following formula;

$$\overline{X} = A + \frac{\Sigma fd}{N}$$

when A = assumed mean;

d = deviations of mid-points from assumed mean, i.e., (m – A);

N = total number of observations.

Steps:

(i) Take an assumed mean.

(ii) From the mid-point of each class deduct the assumed mean.

(iii) Multiply the respective frequencies of each class by these deviations and obtain the total Σfd.

(iv) Apply the formula: $\overline{X} = A + \frac{\Sigma fd}{N}$

Calculate arithmetic mean by the short-cut method.

Solution:

Calculation of Arithmetic Mean

Mark	*Mid-point* *m*	*No. of Students* *f*	*(m – 35)* *d*	*fd*
0–10	5	5	–30	–150
10–20	15	10	–20	–200
20–30	25	25	–10	–250
30–40	35	30	0	0
40–50	45	20	+10	+200
50–60	55	10	+20	+200
		N = 100		Σfd = –200

$$\overline{X} = A + \frac{\Sigma fd}{N} = 35 - \frac{200}{100} = 35 - 2 = 33$$

In order to simplify the calculations, we can divide the deviations by class intervals, i.e., calculate (m – A)/i and then multiply by in the formula for getting mean. The formula becomes;

$$\overline{X} = A + \frac{\Sigma fd}{N} \times i$$

It may be pointed out that when class intervals are unequal we can simplify calculations by taking a common factor. In such a case we should use (m – A)/C instead (m – A)/i while making calculations.

Compute arithmetic mean by step deviation method.

Solution:

Calculation of Arithmetic Mean

Marks	*Mid-point* *m*	*No. of Students* *f*	*(m – 35)* *d*	*(m – 35)/10*	*fd*
0–10	5	5	–30	–3	–15
10–20	15	10	–20	–2	–20
20–30	25	25	–10	–1	–25
30–40	35	30	0	0	0
40–50	45	20	+10	+1	+20
50–60	55	10	+20	+2	+20
		N = 100			Σfd = –20

$$\overline{X} = A + \frac{\Sigma fd}{N} \times i = 35 - \frac{20}{100} \times 10 = 35 - 2 = 33.$$

It is clear from above that all the three methods of finding arithmetic mean in continuous series give us the same answer. The direct method, though the simplest, involves more calculations when mid-points and frequencies are very large in magnitude. For example, observe the following data:

Income in Rs.	***No. of Persons***
400–500	368
500–600	472
600–700	969
700–800	567
800–900	304

In this case step deviation method would be far simpler. In fact, step deviation method should be adopted wherever possible because it minimises the calculations.

While computing mean in continuous series the mid-points of the various classes are taken as representative of that particular class. The reason is that when the data are grouped, the exact frequency with which each value of the variable occurs in the distribution is unknown. We only know the limits within which a certain number of frequencies occur. For example, when we say that the number of persons within the income group 400–500 is 50 we cannot say as to how many persons out of 50 are getting 401, 402, 403, etc. We, therefore, make an assumption while calculating arithmetic mean that the frequencies within each class are spread evenly over the range of the class interval, i.e., there will be as many items below the mid-point as above it. Unless such an assumption is made the value of mean cannot be computed.

This assumption is likely to led to some error. As a result thercof the mean of a number of observations calculated from a frequency distribution will generally be only an approximation to the mean calculated from the original data. However, the possibility of compensating errors must be considered. Some of the mid-points err by being too low and others err by being too high. In general, then mid-points of the classes below the class containing the arithmetic mean tend to be too low and the mid-points of the classes above the class containing the arithmetic mean tend to be too high. It is quite possible, therefore, that when the errors are assumed, those which are too low will offset, in part at least, those which are too high, so that the arithmetic mean for the entire distribution will be approximetely of the same value as is obtained from a list of values.

Example 4:

Calculate arithmetic mean by the short-cut method using frequency distribution.

Solution:

Calculation of Arithmetic Mean

Marks X	*No. of Students* f	*(X – 40)* d	*fd*
20	8	–20	–160
30	12	–10	–120
40	20	0	0
50	10	+10	+100
60	6	+20	+120
70	4	+30	+120
	N = 60		Σfd = 80

$$\overline{X} = \frac{\Sigma fX}{N} = 40 + \frac{60}{60} = 40 + 1 = 41$$

Calculation of Arithmetic Mean—Continuous Series

In continuous series, arithmetic mean may be computed by applying any of the following methods:

(i) Direct method

(ii) Short-cut method.

Direct Method. When direct method is used

$$\overline{X} = \frac{\Sigma fX}{N}$$

where m = mid-point of various classes; f = the frequency of each class; N = the total frequency.

Steps:

(i) Obtain the mid-point of each class and denote it by m.

(ii) Multiply these mid-points by the respective frequency of each class and obtain the total Σfm.

(iii) Divide the total obtained in step (i) by the sum of the frequency, i.e. N.

Example 5:

Calculate arithmetic mean from the following data:

Marks	*0–10*	*10–30*	*30–60*	*60–100*
Nos. of	*5*	*12*	*25*	*8*

Solution:

The class intervals are unequal but still to simplify calculations we can take 5 as the common factor.

Calculation of Mean

Marks	*Mid-point* *m*	*f*	*(m – 45)/5* *d*	*fd*
0–10	5	5	–8	–40
10–30	20	12	–5	–60
30–60	45	25	0	0
60–100	80	8	+7	+56
			N = 50	Σfd = –44

$$\overline{X} = A + \frac{\sum fd}{N} \times C$$

$$A = 45, \Sigma fd = -44, N = 50, C = 5$$

$$\overline{X} = 45 - \frac{44}{50} \times 5 = 45 - 4.4 = 40.6$$

Note. When data are given by the inclusive method it is not necessary to adjust the class for calculating arithmetic mean because the mid-points remain the same whether or not the adjustment is made. However, in case of median and mode adjustment is necessary.

Correcting Incorrect Values. It sometimes happens that due to an oversight or mistake in copying, certain wrong items are taken while calculating mean. The problem is how to find out the correct mean. The process is very simple. From incorrect ΣX deduct wrong items and add correct items and then divide the correct ΣX by the number of observations. The result, so obtained, will give the value of correct mean.

Example 6:

Mean of 100 observations is found to be 40. If at the time of computation two items are wrongly taken as 30 and 27 instead of 3 and 72. Find correct Mean.

Solution:

$$\overline{X} = \frac{\Sigma X}{N} \quad \text{or} \quad \Sigma X = N\overline{X}$$

Here $\overline{X}$ = 40, N = 100

∴ ΣX = 100 × 40	=	4000
Less incorrect items	=	57
	=	3943
Add correct items	=	75
Correct total	=	4018

Correct mean $= \frac{4018}{100} = 40.18.$

Calculation of Arithmetic Mean in Case of Open-end Classes

Open-end classes are those in which lower limit of the first class and the upper limit of the last class are known in such a case we cannot find out the arithmetic mean unless we make an assumption about the unknown limits. The assumption would naturally depend upon the class interval following the first class and preceding the last class. For example, observe the following data:

Marks	No. of students	Marks	No. of students
Below 10	4	30–40	15
10–20	6	40–50	8
20–30	10	Above 50	7

In the above case since the class interval is uniform, the appropriated assumption would be that the lower limit of the first class is zero and the upper limit of the last class is 60. The first class thus would be 0–10 and the last class 50–60. Observe another case:

Marks	No. of students	Marks	No. of students
Below 10	4	60–100	7
10–30	6	Above 100	3
30–60	10		

In the above case since the class interval is 20 in the second class, 20 in the third class, 40 in the fourth class, i.e., it is increasing by 10. The appropriate assumption would be that the lower limit of the first class is zero and the upper limit of the last class 150. In other words, first class is 0–10 and the lase one 100–150.

If the class intervals are of varying width, an effort should not be made to determine the lower limit of the lowest class and upper limit of the highest class. The use of median or mode would be better in such a case. Because of the difficulty of ascertaining lower limit and upper limit in open-end distributions it is suggested that in such distributions arithmetic mean should not be used.

MATHEMATICAL PROPERTIES OF ARITHMETIC MEAN

The following are a few important mathematical properties of the arithmetic mean:

1. The sum of the deviations of the items from the arithmetic mean (taking signs into account) is always zero, i.e., $\Sigma(X - \overline{X}) = 0$. This would be clear from the following example:

X	$(X - \overline{X})$
10	– 20
20	– 10
30	0
40	+ 10
50	+ 20
$\Sigma X = 150$	$\Sigma(X - \overline{X}) = 0$

Here $\overline{X} = \frac{\Sigma X}{N} = \frac{150}{5} = 30$. When the sum of the deviations from the actual mean, i.e., 30, is taken it comes out to be zero. It is because of this property that the mean is characterised as point of balance, i.e., the sum of the positive deviations from it is equal to the sum of the negative deviations from it.

2. The sum of the squared deviations of the items from arithmetic mean is minimum, that is, less than the sum of the squared deviations of the items from any other value. The following example would verify the point:

X	$(X - \overline{X})$	$(X - 4)^2$
2	–2	4
3	–1	1
4	0	0
5	+1	1
6	+2	4
$\Sigma X = 20$	$\Sigma(X - \overline{X}) = 0$	$\Sigma(X - \overline{X})^2 = 10$

The sum of the squared deviations is equal to 10 in the above case. If the deviations are taken from any other value the sum of the squared deviations would be greater than 10. For example, let us calculate the squares of the deviations of item from a value less than the arithmetic mean, say 3.

X	$(X - 3)$	$(X - 3)^2$
2	– 1	1
3	0	0
4	+ 1	1
5	+ 2	4
6	+ 3	9
		$\Sigma(X - 3)^2 = 15$

It is clear that $(\Sigma X - \overline{X})^2$ is greater. This property that the sum of the squares of items is least from the means is of immense use in regression analysis which shall be discussed later.

3. Since $\overline{X} = \frac{\Sigma X}{N}$, $N\overline{X} = \Sigma X$

In other words, if we replace each item in the series by the mean, then the sum of these substitutions will be equal to the sum of the individual items. For example, in the discussion of first property $\Sigma X = 150$ and the arithmetic mean 30. If for each item we substitute 30, we get the same total, i.e., 30 + 30 + 30 + 30 + 30 = 150.

This property is of great practical value. For example, if we know the average wage in a factory, say, Rs. 1,060 and the number of workers employed, say, 200, we can compute total wages bill from the relation $\frac{N\overline{X} = \Sigma X}{X.0}$. The total wage bill in this case would be 200 × 1,060, i.e., Rs. 2,12,000 which is equal to ΣX.

4. If we have the arithmetic mean and number of items of two or more than two related groups, we can compute combined average of these groups by applying the following formula:

$$\overline{X}_{12} = \frac{N_1\overline{X}_1 = \overline{X}_2}{N_1 + N_2}$$

$\overline{X}_{12}$ = combined mean of the two groups

$\overline{X}_1$ = arithmetic mean of first group

$\overline{X}_2$ = arithmetic mean of second group

N_1 = number of items in the first group

N_2 = number of items in the second group

Example 7:

The mean marks of 190 students were found to be 40. Later on it was discovered that a score of 53 was misread as 83. Find the correct mean corresponding to the correct score.

Solution:

We are given $N = 100$, $\overline{X} = 40$

Since $\overline{X} = \frac{\Sigma X}{N}$

$$\Sigma X = N\overline{X} = 100 \times 40 = 4000$$

But this is not correct ΣX

Correct ΣX = Incorrect ΣX – wrong item + correct item

$$= 4000 - 83 + 53 = 3970$$

$$\therefore \text{Correct } \overline{X} = \frac{\text{correct } \Sigma X}{N} = \frac{3970}{100} = 39.7$$

Hence the correct average = 39.7

Example 8:

The mean height of 25 male workers in a factory is 61 cm. and the mean height of 35 female workers in the same factory is 58 cm. Find the combined mean height of 60 workers in the factory.

Solution:

$$\overline{X}_{12} = \frac{N_1 \overline{X}_1 = \overline{X}_2}{N_1 + N_2}$$

$$N_1 = 25, \ \overline{X}_1 = 61, N_2 = 35, \ \overline{X}_2 = 58$$

$$\overline{X}_{12} = \frac{(25 \times 61) + (35 \times 58)}{25 + 35} = \frac{1525 + 2030}{60} = \frac{3555}{60} = 59.25$$

Thus the combined mean height of 60 workers is 59.25 cm.

If we have to find out the combined mean of three sub-groups the above formula can be extended as follows:

$$\overline{X}_{123} = \frac{N_1 \overline{X}_1 + N_2 \overline{X}_2 + N_3 \overline{X}_3}{N_1 + N_2 + N_3}.$$

Merits and Limitations of Arithmetic Mean

Merits

Arithmetic mean is most widely used in practice because of the following reasons:

(a) The mean is typical in the sense that it is the centre of gravity, balancing the values on either side of it.

(b) It is a calculated value, and not based on position in the series.

(c) It is relatively reliable in the sense that it does not vary too much when repeated samples are taken from one and the same population, at least not as much as some other kind of statistical descriptions.

(d) Being determined by a rigid formula, if lends itself to subsequent algebraic treatment better than the median or mode.

(e) It is defined by a rigid mathematical formula with the result that everyone who computes the average gets the same answer.

(f) It is affected by the value of every item in the series.

(g) It is the simplest average to understand and easiest to compute. Neither the arraying of data as required for calculating median for grouping of data as required for calculating mode is needed while calculating mean.

Limitations

Since the value of mean depends upon each and every item of the series, extreme items, i.e. vary small and very large items, unduly affect the value of the average. For example, if in a tutorial group there are 4 students and their marks in a test are 60, 70, 10 and 80 the average marks would be $\frac{60 + 70 + 10 + 80}{4} = \frac{220}{4} = 55$. One single item, i.e., 10, has reduces the average marks considerably. The smaller the number of observations, the greater is likely to be the impact of extreme value.

(a) The arithmetic mean is not always a good measure of central tendency. The mean provides a "characteristic" value, in the sense of indicating where most of the values lie, only when the distribution of the variable is reasonably normal (bell-shaped). In case of a U-shaped distribution the mean is not likely to serve a useful purpose.

(b) In a distribution with open-end classes the value of mean cannot be computed without making assumptions regarding the size of the class interval of the open-end classes. If such classes contain a large proportion of the values, then mean may be subject to substantial error. However, the values of the median and mode cane computed where there open-end classes without making any assumptions about size of class interval.

WEIGHTED ARITHMETIC MEAN

One of the limitations of the arithmetic mean discussed above is that it gives equal importance to all the items. But there re cases where the relative importance of the different items is not the same. When this is so, we compute weighted arithmetic mean. The term 'weight' stands for the relative importance of the different items. The formula for computing weighted arithmetic mean is:

$$\overline{X}_\omega = \frac{\Sigma WX}{\Sigma W}$$

where $\overline{X}_\omega$ represents the weighted arithmetic mean; X represents the variable values, i.e., $X_1, X_2, ..., X_n$.

We represents the weights attached to variable values, i.e. $\omega_1, \omega_2, ... \omega_n$, respectively.

(i) Multiply the weights by the variable X and obtain the total ΣWX.

(ii) Divide this total by the sum of the weights, i.e., ΣW.

In case of frequency distribution, if $f_1, f_2, \ldots f_n$ are the frequencies of the variable values $X_1, X_2, \ldots, X_n$ respectively then the weighted arithmetic mean is given by:

$$\overline{X}_\omega = \frac{\Sigma W(fX)}{\Sigma W}$$

From the expanded form

$$\overline{X}_\omega = \frac{W_1(f_1 X_1) + W_2(f_2 X_2) + \ldots + W_n(f_n X_n)}{W_1 + W_2 + \ldots + W_n}.$$

An important problem that arises while using weighted mean is regarding selection of weights. Weights may be either actual or arbitrary, i.e., estimated. Needless to say, if actual weights are available, nothing like this. However, in the absence of actual weights, arbitrary or imaginary weights may be used. The use of arbitrary weights may lead to some error, but it is better than no weights at all. In practice. It is found that if weights are logically assigned keeping the phenomena in view, the error involved will be so small that it can be easily over looked.

It should be noted that:

(i) Simple arithmetic mean shall be equal to the weighted arithmetic mean if the weights are equal. Symbolically,

$$\overline{X} = \overline{X}_\omega \text{ if } W_1 = W_2.$$

(ii) Simple arithmetic mean shall be less than the weighted arithmetic mean if and only if greater weights are assigned to greater values and smaller weights are assigned to smaller values. Symbolically,

$$\overline{X} < \overline{X}_\omega \text{ if } (\omega_2 - \omega_1)(X_1 - X_2) < 0.$$

(iii) Simple arithmetic mean is greater than the weighted arithmetic mean if and only if smaller weights is attached to the higher values and greater weight is attached to the smaller values. Symbolically.

$$\overline{X} > \overline{X}_\omega \text{ if } (w_2 - w_1)(X_1 - X_2) < 0.$$

It may be noted that weighted arithmetic mean is specially useful in problems relating to:

(i) Construction of index numbers, and

(ii) Standardized birth and death rates.

Example 9:

Comment on the performance of the students of the three universities given below using simple and weighted averages:

University Course study	*Pass %*	*Bombay No. of Students (in hundreds)*	*Pass %*	*Calcutta No. of Students (in hundreds)*	*Pass %*	*Madras No. of Students (in hundreds)*
M.A.	*71*	*3*	*82*	*2*	*81*	*2*
M.Com.	*83*	*4*	*76*	*3*	*76*	*3.5*
B.A.	*73*	*5*	*73*	*6*	*74*	*4.5*
B.Com.	*74*	*2*	*76*	*7*	*58*	*2*
B.Sc.	*65*	*3*	*65*	*3*	*70*	*7*
M.Sc.	*66*	*3*	*60*	*7*	*73*	*2*

Solution:

University of Study	***Pass %*** X	***Bombay No. of Students (in hundreds)*** W	WX	***Pass %*** X	***Calcutta No. of Students (in hundreds)*** W	WX	***Pass %*** X	***Madras No. of Students (in hundreds)*** W	WX
M.A.	71	3	213	82	2	164	81	2.0	162
M.Com.	83	4	332	76	3	228	76	3.5	266
B.A.	73	5	365	73	6	438	74	4.5	333
B.Com.	74	2	148	76	7	532	58	2.0	116
B.Sc.	65	3	195	65	3	195	70	7.0	490
M.Sc.	66	3	198	60	7	420	73	2.0	146
	ΣX = 432	ΣW = 20	ΣWX = 1,451	ΣX = 432	ΣW = 28	ΣWX = 1,977	ΣX = 432	ΣW = 21	ΣWX = 1,51

Simple and Weighted Arithmetic Mean

$$\text{Bombay} \quad \overline{X} = \frac{\Sigma X}{N} = \frac{432}{6} = 72; \qquad \overline{X}_{\omega} = \frac{\Sigma WX}{\Sigma W} = \frac{1{,}451}{20} = 72.55$$

$$\text{Calcutta} \quad \overline{X} = \frac{\Sigma X}{N} = \frac{432}{6} = 72; \qquad \overline{X}_{\omega} = \frac{\Sigma WX}{\Sigma W} = \frac{1{,}977}{28} = 70.61$$

$$\text{Madras} \quad \overline{X} = \frac{\Sigma X}{N} = \frac{432}{6} = 72; \qquad \overline{X}_{\omega} = \frac{\Sigma WX}{\Sigma W} = \frac{1{,}513}{21} = 72.05$$

The arithmetic mean is the same for all the three universities, i.e., 72 and hence, it may be concluded that the performance of students is alike. But this will be a wrong conclusion because what we should compare here is the weighted arithmetic mean. On comparing the weighted arithmetic means we find that for Bombay the mean value is the highest and hence we can say that in Bombay University the performance of students is best.

MEDIAN

The median by definition refers to the middle value in a distribution. In case of median one-half of the items in the distribution have a value the size of the median value or smaller and one-half have a value the size of the median value or larger. The median is just the 50th percentile value below which 50 per cent of the values in the sample fall. It splits the observation into two halves.

As distinct from the arithmetic mean which is calculated from the value of every item in the series, the median is what is called a positional average. The term 'position' refers to the place of a value in a series. The place of the median in a series is such that an equal number of items lie on either side of it.

For example, if the income of five employees is Rs. 900, 950, 1020, 1200 and 1280 the median would be 1020.

900

950

1020 value at middle position of the array

1200

1280

For the above example the calculation of median was simple because of odd number of observations. When an even number of observations are listed, there is no single middle position value and the median is taken to be the arithmetic mean of two middle most items. For example, if in the above case we are given the income of six employees as 900, 950, 1020, 1200, 1280, 1300, the median income would be;

900

950

1020

1200

1280

1300

$$\text{Median} = \frac{1020 + 1200}{2} = \frac{2220}{2} = 1110.$$

Hence, in case of even number of observations median may be found by averaging two middle position values.

Thus, when N is odd, the median is an actual value, with the remainder of the series in two equal parts on either side of it. If N is even, the median is a derived figure, i.e., half the sum of the middle values.

Calculation of Median—Individual Observations

Steps:

(i) Arrange the data in ascending or descending order of magnitude. (Both arrangements would give the same answer.)

(ii) In a group composed of an odd number of values such as 7, add 1 to the total number of values and divide by 2. Thus, 7 + 1 would be 8 which divided by 2 gives 4—the number of the values starting at either and of the numerically arranged groups will be the median value. In a large group the same method may be followed. In a group of 199 items the middle value would be 100th value. This would be determined by $\frac{199 + 1}{2}$. In the form of formula:

$$\text{Med.} = \text{Size of } \frac{N + 1}{2}\text{th item.}$$

Example 10(a)

A contractor employs three types of workers–male, female and children. To a male he pays Rs. 40 per day, to a female worker Rs. 32 per day and to a child worker Rs. 15 per day. What is the average wage per day paid by the contractor?

Solution:

The average wage is not the simple arithmetic mean, i.e, $\frac{40 + 32 + 15}{3}$ = Rs. 29 per day. It we assume that the number of male, female and child workers is the same, this answer would be correct. For example, if we take 10 workers in each case then the mean wage would be

$$\frac{(10 \times 40) + (10 \times 32) + (10 \times 15)}{10 + 10 + 10} = \frac{400 + 320 + 150}{30} = \text{Rs. } 29$$

However, the number of male, female and child workers employed is generally different, it we know how many workers of each type are employed by the contractor in question, nothing like this. However, in the absence of this we take assumed weights. Let as assume that the number of male, female and child workers employed is 20, 15 and 5 respectively. The average wage would be the weighted mean calculated as follows:

Wages per day (Rs.)	***No. of workers***	
XW	***W***	***WX***
40	20	800
32	15	480
15	5	75
	SW = 40	ΣWX = 1,355

$$\overline{X}_{\omega} = \frac{\Sigma WX}{\Sigma W} = \frac{1{,}355}{40} = 33.875 \text{ or } 33.88$$

Example 10(b):

A train runs 25 miles at a speed of 30 m.p.h., another 50 miles at a speed of 40 m.p.h., then due to repairs of the track travels for 6 minutes at a speed of 10 m.p.h. and finally covers the remaining distance of 24 miles at a speed of 24 m.p.h. What is the average speed in miles per hour?

Solution:

Time taken in covering 25 miles at a speed of 30 m.p.h. = 50 minutes. Time taken in covering 50 miles at a speed of 40 m.p.h. = 75 minutes. Distance covered in 6 minutes at a speed of 10 m.p.h. = 1 mile. Time taken in covering 24 miles at a speed of 24 m.p.h. = 60 minutes.

Therefore, taking the time taken as weights we have the weighted mean as

	Speed in m.p.h.	***Time taken***
XW	***W***	***WX***
30	50	1,500
40	75	3,000
10	6	60
24	60	1,440
		$\Sigma WX = 6{,}000$

$$\therefore \text{Average speed} = \frac{6{,}000}{191} = 31.41 \text{ m.p.h.}$$

Example 11:

Obtain the value of median from the following data:

391 384 591 407 672

522 777 753 2,488 1,490

Solution:

Calculation of Median

Sl. No.	***Data arranged in ascending order (X)***	***Sl. No.***	***Data arranged in ascending order (X)***
1	384	6	672
2	391	7	753
3	405	8	777
4	522	9	1,490
5	591	10	2,488

$$\text{Median} = \text{size of } \frac{N+1}{2}\text{th item} = \frac{11}{2} = 5.5\text{th item.}$$

$$\text{Size of 5.5th item} = \frac{\text{5th item + 6th item}}{2} = \frac{591 + 672}{2} = \frac{1{,}263}{2} = 631.5$$

Computation of Median—Discrete Series

Steps:

(i) Arrange the data in ascending or descending order of magnitude.

(ii) Find out the cumulative frequencies.

(iii) Apply the formula: Median = size of $\frac{N+1}{2}$.

(iv) Now look at the cumulative frequency column and find that total which is either equal to $\frac{N+1}{2}$ or next higher to that and determine the value of the variable corresponding to it. That gives the value of median.

Example 12:

Calculate the median for the following frequency distribution:

Marks	*No. of Students*	*Marks*	*No. of Students*
45–50	*10*	*20–25*	*31*
40–45	*15*	*15–20*	*24*
35–40	*26*	*10–15*	*15*
30–35	*30*	*5–10*	*17*
25–30	*42*		

Solution:

First arrange the data in ascending order and then find out median.

Calculation of Median

Marks	*f*	*c.f.*	*Marks*	*f*	*c.f.*
5–10	7	7	30–35	30	149
10–15	15	22	35–40	26	175
15–20	24	46	40–45	15	190
20–25	31	77	45–50	10	200
25–30	42	119			

$$\text{Med.} = \text{size of } \frac{N}{2} \text{ item} = \frac{200}{2} = 100\text{th item}$$

Median lies in the class 25–30

$$\text{Med.} = L + \frac{N/2 - c.f.}{f} \times i$$

$L = 25,\ N/2 = 100,$

$c.f. = 77,\ f = 42,\ i = 5$

$$\text{Med. } 25 + \frac{100 - 77}{42} \times 5 = 25 + 2.74 = 27.74.$$

Example 13:

From the following data of the wages of 7 workers compute the median wage:

Wages (in Rs.) 1100 1150 1080 1120 1200 1160 1400

Solution:

Calculation of Median

Sl. No.	*Wages arranged in ascending order*	*Sl. No.*	*Wages arranged in ascending order*
1	1080	5	1160
2	1100	6	1200
3	1120	7	1400
4	1150		

$$\text{Median} = \text{size of } \frac{N+1}{2}\text{th item} = \frac{7+1}{2} = 4\text{th item.}$$

Size of 4th item = 1150. Hence the median wage = Rs. 1150.

We thus find that median is the middlemost item : 3 persons get a wage less than Rs. 1150 and equal number, i.e., 3, get more than Rs. 1150.

The procedure for determining the median of an even-numbered group of items is not as obvious as above. If there were, for instance, different values in a group, the median is really not determinable since both the 5th and 6th values are in the centre. In practice, the median value for group composed of an even number of items is estimated by finding the arithmetic mean of the two middle values—that is, adding the two values in the middle and dividing by two. Expressed in the form of formula, it amounts to:

$$\text{Median} = \text{Size of } \frac{N+1}{2}\text{th item}$$

Thus we find that it is both when N is odd as well as even that 1 (one) has to be added to determine median value.

Example 14:

From the following data find the value of median:

Income (Rs.)	*1000*	*1500*	*800*	*2000*	*2500*	*1800*
No. of persons	*24*	*26*	*16*	*20*	*6*	*30*

Solution:

Calculation of Median

Income arranged in ascending order	*No. of persons f*	*c.f.*	*Income arranged in ascending order*	*No. of persons f*	*c.f.*
800	16	16	1800	30	96
1000	24	40	2000	20	116
1500	26	66	2500	6	122

Median = Size of $\frac{N+1}{2}$th item = $\frac{122+1}{2}$ = 61.5th item.

Size of 61.5th item = 1500.

Calculation of Median—Continuous Series

Steps. Determine the particular class in which the value of median lies. Use N/2 as the rank of the median and not (N + 1)/2. Some writers have suggested that while calculating median in continuous series 1 should be added to total frequency if it is odd (say 99) and should not be added if it is even figure (say, 100). However, 1 is to be added in case of individual and discrete series because specific items and individual values are involved. In a continuous frequency distribution all the frequencies lose their individuality. The effort now is not to find the value of one specific item but to find a particular point on a curve—that one value which will have 50 percent of frequencies on one side of it and 50 percent of the frequencies on the other. It will be wrong to use the above rule. Hence it is N/2 which will divide the area of curve into two equal parts and as such we should use N/2 instead of (N + 1)/2, in continuous series. After ascertaining the class in which median lies, the following formula is used for determining the exact value of median.

$$\text{Median} = L + \frac{N/2 - c.f.}{f} \times i$$

L = Lower limit of the median class, i.e., the class in which the middle item of the distribution lies.

c.f. = Cumulative frequency of the class preceding the median class or sum of the frequencies of all classes lower than the median class.

f = Simple frequency of the median class.

i = The class interval of the median class.

It should be remembered that while interpolating the median value in a frequency distribution it is assumed that the variable is continuous and that there is an orderly and even distribution of items within each class.

Example 15:

Calculate the median from the following data:

Weight (in gms.)	*No. of Apples*	*Weight (in gms.)*	*No. of Apples*
410–419	*14*	*450–459*	*45*
420–429	*20*	*460–469*	*18*
430–439	*42*	*470–479*	*7*
440–449	*54*		

Solution:

Since we are given inclusive class intervals, we should convert in to the exclusive the by deducting 0.5 from the lower limits and adding 0.5 to the upper limits.

Weight	**f**	**c.f.**
409.5–419.5	14	14
419.5–429.5	20	34
429.5–439.5	42	76
439.5–449.5	54	130
449.5–459.5	45	175
459.5–469.5	18	193
469.5–479.5	7	200
	N = 200	

$$\text{Med.} = \text{Size of } \frac{N}{2}\text{th item} = \frac{200}{2} = 100\text{th item}$$

Median lies in the class 439.5 – 449.5

$$\text{Med.} = L + \frac{N/2 - c.f.}{f} \times i$$

$L = 439.5$, $N/2 = 100$,

$c.f. = 76$, $f = 54$, $i = 10$

$$\text{Med.} = 439.5 + \frac{100 - 76}{54} \times 10$$

$$= 439.5 + 4.44 = 443.94.$$

Example 16:

Compute median from the following data:

Mid-value	*Frequency*	*Mid-value*	*Frequency*
115	*6*	*165*	*60*
125	*25*	*175*	*38*
135	*48*	*185*	*22*
145	*72*	*195*	*3*
155	*116*		

Solution:

Since we are given the mid-values, we should find out the upper and lower limits the various classes.

Calculation of Median

Class group	*f*	*c.f.*	*Class group*	*f*	*c.f.*
100–120	6	6	160–170	60	327
120–130	25	31	170–180	38	365
130–140	48	79	180–190	22	387
140–150	72	151	190–200	3	390
150–160	116	267			

$$\text{Med.} = \text{Size of } \frac{N}{2}\text{th item} = \text{Size of } \frac{390}{2} = 195\text{th item}$$

Median lies in the class 150–160

$$\text{Median} = L + \frac{N/2 - c.f.}{f} \times i$$

$$L = 150,\ N/2 = 195,$$
$$c.f. = 151,\ f = 116,\ i = 10$$

$$\text{Median} = 150 + \frac{195 - 151}{116} \times 10 = 150 + 3.79 = 153.79.$$

Example 17:

From the following data calculate median:

Marks	*No. of*	*Marks*	*No. of*
Less than 5	***Students***	***Less than 5***	***Students***
	29	*30*	*644*
" 10	*224*	*" 35*	*650*
" 15	*465*	*" 40*	*653*
" 20	*582*	*" 45*	*655*
" 25	*634*		

Solution:

Since we are given cumulative frequencies; first find simple frequencies and then calculate median.

Marks	*No. of Students* *f*	*c.f.*
0–5	29	29
5–10	195	224
10–15	241	465
15–20	117	582
20–25	52	634
25–30	10	644
30–35	6	650
35–40	3	653
40–45	2	655

$$\text{Med.} = \text{Size of } \frac{N}{2}\text{th item} = \text{Size of } \frac{655}{2} = 327.5\text{th item}$$

Median lies in the class 10–15

$$\text{Median} = L + \frac{N/2 - c.f.}{f} \times i$$

$L = 10$, $N/2 = 327.5$,

c.f. $= 224$, $f = 245$, $i = 10$

$$\text{Median} = 10 + \frac{327.5 - 224}{241} \times 10 = 10 + 4.27 = 14.27.$$

Example 18:

An incomplete distribution is given below:

Variable:	*0–10*	*10–20*	*20–30*	*30–40*	*40–50*	*50–60*	*60–70*
Frequency:	*10*	*20*	*?*	*40*	*?*	*25*	*15*

(i) You are given that the median value is 35. Find out missing frequency (given the total frequency = 170)

(ii) Calculate the arithmetic mean of the completed table.

Solution:

Let the missing frequency of the class 20–30 be denoted by f_1 and that of 40–50 by f.

The total frequency = 170

The frequencies of the classes other than the missing ones are

$(10 + 20 + 40 + 25 + 15) = 110.$

$$110 + f_1 + f_2 = 170$$

Hence $f_1 + f_2 = (170 - 110) = 60$

$$\text{Med.} = L + \frac{N/2 - c.f.}{f} \times i$$

$$\text{Med.} = \text{Size of } \frac{N}{2} \text{th item} = \frac{170}{2} = 85\text{th item}$$

We are given median = 35

Hence it must lie in the class 30–40.

Thus the various values known to us are

Med. = 35, L = 30, N/2 = 85, c.f. = $(10 + 20 + f_1)$, i = 10, $f_2 = 40$

Substituting the values in the median formula

$$35 = 30 + \frac{85 - (10 + 20 + f_1)}{40} \times 10$$

$$35 = 30 + \frac{85 - 10 + 20 + f_1}{40} \times 10$$

$$35 = 30 + \frac{55 - f_1}{4}$$

$$30 + \frac{55 - f_1}{4} = 5 \quad \text{or} \quad 55 - f_1 = 20 \quad \text{or} \quad f_1 = 35$$

Since $f_1 + f_2 = 60$, f_2 shall be 60–35 = 35. Thus the missing frequencies are $f_1 = 35$, $f_2 = 25$.

Calculation of Arithmetic Mean

Variable	*f*	*m.p.*	*(m – 35)/10* *d*	*fd*
0–10	10	5	– 3	– 30
10–20	20	15	– 2	– 40
20–30	35	25	– 1	– 35
30–40	40	35	0	0
40–50	25	45	+ 1	+ 25
50–60	25	55	+ 2	+ 50
60–70	15	65	+ 3	+ 45
	N = 170			Σfd = 185

$$\overline{X} = A + \frac{\Sigma fd}{N} \times i$$

A = 35, Σfd = 15, N = 170, i = 10

$$\overline{X} = 35 + \frac{15}{170} \times 10$$
$$= 35 + .882 = 35.882$$

Hence the arithmetic mean of the completed table is 35.882.

Calculation of Median when Class Intervals are Unequal

When the class intervals are unequal, the frequencies need not be adjusted to make the class intervals equal and the same formula for interpolation can be applied as discussed above.

Example 19:

Calculate the lower and upper quartiles, third decile and 20th percent from the following data:

Variable:	*2.5*	*7.5*	*12.5*	*17.5*	*22.5*
Frequency:	*7*	*18*	*25*	*30*	*20*

Solution:

Since we are given mid-points, we will first find the lower and upper limits of the various classes. The method for finding these limits is to take the difference between the two central values, divide it by 2, deduct the values so obtained from the lower limit and add it to the upper limit, in the given cases $\frac{7.5 - 2.5}{2} = \frac{5}{2} = 2.5$. The first class shall be 0.5, second 5–10, etc.

Calculation of Q_1, Q_2, Q_3, P_{20}

Class group	*f*	*c.f.*
0–5	7	7
5–10	18	25
10–15	25	50
15–20	30	80
20–25	20	100
	N = 100	

Lower Quartile Q_1 = Size of $\frac{N}{4}$th item $= \frac{100}{4} = 25$th item

Q_1 lies in the class 5–10.

$$Q_1 = L + \frac{N/4 - c.f.}{f} \times i$$

L = 5, N/4 = 25,

c.f. = 7, f = 18, i = 5

$$Q_1 = 55 + \frac{25 - 7}{18} \times = 5 + 5 = 10$$

Upper Quartile Q_3 = Size of $\frac{3N}{4}$th item = $\frac{3 \times 100}{4}$ = 75th item

Q_3 lies in the class 15–20

$$Q_3 = L + \frac{3N/4 - c.f.}{f} \times i$$

L = 15, 3N/4 = 75,

c.f. = 50, f = 30, i = 5

$$\therefore \quad Q_3 = 15 + \frac{75 - 50}{30} \times 5 = 15 + 4.17 = 19.17$$

Third Quartile D_3 = Size of $\frac{3N}{10}$th item = $\frac{3 \times 100}{10}$ = 30th item

D_3 lies in the class 10–15

$$D_3 = L + \frac{3N/10 - c.f.}{f} \times i$$

L = 10, 3N/10 = 30, c.f. = 25, f = 25, i = 5

$$\therefore \quad D_3 = 10 + \frac{30 - 25}{25} \times 5 = 10 + 1 = 11$$

Twentieth Percentile P_{20} = Size of $\frac{20N}{100}$th item

$$= \frac{20 \times 100}{100} = 20\text{th item}$$

$$P_{20} = L + \frac{20N/100 - c.f.}{f} \times i$$

L = 5, 20N/100 = 20, c.f. = 7, f = 18, i = 5

$$\therefore \quad P_{20} = 5 + \frac{20 - 7}{18} \times 5 = 5 + 3.61 = 8.61.$$

Determination of Median, Quartiles, etc., Graphically

Median can be determined graphically by applying any of the following two methods:

(a) Draw two ogives—one by 'less than' method and the other by 'more than' method. From the point where both these curves intersect each other draw a perpendicular on the X-axis. The point where this perpendicular touches the X-axis gives the value of median.

(b) Draw only one ogive by 'less than' method. Take the variable on the X-axis and frequency on the Y-axis. Determine the median value by the formula: median = size of $\frac{N}{2}$th item. Locate this value on the Y-axis and from it draw a perpendicular on the cumulative frequency curve. From the point where it meets the ogive draw another perpendicular on the X-axis and the point where it meets the X-axis is the median.

The other partition values like quartiles, deciles, etc., can also be determined graphically by following method No. 2.

Example 20:

Calculate median from the following data:

Marks	*0–10*	*10–30*	*30–60*	*60–80*	*80–90*
No. of Students	*5*	*15*	*30*	*8*	*2*

Solution:

Since class intervals are unequal, let us first convert it to a distribution with equal class intervals on the assumption that the frequencies are equally distributed throughout a class.

Calculation of Median

Marks	***f***	***c.f.***	***Marks***	***f***	***c.f.***
0–10	5	5	60–80	8	58
10–30	15	20	80–90	2	60
30–60	30	50			

$$\text{Med.} = \text{Size of } \frac{N}{2}\text{th item} = \text{Size of } \frac{60}{2} = 30\text{th item}$$

Median lies in the class 30–60

$$\text{Median} = L + \frac{N/2 - c.f.}{f} \times i$$

$$L = 300,\ N/2 = 30,\ c.f. = 20,\ f = 30,\ i = 30$$

$$\text{Median} = 30 + \frac{30 - 20}{30} \times 30 = 30 + 10 = 40.$$

Mathematical Property of Median

The sum of the deviations of the items from median, ignoring sign, is the least. For example, the median of 4, 6, 8, 10, 12 is 8. The deviations from 8 ignoring signs are 4, 2, 0, 2, 4 and the total is 12. This total is smaller than the one obtained if deviations are taken from any other value. Thus if deviations are taken from 7, values ignoring signs would be 3, 1, 1, 3, 5, and the total 13.

Merits and Limitations of Median

Merits

(a) Perhaps the greatest advantage of median is, however, the fact that the median actually does indicate what many people incorrectly believe the arithmetic mean indicates. The median indicates the value of the middle item in the distribution. This is a clear-cut meaning and makes the median a measure that can be easily explained.

(b) The value of median can be determined graphically whereas the value of mean cannot be graphically ascertained.

(c) It is the most appropriate average in dealing with qualitative data, i.e., where ranks are given or there are other types of items that are not counted or measured but are scored.

(d) In markedly skewed distributions such as income distributions or price distributions where the arithmetic mean would be distorted by extreme values, the median is especially useful. Consequently, the median income for some purposes be regarded as a more representative figure, for half the income earners must be receiving at least the median income. One can say as many receive the median income and as many do not.

(e) It is especially useful in case of open-end classes since only the position and not the values of items must be known. The median is also recommended if the distribution has unequal classes, since it is easier to compute than the mean.

(f) Extreme values do not affect the median as strongly as they do the mean. For example, the median of 10, 20, 30, 40 and 150 would be 30 whereas the mean 50. Hence very often when extreme values are present in a set of observations, the median is a more satisfactory measure of the central tendency than the mean.

Limitations

(a) For calculating median it is necessary to arrange the data; other averages do not need any arrangement.

(b) The value of median is affected more by sampling fluctuations than the value of the arithmetic mean.

(c) Since it is a positional average, its value is not determined by each and every observation.

(d) It is not capable of algebraic treatment. For example, median cannot be used for determining the combined median of two or more groups as is possible in case of mean. Similarly, the median wage of a skewed distribution times the number of workers will not give the total payroll. Because of this limitation the median is much less popular as compared to the arithmetic mean.

(e) It is erratic if the number of items is small.

(f) The median, in some cases, cannot be computed exactly as the mean. When the number of items included in a series of data is even, the median is determined approximately as the mid-point of the two middle items.

Usefulness. The median is useful for distributions containing open-end intervals since these intervals do not enter its computation. Also since the median is affected by the number rather than the size of items, it is frequently used instead of the mean as a measure of central tendency in cases where such values are likely to distort the mean.

Related Positional Measure

Besides median, there are other Measure which divide a series into equal parts. Important amongst these are quartiles, deciles and percentiles. Quartiles are those values of the variate which divide the total frequency into four equal parts, deciles divide the total frequency into 10 equal parts and the percentiles divide the total frequency into 100 equal parts. Just as one point divides a series into two parts, three points would divide it into four parts, 9 points into 10 parts and 99 points into 100 parts. Consequently, there re only 3 quartiles, 9 deciles and 99 percentiles for a series. The quartiles are denoted by symbol Q, deciles by D and percentiles by P. The subscripts 1, 2, 3, etc., beneath Q, D, etc., would refer to the particular value that we want to compute. Thus Q_1 would denote first quartile, Q_2 second quartile, Q_3 third quartile, D_1 first decile, D_8 8th decile, P1 first percentile and P_{60} 60th percentile, etc.

Graphically any set of these partition values divides the area of the frequency curve or histogram into equal parts. If vertical lines are drawn as third quartiles, for example, the area of the histogram will be divided by these lines into four equal parts. The 9 deciles divide the area of the histogram or frequency curve into 10 equal parts and the 99 percentiles divide the area into 100 equal parts.

In economics and business statistics quartiles are more widely used than deciles and percentiles. The quartiles are the points on the X-scale that divide the distribution into four equal parts. Obviously, there are three quartiles, the second coinciding with the median. More precisely stated, the lower quartile Q_1 is that point on the X-scale such that one-fourth of the total frequency is less than Q_1 and three-fourths is greater than Q_1. The upper quartile Q_3, is that point on the X-scale such that three-fourths of the total frequency is below Q_3 and one-fourth is above it.

The deciles and percentiles are important in psychological and educational statistics concerning grades, rates, ranks, etc. they are of use in economics and business statistics in personnel work, productivity ratings and other such situations.

It should be noted that quartiles, deciles, etc., are not averages. They are Measure of dispersion and as such shall be discussed in detail in the next chapter. Here only a passing reference is made. The method of computing these partition values is the same as discussed for median.

Just as quartiles divide the series into 4 equal parts, pantiles divide into 5 equal parts, septiles into 7 equal parts an doctiles into 8 equal parts. However, these partition values are rarely used in practice.

Computation of Quartiles, Percentiles, etc.

The procedure for computing quartiles deciles, etc., is the same as the median. While computing these values in individual and discrete series we add 1 to N whereas in continuous series we do not add 1.

Thus $Q_1 = \text{Size of } \frac{N+1}{4} \text{ item}$

(individual observations and discrete series)

$Q_1 = \text{Size of } \frac{N}{4}$ (in continuous series)

$Q_3 = \text{Size of } \frac{3(N+1)}{4}$ th item (in individual and discrete series)

$Q_3 = \text{Size of } \frac{3N}{4}$ th item (in continuous series)

$D_4 = \text{Size of } \frac{4(N+1)}{10}$ th item (in individual and discrete series)

$Q_3 = \text{Size of } \frac{4N}{10}$ th item (in continuous series)

$P_{60} = \text{Size of } \frac{60(N+1)}{100}$ th item (in individual and discrete series)

$Q_3 = \text{Size of } \frac{60N}{100}$ th item (in continuous series)

Example 21:

Calculate the mode from the following data of the marks obtained by 10 students:

Sl. No.	*Marks obtained*	*Sl. No.*	*Marks obtained*
1	*10*	*6*	*27*
2	*27*	*7*	*20*
3	*24*	*8*	*18*
4	*12*	*9*	*15*
5	*27*	*10*	*30*

Solution:

Calculation of Mode

Size of item	*Number of time it occurs*	*Size of item*	*Number of time it occurs*
10	1	20	1
12	1	24	1
15	1	27	3
18	1	30	1
			Total 10

Since the item 27 occurs the maximum number of times, i.e., 3, hence the modal marks are 27.

Note. Thus the process of determining mode in case of individual observations essentially involves grouping of data.

When there are two or more values having the same maximum frequency, one cannot say which is the modal value of hence mode is said to be ill-defined. Such a series is also shown as bimodal or multimodal. For example, observe the following data:

Income (in Rs.)	110	120	130	120	110	140	130	120	130	140

Size of items	110	120	130	140
No. of times it occurs	2	3	3	2

Since 120 and 130 have the same maximum frequency, i.e., 3, mode is ill defined in the case.

Calculation of Mode—Discrete Series

In discrete series quite often mode can be determined just by inspection, i.e., by looking to that value of the variable around which the items are most heavily concentrated. For example, observe the following data:

Size of garment	28	29	30	31	32	33
No. of persons wearing:	10	20	40	65	50	15

From the above data we can clearly say that the modal size is 31 because the value 31 has occurred the maximum number of times, i.e. 65. However, where the mode is determined just by inspection, an error of maximum frequency and the frequency preceding it or succeeding it is very small and the items are heavily concentrated on either side. In such cases it is desirable to prepare a grouping table and an analysis table. These tables help us in ascertaining the modal class.

A grouping table has six columns. In column 1 the maximum frequency is marked or put in a circle; in column 2 frequencies are grouped in two's; in column 3 leave the first frequency and then group the remaining in two's; in column 4 group the frequencies in three's; in column 5 leave the first frequency and group the frequencies in three's; and in column 6 leave the first two frequencies and then group the remaining in three's. In each of these cases take the maximum total and mark it in a circle or by bold type.

After preparing the grouping table, prepare an analysis table. While preparing the table put column number on the left-hand side and the various probable values of mode on the right-hand side. The values against which frequencies are the highest are marked in the grouping table and then entered by means of a bar in the relevant 'box' corresponding to the values they represent.

The procedure of preparing grouping table and analysis table shall be clear from the following example:

Example 22:

Calculate mode from the following data:

Marks	*No. of students*	*Marks*	*No. of students*
Above 0	*80*	*Above 60*	*25*
Above 10	*77*	*Above 70*	*16*
Above 20	*72*	*Above 80*	*10*
Above 30	*65*	*Above 90*	*8*
Above 40	*55*	*Above 100*	*0*
Above 50	*43*		

Solution:

Since this is cumulative frequency distribution, we first convert it into a simple frequency distribution.

Marks	*No. of students*	*Marks*	*No. os students*
0–10	3	50–60	15
10–20	5	60–70	12
20–30	7	70–80	6
30–40	10	80–90	2
40–50	12	90–100	8

By inspection the modal class is 50–60.

$$M_o = L + \frac{\Delta_1}{\Delta_1 + \Delta_2} \times i$$

$$L = 50;\ \Delta_1 = (15 - 12) = 3,$$

$$\Delta_2 = (15 - 12) = 3;\quad i = 10$$

$$M_o = 50 + \frac{3}{3 + 30} \times 10 = 50 + 5 = 55.$$

Merits and Limitations of Mode

Merit

The main merits of mode are:

(a) It can be used to describe qualitative phenomenon. For example, if we want to compare the consumer preferences for different types of products, say, soap, toothpaste, etc., or different media of advertising we should complete the modal preferences expressed by different groups of people.

(b) The value of mode can also be determined graphically whereas the value of mean cannot be graphically ascertained.

(c) By definition mode is the most typical or representative value of a distribution. Hence, when we talk of modal wage, modal size of shoe or modal size of family it is this average that we refer to. The mode is a measure which actually does indicate what many people incorrectly believe the arithmetic mean indicates. The mode is the most frequently occurring value. If the modal wage in a factory is Rs. 916 then more workers receive Rs. 916 than any other wage. This is what many believe the "average" wage always indicates, but actually such a meaning is indicated only if the average used is the mode.

(d) Like median, the mode is not unduly affected by extreme values. Even if the high values are very high and the low values are very low we choose the most frequent value of the data to the modl value: for example, the mode of 10, 2, 5, 10, 5, 60, 5, 10, 60 is 10 as this value, i.e., 10 has occurred most often in the data set.

(e) Its value can be determined in open-end distributions without ascertaining the class limits.

Limitations

The important limitations of this average are:

(a) It is not a rigidly defined measure. There are several formulae for calculating the mode, all of which usually give somewhat different answers in fact, mode is the mot unstable average and its value is difficult to determine.

(b) While dealing with quantitative data, the disadvantages of the mode outweigh its good features and hence it is seldom used.

(c) The value of mode cannot always be determined. In some cases we may have a bimodal series.

(d) It is not capable of algebraic manipulations. For example, from the modes of two sets of data we cannot calculate the overall mode of the combined data. Similarly, the modal wage times the number of workers will not give the total payroll—except, of course, when the distribution is normal and then the mean, median and mode are all equal.

(e) The value of mode is not based on each and every item of the series.

Usefulness

The mode is employed when the most typical value of a distribution is desired. It is the most meaningful measure of central tendency in case of highly skewed or non-normal distributions, as it provides the best indication of the point of maximum concentration.

RELATIONSHIP AMONG MEAN, MEDIAN AND MODE

A distribution in which the values of mean, median and mode coincide (i.e., mean = median = mode) is known as a symmetrical distribution. Conversely stated, when the values of mean, median and mode are not equal to distribution is known as asymmetrical or skewed. In moderately skewed or asymmetrical distributions a very important relationship exists among mean, median and mode. In such distributions the distance between the mean and the median is about one-third the distance between the mean and the mode as will be clear from the diagram given below.

Karl Pearson has expressed this relationship as follows:

$$\text{Mode} = \text{Mean} - 3\ [\text{Mean} - \text{Median}]$$

$$\text{Mode} = 3\ \text{Median} - 2\ \text{Mean}$$

and
$$\text{Median} = \text{Mode} + \frac{2}{3}\ [\text{Mean} - \text{Mode}]$$

If we know any of the two values out of the three, we can compute the third from these relationships. The following example will illustrate this point:

Example 23:

Daily income of ten families of a particular place is given below.

85 70 15 75 500 8 45 250 40 36

Solution:

Calculation of Geometric Mean

X	*Log X*	*X*	*Log X*
85	1.9294	8	0.9031
70	1.8451	45	1.6532
15	1.1761	250	2.3979
75	1.8751	40	1.6021
500	2.6990	36	1.5563

$$\text{G.M.} = \text{AL}\left(\frac{\sum \log x}{N}\right) = \text{AL}\frac{17.6373}{10} = \text{AL } 1.7637 = 58.03.$$

Example 24:

From the following data of weight of 122 persons determine the modal weight.

Solution:

By inspection it is difficult the modal weight:

Weight (in ibs.)	*No. of persons*	*Weight (in ibs)*	*No. of persons*
100-110	4	140-150	33
110-120	6	150-160	17
120-130	20	160-170	8
130-140	32	170-180	2

By inspection it is difficult to say which is the modal class. Hence we prepare a grouping table and an analysis table.

Grouping Table

Weight (in Ibs.)	*No. of persons* *I*	*II*	*III*	*IV*	*V*	*VI*
100-110	4					
		10				
110-120	6			30		
			26			
120-130	20				58	
		52				
130-140	32					85
			65			
140-150	33			82		

		50			58	
150-160	17					27
			25			
160-170	8					
		10				
170-180	2					

Class in which mode is expected to lie

Col. No.	*120-130*	*130-140*	*140-150*
1			1
2	1	1	
3		1	1
4		1	1
5	1	1	1
6	1	1	1
	Total 3	5	5

This is a bi-modal series. Hence mode has to be determined by plying the formula:

Mode = 3 Median – 2 Mean

Weight in ib.	*m*	*No. of persons* *f*	*c.f.*	*(m-135)/10* *d*	*fd*
100-110	105	4	4	–3	–12
110-120	115	6	10	–2	–12
120-130	125	20	30	–1	–20
130-140	135	32	62	0	0
140-150	145	33	95	+1	+33
150-160	155	17	112	+2	–34
160-170	165	8	122	+3	–24
170-180	175	2	122	+4	+8
		N = 122			Σfd = 55

$$\overline{X} = A + \frac{\sum fd}{N} xi$$

A = 135, Σfd = 55,

N = 122, i = 10

$$\overline{X} = 135 + \frac{55}{122} \times 135 + 4.51 = 139.51$$

$$\text{Med.} = \text{size of } \frac{N}{2}\text{th item} = \text{size of } \frac{122}{2} = 61\text{st item.}$$

Hence median lies in the class 130–140.

$$\text{Median} = L + \frac{N/2 - \text{c.f.}}{2f} \times i$$

L = 130, N/2 = 61, c.f. = 30,
f = 32, i = 10

$$\text{Median} = 130 + \frac{(61 - 30)}{32} \times 10 = 130 + \frac{310}{32} = 139.69$$

Mode = 3 Median = 2 Mean

Mode = (3 × 139.69) – (2 × 139.51) = 419.07 – 279.02 = 140.05.

Hence model weigh is

Mode when Class Intervals are Unequal

The formula for calculating the value of mode given above is applicable only where there are equal class intervals. If the class intervals are unequal, then we must make them equal before we start computing the value of mode. The class interval should be made equal and frequencies adjusted on the assumption that they are equally distributed thought the class.

Locating Mode Graphically

In a frequency distribution the value of mode can also be determined graphically. The steps in calculation are:

(a) Draw a histrogram of the given data.

(b) Draw two lines diagonally in the inside of the modal class bar, starting from each upper corner of the bar to the upper corner of the adjacent bar.

(c) Draw a perpendicular line from the intersection of the two diagonal lines to the X-axis (horizontal scale) which gives us the modal value.

Example 25:

Calculate geometric mean from the following data:

125 1462 38 7 0.22 0.08 12.75 0.5

Solution:

Calculation of G.M.

X	*Log X*
125	2,0969
1462	3,1650
38	2,5798
7	0.8451
0.22	.3424
0.08	.9031
12.75	1.1055
0.5	.6990
	Σ log X = 6.7360

$$\text{G.M.} = \text{AL}\left(\frac{\sum \log X}{N}\right) = \text{AL}\left(\frac{6.7360}{8}\right)$$

$$= \text{Al } (0.8421 = 6.952).$$

Calculation of Geometric Mean – Discrete Series

$$\text{G.M.} = \text{Antilog}\left(\frac{\sum f \log X}{N}\right)$$

Steps : Find the logarithms of the variable X.

(i) Multiply these logarithms with the respective frequencies and obtain the total Σf log X.

(ii) Divide Σf log X by the total frequency and take the antilog of the value so obtained.

Calculation of Geometric Mean–Continuous Series

$$\text{G.M.} = \text{Antilog}\left(\frac{\sum f \log X}{N}\right)$$

Steps :

(i) Find out the mid-points of teh classes and take their logarithms.

(ii) Multiply these logarithms with the respective frequencies of each class and obtain the total Σf log m.

(iii) Divide the total obtained in step (ii) by the total frequency and take the antilog of the value so obtained.

Example 26:

(i) In a moderately asymmetrical distribution, the mode and mean are 32.1 and 35.4 respectively. Find out the value of Median.

Solution:

(i) Mode = 3 Median – 2 Mean

Given mean = 35.4, mode = 32.1

32.1 = 3 median – 2 × 35.4

3 median = 32.1 + 70.8 = 102.9 or median = 102.9/3 = 34.3

Given median = 20.6, mode = 26, find mean.

(ii) Mode = 3 Median – 2 Mean

Given mean = 20.6, mode = 26

26 = 3 × 20.6 – 2 mean

26 = 61.8 – 2 mean.

2 mean = 61.8 – 26 = 35.8 or mean = 17.9

GEOMETRIC MEAN

Geometric mean is defined as the Nth root of the product of N items or values. If there are two items, we take the square root: if there are three items, the cube root; and so on. Symbolically.

$$G.M. = \sqrt{(X_1) \times (X_2) \times (X_3) \times \ldots\ldots.. X_n}$$

here X_1, X_2, X_3, etc. refer to the various items of the series.

Thus the geometric mean of 3 values 2, 3 4, would be:

$$G.M. = \sqrt[3]{2 \times 3 \times 4} = \sqrt{24} = 2.885$$

When the number of items is three or more the task of multiplying the numbers and of extracting the root becomes excessively difficult. To amplify calculations logarithms are used. Geometric mean then is calculated as follows:

$$\log G.M. = \frac{\log X_1 + \log X_2 + \ldots + \log X_n}{N}$$

$$\text{or } \log G.M. = \left(\frac{\sum \log X}{N}\right) \quad \therefore G.M. = \text{Anti}\log\left(\frac{\sum \log X}{N}\right)$$

$$\text{In discrete series G.M.} = \text{Antilog}\left(\frac{\sum f \log X}{N}\right)$$

In continuous series G.M. = Antilog $\left(\frac{\sum f \log m}{N}\right)$

Properties of Geometric Mean

The following are two important mathematical properties of geometric mean:

(a) The product of the value of series will remain unchanged when the value of geometric mean is substituted for each individual value. For example, the geometric mean for series 2, 4, 8 is 4 : therefore, we have

$$2 \times 4/ \times 8 = 64 = 4 \times 4 \times 4$$

(b) The sum of the deviations of the logarithms of the original obwervations above or below the logarithm of the geometric mean is equal. this also means that the value of the geometric mean is such as to balance the ratio deviations of the observations from it. Thus, using the same pervious numbers, we find that

$$\left(\frac{4}{2}\right)\left(\frac{4}{4}\right) = 2 = \left(\frac{8}{4}\right)$$

Because of this property this Measures of Central Tendency is especially adapted to average ratios, rates of change, and logarithmically distributed series.

Calculations of Geometric Mean—Individual Observations

$$\text{G.M.} = \text{Antilog}\left(\frac{\sum \log X}{N}\right)$$

Steps:

(a) Take the logarithms of the variable X and obtain the total Σ log X.

(b) Divides S log X by n and take the antilog of the value so obtained. This gives the value of geometric mean.

Σ log X

If $f_1, f_2, f_3, \ldots f_n$—represents frequencies of

$X_1, X_2, X_3, \ldots X_n$ respectively then

$$\text{G.M.} = \sqrt[3]{\frac{(X_1 . X_1 . X_1 \ldots . f_2 \text{ times}) (X_2 . X_2 . X_3 \ldots . f_2 \text{ times})}{(X_2 . X_3 . X_4 \ldots . f_2 \text{ times}) (X_n . X_n . X_n \ldots . \text{ times})}}$$

$$X_1^{f_1} . X_2^{f_2} . X_n^{f_n}$$

Taking logarithms of both the sides to simplify calculation

$$\log G.M. = \frac{\log (X_1 f_1 \times X_2 f_2 \times X_3 \ldots. X_n f_n}{N} \qquad [\because N = \Sigma f]$$

$$= \frac{\log X_1^{f2} \log X_2^{f^2} \log X_3^{f^3} \ldots. \log X_{f_n}}{N}$$

$$= \frac{f_1 \log X_1 + f_2 \log X_2 + f_2 \log X_2 \ldots. f^n \log X_n}{N}$$

$$= \frac{\sum f \log X}{N}$$

$$G.M. = \text{Antilog}\left(\frac{\sum \log X}{N}\right)$$

Example 27:

Find the Geometric mean for the data given below:

Marks	*Frequency*	*Marks*	*Frequency*
4-8	*6*	*24-28*	*12*
8-12	*10*	*28-32*	*10*
12-16	*18*	*32-36*	*6*
16-20	*30*	*36-40*	*2*
20-24	*15*		

Solution:

Calculation of Geometric Mean

Marks	*m.p.m*	*t*	*log m*	*l × log m*
4-8	6	6	0.7782	4.6692
8-12	10	10	1.0000	10.0000
12-16	14	18	1.1461	20.6298
16-20	18	30	1.2553	37.6590
20-24	22	15	1.3424	20.1360
24-28	26	12	1.4150	16.9800
28-32	30	10	1.4771	14.7710
32-36	34	6	1.5315	9.1890
36-40	38	2	1.5798	3.1596
		N = 109	$\Sigma f \times \log m$ = 137.1936	

$$G.M. = AL\left(\frac{\sum f \log X}{N}\right) = AL\left(\frac{137.1936}{109}\right) = A.L.\ 1.2587 = 18.14$$

Uses of Geometric Mean

Geometric mean is specially useful in teh following cases:

(a) The geometric mean is used to find the average per cent increase in sales, production, population or other economic or business series. For example, from 1996 to 1998 prices increased by 5, 10 and 18 per cent respectively. The average annual increase is not 11 per cent $\left(\frac{5+10+18}{3} = 11\right)$ as given by the arithmetic average but 10.9 per cent as obtained by the geometric mean.

(b) Geometric mean is theoretically considered to be the best average in the construction of index numbers. It satisfies the time reversal test and gives equal weight to equal ratio of change.

(c) This average is most suitable when large weights have to be given to small items and small weights to large items situations which we usually come across in social and economic fields.

The following examples illustrate the use of geometric mean:

Example 28:

Find the average rate of increase in population which in the first decade has increased by 20%, in the second decade by 30% and in the third decade by 40%.

Solution:

Calculation of Geometric Mean

Decade	*% Rise*	*X Population at the end of the decade taking population of the previous decade as 100*	*log X*
1st	*20*	*120*	*2.0792*
2nd	*30*	*130*	*2.1139*
3rd	*40*	*140*	*2.1461*
			$\Sigma \log X = 6.3292$

$$G.M = A.L. \left(\frac{\sum \log X}{N}\right) = AL\left(\frac{6.3392}{3}\right) = A.L\ (2.1131) = 129.7$$

Thus the average rate of increase in population is (129.7 – 100) 29.7 per cent predicate:

Example 34:

The population of a country has increased from 84 million in 1988 to 108 million in 1998. Find the annual rate of growth of population.

Solution:

Let r be the rate of growth. Applying the compound interest formula:

$$P_n = P_0(1 + r)^n$$

$$84(1 + r)^{10} = 108$$

Taking

logarithms, $\log (1 + r) = \dfrac{\log 108 - \log 84}{10} = \dfrac{2.0334 - 1.9243}{10} = 0.0109$

$$1 + r = 1.026;\ r = 0.026 = 2.6\%$$

Example 29:

The price of a commodity increased by 5% from 1995 to 1996, 8% from 1996 to 1997 and 77% from 1997 to 1998. The average increase from 1996 to 1998 is quoted as 26% and not 30%. Explain and verify the result.

Solution:

The appropriate average here is the geometric mean and not the arithmetic mean. The arithmetic mean of 5, 8, 77 is 30 but this is not the correct answer. Correct answer shall be obtained if we calculate geometic mean.

% Rise	*X* *Price at the end of the year taking preceding year as 100*	*log X*
5	105	2.0212
8	108	2.0334
77	177	2.2480
		$\Sigma \log X = 6.3026$

$$\text{G.M.} = \text{A.L.} \left(\frac{\sum \log X}{N}\right) = \text{AL}\left(\frac{6.3026}{3}\right) = \text{A.L. } (2.1009) = 126.2$$

The average increase from 1996 to 1998 = 126.2 – 100 = 26.2% or approx. 26%. Verification. When the average rise is 30%.

Year		*Rate of change*	*Total change*	*Price at the end of each year*
I	year	30% on 100	30	130.0
II	year	30% on 130	39	169.0
III	year	30% on 169	50.7	219.7

when the average rise is 26%

I	year	26% on 100.00	26.00	126.00
II	year	26% on 126.00	32.76	158.76
III	year	26% on 158.76	41.28	200.04

When the rise is of 5, 8 and 77% the changed price at the end of each year:

I	year	5% on 100.00	5.00	105.00
II	year	8% on 126.00	8.40	113.40
III	year	77% on 158.76	87.318	200.00

The above calculations make it clear that in the second and third cases the price at the end of the third year is almost the same, the slight difference being due to approximation of 26.2 to 26. Hence, the average increase is 26%.

Example 30:

The geometric mean of 10 observations was calculated as 28.6. It was later discovered that one of the observations was recorded as 23.4 instead of 32.4. Apply appropriate correction and calculate the correct geometric mean.

Solution:

Geometric mean of n observations is given by:

$$\text{G.M.} = (X_1, X_2, X_3,..., X_{n)})^{1/n}$$

$$\text{or G.M.}^n = X_1, X_2, X_3,...X_n$$

Thus the product of the number is given by G.M^n or $(28.6)^{10}$ since in the given case n = 10 and G.M. = 28.6.

If the wrong observation 23.4 is replaced by the correct value 32.4, the correct value of the product of 10 numbers is obtained by dividing $(28.6)^{10}$ by wrong observation and multiplying by the correct observation. Hence

$$\text{Corrected product } (X_1, X_2,..., X_n) = \frac{(26.6)^{10} \times 32.4}{23.4}$$

Correct value of geometric mean is given by:

$$\text{G.M.c} = \left[\frac{(28.6)^{10} \times 32.4}{23.4}\right]^{1.10}$$

$$\log \text{G.M.}_c = \frac{1}{10}[10 \log 28.6 + 32.4 - \log 23.4]$$

$$= \frac{1}{10}[10 \times 1.4564 + 1.5105 - 1.3692]$$

$$= \frac{1}{10}[14.564 + 1.5105 - 1.3692]$$
$$= 1.470\text{-}53$$
$$G.M._c = A.L.\ 1.47053 = 29.54.$$

Weighted Geometric Mean

Like weighted arithmetic mean, we can also calculate weighted geometric mean with the help of the following formula:

$$G.M.\ 10 = A.L. \left[\frac{(\log X_1 \times W_1) + (\log X_2 \times W_2) + ... + (\log X_n \times W_n)}{W_1 + W_2 + W_3 + ... + W_n}\right]$$

$$= A.L. \left[\frac{\sum (\log X \times W)}{\sum W}\right]$$

Symbolically, $G.M._w = \sqrt{X_1^{w1} \times X_2^{w2} \times X_3^{w3} \times ... \times X_2^{wn}}$

If $W_1, W_2, W_3, ..., W_n$ are weights assigned to different values of $X_1, X_2, X_3, ..., X_n$

$$G.M._w = \sqrt{X_1^{w1}, X_2^{w2}; X_3^{w3} ... X_n^{wn}}$$

Taking logaritms of both sides

$$\log G.M._w = \frac{\log\left(X_1^{w1}, X_2^{w2}, X_3^{w3} ... X_n^{wn}\right)}{\sum W}$$

where $N = \sum W = W_1, + W_2, + W_3 + ..., + X_n$

$$= \frac{\log\left(X_1^{w1} + \log X_2^{w2} + \log X_3^{w3} + ... \log X_n^{wn}\right)}{\sum W} = \frac{\sum W \log X}{\sum W}$$

$$G.M._w = \text{Antilog}\ \frac{\sum W \log X}{\sum W}$$

Since it is difficult to find nth root, the Geometric Mean can be calculated with the help of Logarithms.

Example 31:

The annual rates of growth of an economy over the last five years were 1.5, 2.7, 3.0, 4.5 and 6.2 percent respectively. What is the compound rate of growth perannum of the economy for the period?

Solution:

Apply the geometric mean.

Annual rate of growth	*Growth relatives at the end of the year X*	*log*
1.5	101.5	20.0064
2.7	102.7	2.0116
3.0	103.0	2.0128
4.5	104.5	2.0191
6.2	106.2	2.0261
		Σ log X = 10.076

$$\text{G.M.} = \text{AL.}\left(\frac{\sum \log X}{N}\right) = \text{AL}\left(\frac{10.076}{5}\right) = \text{A.L. } (2.0152) = 103.5$$

The compound rate of growth per annum = 103 – 100 = 3.5

Note. The same result shall be obtained by applying the compound interest formula which is discussed below.

COMPOUND INTEREST FORMULA

Geometric mean is most frequently used in the determination of average per cent of change. For example, if a city had a population of 2,00,000 in a given year and 2.40.000 ten years later, we may be interested in finding out the annual per cent of change. The increase is (2,40,000 – 2,00,000), i.e., 40,000 over a period of 10 years and so one may say that the annual per cent increase is 2. However, if we compute 2 per cent increase each year over the preceding year the population figure turns out to be 2,43, 800. This means that the correct figure is little less than 2 per cent because we are actually compounding. The average annual per cent increase may be computed by applying the formula:

$$P_n = P_0(1 + r)_n$$

where P_0 = The value at the beginning of the period;

P_n = The value at the end of the period n;

r = Rate of change;

n = length of time period.

It follows from the above formula that $r = n\sqrt{\frac{P_n}{P_0}} - 1;$

For the above data 2,40,000 = 2,00,000 $(1 + r)^{10}$

Taking logarithms 5.3802 = 5.301 + 10 log (1 + r)

log (1 + r) = 0.00792

(1 + r) = Antilog 0,0079

1 + r = 0.0184 or r = 0.0184 = 1.84 per cent.

The expression $P_n = P_0 (1 + r)^n$ is called compound interest formula because it is extremely useful in problems involving compound interest. In the above case we have used it to determine average annual per cent of growth. However, if we know any three values of the four used in the formula we can find out the fourth one. Thus, we may determine:

(a) Average annual per cent of change r.

(b) Population the given number of years later Pn assuming the constant relative change.

(c) Number of years, n, after which the given population will be attained, again assuming a constant relative change.

(d) Population the given number of years earlier P_0, if the per cent of change was constant.

It may by pointed out that the assumption of a constant relative change for population is not valid over extended period for any country except possibly "new" countries.

Example 32:

If arithmetic mean and geometric mean of two values are 10 and 8 respectively, find values.

Solution:

It two values are a and b then $\frac{a + b}{2} = 10 \sqrt{ab} = 8$ or ab = 64

a + b = 20, ab = 64

$(a - b) = \sqrt{(a + b)^2 - 4ab} = \pm 12$ a + b = 20, a − b = ± 12

Hence a = 16 and b = 4. The values are 16,4.

Example 33:

The weighted geometric mean of the four numbers 8, 25, 17 and 30 is 15.3. If the weights of the first three numbers are 5, 3 and 4 respectively, find the weight of the fourth number.

Solution:

Let the weight of the fourth number be W_1.

Calculation of Geometric Mean

X	*W*	*Log X*	*W. Log X*
8	5	0.9031	4.5155
25	3	1.3979	4.1937
17	4	1.2304	4.9216
30	W1	1.4771	$1.4771 + W_1$
	$\Sigma W = 12 + W_1$		ΣW. Lot X = 13.6308 $1.4771 + W_1$

$$\text{Log G.M.}_w \left[\frac{\sum W \log X}{\sum W}\right] \log 15.3 = \frac{13.63.08 + 1.4771\ W_1}{12 + W_1}$$

$$1.1847\ (12 + W_1) = 13.6308 + 1.4771\ W_1$$

$$14.2164 + 1.1847\ W_1 = 13.6308 + 1.4771\ W_1$$

$$1.1847\ W_1 - 1.471\ W_1 = 13.6308 - 14.2164 = -0.5856$$

$$W_1 = \frac{0.5856}{0.2924} = 2.003 \text{ or } 2 \text{ app.}$$

Thus the weight of the fourth number is 2.

Merits and Limitations of Geometric Mean

Merits It is based on each and every item of the series.

(a) It gives less weight to large items and more to small ones than does the arithmetic average. It is because of this reason that geometric mean is never larger than the arithmetic mean. On occasions it may turn out to be same as the arithmetic mean, but usually it is smaller.

(b) It is useful in average ration and percentages and in determining rates of increase and decrease.

(c) It is rigidly defined.

(d) It is capable of alebratic manipulation. For example, if the geometric average of two or more series and their number of items is known, a combined G.M. can be easily calculated by applying the formula.

$$\text{G.M}_{12} = \text{antilog}\left[\frac{N_1 \log \text{G.M.}_1 + N_2 \log \text{G.M.}_2}{N_1 + N_2}\right]$$

For example if there are 2 sets of two figures each and their geometric neans are 8 and 12 respectively, we can calculate the combined geometric nean as follows:

$$\text{G.M.}_{12} = \text{A.L.}\left[\frac{(2 \times \log 8)\ (2 \times \log 12)}{2 + 2}\right]$$

$$= \text{A.L.}\left[\frac{(2 \times 9031)\ (2 \times 1.0792)}{4}\right]$$

$$= \text{A.L.}\left[\frac{1.8062 + 2.1584}{4}\right] = \text{A.L. } 0.99115 = 9.8$$

Limitations It is difficult to understand.

(a) It cannot be computed when there are both negative and positive values in a series or one or more of the values are zero.

(b) It is difficult to compute and to interpret and so has restricted application.

HARMONIC MEAN

The harmonic mean is based on the reciprocals of numbers averaged. It is defined as the reciprocal of the arithmetic mean of the reciprocal of the individual observations. Thus, by definition

$$\text{H.M.} = \frac{N}{\left(\frac{1}{X_1} + \frac{1}{X_2} + \frac{1}{X_3} + \ldots + \frac{1}{X_n}\right)}$$

When the number of items is large, the computation of harmonic mean in the above manner becomes tedious. To simplify calculations we obtain reciprocal of the various items from the table and apply the following formulae:

In individual observations. $\text{H.M.} = \frac{N}{\sum(1/X)}$

In discrete series, $\text{H.M.} = \frac{N}{\sum\left(f \times \frac{1}{X}\right)}$

In continuous series $\text{H.M.} = \frac{N}{\sum\left(f \times \frac{1}{X}\right)} = \frac{N}{\sum(f/m)}$

Calculation of Harmonic Mean–Individual Observations

In individual series harmonic mean is computed by applying the following formula:

$$\text{H.M.} = \frac{N}{\left(\frac{1}{X_1} + \frac{1}{X_2} + \frac{1}{X_3} + \ldots + \frac{1}{X_n}\right)}$$

X_1, X_2, X_3, etc, refer to the various items of the variable.

Example 34:

An automobile driver travels from plain to hill station 1000 km, distance at an average speed of 30 km. per hour. He then makes the return trip at average speed of 20 km. per hour. What is his average speed over the entire distance (200 km)?

Solution:

If the problem is given to a layman he is most likely to compute the arithmetic mean of two speeds i.e.,

$$\overline{X} = \frac{30 \text{ km} + 20 \text{ km.}}{2} = \text{km. ph.}$$

But this is not the correct average. Harmonic mean would be more suitable in this situation. Harmonic mean of 30 and 20 is

$$\text{H.M.} = \frac{2}{\frac{1}{20} + \frac{1}{30}} = \frac{2}{\frac{10}{120}} = \frac{2 \times 120}{20} = 24 \text{ km.p.h.}$$

It can be proved that harmonic mean is the appropriate average in this case by tabulating the time and distance for each trip separately as follows:

	Distance (km)	*Average speed km. p.h.*	*Time taken*
Ging	100	30	3 hours 20 minutes
Returning	100	20	5 hours
Total	200		8 hours 20 minutes

Thus the total time required for covering a distance of 200 km. is 8 hours 20 minutes which gives an average speed of 24 km. p.h. and not 25 km. p.h.

The above problem can be changed in such a manner that arithmetic mean is the appropriate average. Suppose the driver makes the same trip but it is given that he travels at 30 km. per hour for half of the time and at 20 km. per hour for other half of the time Now the correct answer about the average speed would be given by the arithmetic mear i.e. average speed = (30 + 20)/2 = 25 km. per cent To verify the result we again prepare a table of time and distance at each speed:

Speed km. p.h	*Distance*	*Time required*
30	120 km.	120/30 = 4 hours
20	80 Km.	80/20 = 4 hours
Total	200 Km.	8 hours

Thus he has covered 200 km. in 8 hours. Hence the average speed is 25 km. per hour.

The above example clearly shows that when distances are the same for the two speeds harmonic mean gives the correct answer but when times are same, the arithmetic mean of the rates of speed gives the correct answer.

Example 35:

From the following data compute the value of harmonic mean:

Class interval	*10-20*	*20-30*	*30-40*	*40-50*	*50-60*
Frequency	*4*	*6*	*10*	*7*	*3*

Solution:

Calculation of Harmonic Mean

Class interval	*Mid-points* *m*	*Frequency* *f*	*f/m*
10-20	15	4	0.267
20-30	25	6	0.240
30-40	35	10	0.286
40-50	45	7	0.156
50-60	55	3	0.055
		N = 30	Σ(f/m) = 1.004

$$\text{H.M.} = \frac{N}{\Sigma(1/X)} = \frac{30}{1.004} = 29.88.$$

Uses of Harmonic Mean

The harmonic mean is restricted in its field of usefulness. It is useful for computing the average rate of increase in profits of a concern or average speed at which a journey has been performed or the average price at which an article has been sold. The rate usually indicates the relation between two different types of measuring units that can be expressed reciprocally. For example if a man walked 20 km. in 5 hours the rate of his walking speed be expressed as

$$\frac{20 \text{ km.}}{5 \text{ hours}} = 4 \text{ km. per hour}$$

where the unit of the first term is a km. and the unit of the second term is an hour. Or reciprocally,

$$\frac{20 \text{ km.}}{5 \text{ hours}} = \frac{1}{4} \text{ hours per km.}$$

where the unit of the first term is an hour and the unit of the second is term is km.

Example 36:

From the following data compute the value of harmonic mean:

Marks	*10*	*20*	*25*	*40*	*50*
No. of students	*20*	*30*	*50*	*15*	*5*

Solution:

Calculation of Harmonic Mean

Marks X	*f*	*(f/X)*
10	20	2.000
20	30	1.500
25	50	2.000
40	15	0.375
50	5	0.100
	N = 120	Σ(f/X) = 5.975

$$\text{H.M.} = \frac{N}{\sum(1/X)} = \frac{120}{5.975} = 20.08.$$

Calculation of Harmonic Mean–Continuous Series

For calculating harmonic mean in continuous series the procedure is the same as applied to discrete series. The only difference is that here we take the reciprocal of the mid-points.

Example 37:

An aeroplane covers the four sides of a square at speeds of 1,000 2,000 3,000 and 4,000 km. per hour respectively. What is the average speed of the plane in its flight around the square?

Solution:

If we compute the arithmetic mean we get the following answer;

$$\overline{X} = \frac{1{,}000 + 2{,}000 + 3{,}000 + 4{,}000}{4} = \frac{10{,}000}{4} = 2{,}500 \text{ km. per hour.}$$

However that is not the correct answer. In such a problem harmonic mean is an appropriate average

$$\text{H.M.} = \frac{4}{\frac{1}{1{,}000} + \frac{1}{2{,}000}\ \frac{1}{3{,}000} + \frac{1}{4{,}000}}$$

$$= \frac{4}{\frac{12 + 6 + 4 + 3}{12{,}000}} = \frac{4}{\frac{25}{12{,}000}} = \frac{4 \times 12{,}000}{25} = 1{,}920 \text{ km. per hour.}$$

Verification. Suppose one side of square is 1000 km. Each side i.e. 1,000 km. it covers at average speeds of 1,000 2,000 3,000 4,000 kms respectively. From this we can calculate the time taken in covering the entire distance.

Distance	*Speed (km. p.h)*	*Time taken*
1,000	1,000	60 minutes
1,000	2,000	30 minutes
1,000	3,000	20 minutes
1,000	4,000	15 minutes
Total		125 minutes

In 125 minutes it covers 4,000 km.

In 60 minutes is would cover $\frac{4,000}{125} = 60 = 1,920$ km.

Thus the average speed over the entire distance is 1,920 and not 2,500 km. per hour.

Weighted Harmonic Mean

At times it may be necessary to calculate the weighted harmonic mean. For example if we are given not only the speed of the aeroplane but the distances travelled also simple harmonic mean cannot be used. Weighted harmonic mean is calculated with the help of the following formula:

$$H.M._{\omega} = \frac{\sum \omega}{\left(\frac{1}{a} \times \omega_1\right) + \left(\frac{1}{b} \times \omega_2\right) + \left(\frac{1}{c} \times \omega_3\right)} \text{ or } \frac{\sum \omega}{\sum (\omega / x)}.$$

Example 38:

Find the weighted geometric mean from the following data:

Group	*Index Number*	*Weights*
Food	*260*	*46*
Fuel & Lighting	*180*	*10*
Clothing	*220*	*8*
House Rent	*230*	*20*
Education	*120*	*12*
Misc.	*200*	*4*

Solution:

Calculation of Weighted Geometric Mean

Group	*Index No. X*	*Weights W*	*Log X*	*W log X*
Food	260	46	2.4150	111.0900
Fuel & Lighting	180	10	2.2553	22.5530
Clothing	220	8	2.3424	18.7392
House Rent	230	20	2.3617	47.2340
Education	120	12	2.0792	24.9504
Misc.	200	4	2.3010	9.2040
		$\Sigma W = 100$		$\Sigma W \text{ Log } X = 233.7706$

$$G.W._{w} = A.L.\left[\frac{\Sigma W \log X}{\Sigma W}\right] = A.L.\left[\frac{233.7706}{100}\right]$$

$$= A.L\ 2.3377 = 217.6$$

Example 39:

(a) Find the harmonic mean from the following:

2574 475 75 5 0.8 0.08 0.005 0.0009.

Solution:

Calculation of Harmonic Mean

X	*(1/X)*	*X*	*(1/X)*
2574	0.0004	0.8	1.2500
475	0.0021	0.08	12.5000
75	0.0133	0.005	200.0000
5	0.2000	0.0009	1111.1111
			$\Sigma(1/X) = 1325.0769$

$$H.M. = \frac{N}{\Sigma(1/X)} = \frac{8}{1325.0769} = 0.006.$$

Example 40:

(b) Calculate the harmonic mean from the following data:

3834 382 63 8 0.4 0.03 0.009 0.005

Solution:

Calculation of Harmonic Mean

X	(f/X)	X	(f/X)
3834	0.0003	0.4	2.5000
382	0.0027	0.03	33.3333
63	0.0159	0.009	111.1111
8	0.1250	0.0005	2000.0000
	N = 120		Σ(1/X) = 5.975

$$\text{H.M.} = \frac{N}{\Sigma(1/X)} = \frac{8}{2147.0883} = 0.003726.$$

Calculation of Harmonic Mean-Discrete Series

In discrete series, harmonic mean is computed by applying the following formula:

$$\text{H.M.} = \frac{N}{\Sigma\left(f \times \frac{1}{X}\right)} = \frac{N}{\Sigma(f/m)}$$

Steps:

(i) Take the reciprocal of the various items of the variable X.

(ii) Multiply the reciprocal by frequents and obtain the total $\Sigma\left(f \times \frac{1}{X}\right)$.

(iii) substitute the values of N and $\Sigma\left(f \times \frac{1}{X}\right)$. in the above formula.

Note. Instead of first finding out the reciprocals and then multiplying them by frequencies it will be far more easider to divide each frequency by the respective value of the variable.

Example 41:

The following figures show the number of passengers carried on each of 50 journeys by an aircraft with a seating capacity of 100. Calculate:

(i) the average capacity used and (ii) it 65 passengers is the smallest profitable load the proportion of lights which were unprofitable.

10	18	61	63	72	71	82	25	45	66	68	95	92
31	41	56	33	78	65	72	74	89	67	68	32	46
49	35	37	43	55	57	69	72	84	69	92	75	83
68	75	42	45	39	11	37	38	62	89	72		

Solution:

(i) Let us obtain the total of all the items given and find out the arithmetic mean

$$\overline{X} = \frac{\Sigma X}{N} = \frac{2918}{50} = 58.36$$

Thus the average capacity used is 58.36 passengers.

(ii) The number of flights that carried passengers below 65 is 23. Hence the proportion of flights which were unprolitable would be 23/50 × 100 = 46 per cent.

Example 42:

A cycilist covers his first five km. at an average speed of 10 km. p.h. another km. at 8km. p.h. and the last two km. at 5 km. p.h. Find average spped of the entire journey and verify your answer.

Solution:

Weighted harmonic mean would be appropriate here

$$H.M._w = \frac{\Sigma w}{\left(\frac{1}{a} \times w_1\right) + \left(\frac{1}{b} \times w_2\right) + \left(\frac{1}{c} \times w_3\right)}$$

The various speeds are 10 km. ph. 8 km. p.h. and 5 km. henceab and c respectively are 108 and 5. The distances covered are respectively 53 and 2 km. Hence w_1, w_2 and w_3 3 are 53 and 2

$$H.M._w = \frac{10}{\left(\frac{1}{10} \times 5\right) + \left(\frac{1}{8} \times 3\right) + \left(\frac{1}{2} \times 2\right)}$$

$$= \frac{10}{\frac{1}{2} + \frac{3}{8} + \frac{2}{5}} = \frac{10}{\frac{51}{40}} \; \frac{10 \times 40}{51} = 7.84. \text{ km. p.h.}$$

Thus the average speed for the entire journey is 7.84 km. p.h.

Verification:

Speed (km. p.h.)	***Distance (km.)***	***Time taken (in minutes)***
10	5	30.0
8	3	22.5
5	2	24.0
Total	10	76.5

In 76.5 minutes he covers 10 km.

In 1 minute the would cover $\frac{10}{76.5} \times 60 = 7.84.$

In 60 minutes he would cover $\frac{10}{76.5} \times 60 = 7.84$ km.

Merits and Limitations of Harmonic Mean

Merits. Its value is based on every item of the series.

(a) In problems relating to time and rates it gives better results than other averages.

(b) It tends itself to algebraic manipulation.

Limitations. It is not easily understood.

(a) It gives largest weight to smallest items. This is generally not a destrable feature and as such this average is not very useful for the analysis of economic data.

(b) It is difficult to compute.

(c) Its value cannot be computed when there are both positive and negative items in a series or when one or more items are zero.

Because of these limitations the harmonic mean has little practical application and is not a good representation of a statistical seriousness the phenomenon is such where small items need to be given a very high weightage.

RELATIONSHIP AMONG THE AVERAGES

In any distribution when the origina' ·items differ in size the value of A.M. G.M. and H.M. would also differ and will be in the following orr:

$$\text{A.M.} \geq \text{G.M.} \geq \text{H.M.}$$

i.e. arithmetic mean is greater than geometric mean and geometric mean is greater than harmonic mean. The equality signs hold only if all the numbers $X_1, X_2, \ldots X_n$ are identical.

Proof:

Prove that if a and b are two positive numbers their A.M. ≥ G.M. ≥ H.M.

Solution:

Let a and b be two positive quantities such that a 1 b.

The A.M and H.M. of these two quantities are

$$\overline{X} = \frac{a+b}{2}; \text{ G.M.} = \sqrt{a \times b}; \text{ H.M.} = \frac{2}{\frac{1}{a}+\frac{1}{b}} = \frac{2ab}{a+b}$$

We have to prove that A.M. > G.M > H.M. Let us first prove that A.M. > G.M. or $\frac{a+b}{2} > \sqrt{a+b}$.

$$\frac{a+b}{2} > \sqrt{a+b}; \; a+b > \sqrt{ab}$$

$a + b - 2\sqrt{ab} > 0$ [Since $a + b - 2\sqrt{ab} = (\sqrt{a} - \sqrt{b})^2$]

$(\sqrt{a} - \sqrt{b})^2 > 0.$

But the square of any real quantity is positive.

Hence $(\sqrt{a} - \sqrt{b})^2$ will be positive. Hence $\frac{a + b}{2} > \sqrt{ab}$

Let us now prove that G.M. > H.M.

$$\Rightarrow \qquad \sqrt{ab} > \frac{2ab}{a + b} \; 1 > \frac{\sqrt{ab}}{a + b} \text{ or } a + b > 2\sqrt{ab}.$$

This has slready been proved above. Hence G.M. > H.M.

Since we have shown that A.M. > G.M. and G.M. > H.M. it is automatically proved that A.M. > G.M. > H.M.

If a and b are equal in that case A.M. = G.M. = H.M. Thus,

A.M. > G.M. > H.M.

Example 43:

The following are the weekly wages in rupees of 30 workers of a firm:

140	*139*	*126*	*114*	*100*	*88*	*62*	*77*	*99*	*103*	*108*
129	*144*	*148*	*134*	*63*	*69*	*148*	*132*	*118*	*142*	*116*
123	*104*	*95*	*80*	*85*	*106*	*123*	*133*			

The firm gave bonus of Rs. 10 15 20 25 30 and 35 for individuals in the respective salary: exceeding 60 but not exceeding 75 exceedings but not exceeding 90 and so on up to exceeding 135 and not exceeding 150.

Find the average bonus paid.

Solution:

Let us first prepare a frequency distribution of the given data and then calculate the average bonus paid.

Weekly wages (Rs.)	***Tally Bars***	***Frequency f***	***Bonus Paid X***	***fx***
61-75	|||	3	10	30
76-90	||||	4	15	60
91-105	||||	5	20	100
106-120	||||	5	25	125
121-135	|||| ||	7	30	210
136-150	|||| |	6	35	210
		n = 30		$\Sigma fX = 735$

$$\text{Average bonus paid} = \frac{\Sigma X}{N} = \frac{735}{30} \text{ Rs. } 24.5$$

Example 44:

Calculate arithmetic mean median and mode from the following frequency distribution:

Variable	*Frequency*	*Variable*	*Frequency*
10-13	*8*	*25-28*	*54*
13-16	*15*	*28-31*	*36*
16-19	*27*	*31-34*	*18*
19-22	*51*	*34-37*	*9*
22-25	*75*	*37-40*	*7*

Solution:

Calculation of Mean Median & Mode

Variable	*m.p. m*	*f*	*(m–23.5)/3 d*	*fd*	*c.f*
10-13	11.5	8	–4	–32	8
13-16	14.5	15	–3	–45	23
16-19	17.5	27	–2	–54	50
19-22	20.5	51	–1	–51	101
22-25	23.5	75	0	0	176
25-28	26.5	54	+1	+54	230
28-31	29.5	36	+2	+72	266
31-34	32.5	18	+3	+54	284
34-37	35.5	9	+4	+36	293
37-40	38.5	7	+5	+35	300
		N = 300		Σfd = 69	

$$\overline{X} = A + \frac{\Sigma X}{N} i = 23.5 + \frac{69}{300} \times 3 = 24.19$$

Med. = Size of N/2th item = 300/2 = 15th item

Median lies in the class 22 – 25.

$$\text{Med.} = L + \frac{N/2 - c.f.}{f} \times i = 22 + \frac{150 - 101}{75} \times 3 = = 22 + 1.96 = 23.96$$

Mode. By inspection mode lies in the class 22 – 25.

$$M_0 = L + \frac{\Delta_1}{\Delta_1 + \Delta_2} \times i$$

$L = 22 = \Delta_1 = (75 - 51) = 24$ $\Delta_2 = (75 - 54) = 21$, $i = 3$

$$M_0 = 22 + \frac{24}{24 + 21} \times 3 = 22 + 1.6 = 23.6$$

Example 45:

In 500 small-scale industrial units the return on investment ranged from 0 to 30 per cent, no unit sustaining any loss, 5 per cent of the units had returns ranging from 0 per cent up to (and including) 5 per cent and 15 per cent of the units earned returns cent and the upper quartile 20 per cent. The uppermost layer of the returns exceeding 25 per cent was earned by 50 units.

Present this information in the form of a frequency table with intervals of 5 per centas follows:

Exceeding 0 per cent but not exceeding 5 per cent

" 5 " " 10 "

" 10 " " 15 "

" 15 " " 20 "

" 20 " " 25 "

" 25 " " 30 "

Use $\frac{N}{4}, \frac{2N}{4}, \frac{3N}{4}$ *as the ranks of the tower middle and upper quartiles respectively. Find the rate of return round which there is maximum concentration of the units.*

Solution:

The information given above can be summarized as follows:

Table Showing the Distribution of Small-Scale Industrial Units According to the Table of Returns of Investment

Rate of Return on Investment Number of		***Firms % of***			
		Total			**Firms**
Exceeding 0 but not exceeding	5		5	(a)	25
5	10		15	(b)	75
10	15	(b)	30	(d)	150
15	20	(c)	25	(f)	125
20	25	(b)	15	(h)	75
25	30		10		50
			100		500

On the basis of information provided the computations are shown below:

(a) 5% of 500 = 25,

(b) 15% of 500 = 75

(c) The rate of return of 15% being the median $\frac{3N}{2}$ would represent 50% of the firms 20% of the data is comprised in previous class. Tusclasses would represent 50% of the firms 20% of the data is comprised in previous calss. Thus classes would represent 30% of the firms.

(d) In the consequence 30% of 500 = 150.

(e) Firms having 20% return constitute the quartile $\frac{3N}{4}$ This shall cover 75% of the data comprised in the preceding calss being representation of 50% firms this class will cover 25% of data.

(f) In consequence 25% of 500 = 125.

(g) The residual balance of given data equals 15%.

(h) 15% being the residual balance it represents 75% firms.

The rate of return around which there is maximum concentration is the modal calss. The modal lies in the class 10-15.

$$\text{Mode} = L + \frac{\Delta_1}{\Delta_1 + \Delta_2} \times i$$

$$L = 10\ \Delta_1 = 150 - 75 = 75 = \Delta_2 = 150 - 125 = 25\ i = 5$$

$$\therefore M_0 = 10 + \frac{75}{75 + 25} \times 5 + 10 + 3.75 = 13.75$$

Hence the rate of return around which there is maximum concentration of units is 13.75%.

Example 46:

The mean annual salaries paid to 1,000 employees of a company was Rs. 5,000. The mean annual salaries paid to male and female employees were Rs. 5,200 and Rs. 4,200 respectively. Determine the percentage of males and females employed by the company.

Solution:

Let N_1 represent percentage of males and N_2 percentage of females so that $N_1 + N_2 = 100$

We are given $\overline{X}_{12} = 5{,}000$ $\overline{X}_1 = 5{,}200$, $\overline{X}_2 = 4{,}200$.

Substituting the values in the formula:

$$\overline{X}_{12} = \frac{N_1\overline{X}_1 + N_2\overline{X}_2}{N_1 + N_2};\ 5{,}000 = \frac{N_1(5{,}200) + N_2(4{,}200)}{100}$$

$$5{,}00{,}000 = 5{,}200\ N_1 + [(100 - N_1)\ (4{,}200)]$$

$$= \left[\begin{array}{l} \text{Since } N_1 + N_2 = 100 \\ N_2 = 100 - N_1 \end{array}\right]$$

$$\Rightarrow \quad 5{,}00{,}000 = 5{,}200\ N_1 + 4{,}20{,}000 - 4{,}200\ N_1$$

$$1{,}000\ N1 = 80{,}000$$

$$N_1 = 80 \text{ and } N_2$$

$$= 100 - N_1\ (100 - 80) = 20$$

Thus, the percentage of males and females employed is 80 and 20 respectively.

Example 47:

The arithmetic mean the mode and the medium of a group of 75 observations were calculated to be 27, 34 and 29 respectively. It was later discovered that one observation was wrongly read as 43 instead of the correct value 53. Examine to what extent the calculated values of the three averages will be affected by error.

Solution:

Correct mean $\quad \overline{X} = \frac{\Sigma X}{N} \quad \text{or} \quad \Sigma X = N\overline{X}$

$$N = 75\ \overline{X} = 27$$

$\therefore \quad \Sigma X = 75 \times 27 = 2{,}025$

Correct $\quad \Sigma X = 2{,}025 - 43 + 53 = 2{,}035$

Correct $\quad \overline{X} = \frac{2{,}035}{75} = 27.13$

The values of median and mode will not be affected by this error because these are positional averages and the median value is 29 which is far away from the values 43 and 53. Similarly the value of mode would not be affected by the error since the modal value 34 is far away from the value 43 and 53.

Example 48:

Calculate Median and Mode of the data given below. Using them find arithmetic mean:

Marks:	*10*	*20*	*30*	*40*	*50*	*60*
No. of students:	*8*	*23*	*45*	*65*	*75*	*80*

Solution:

Calculation of Median, Mode and Arithmetic Mean

Marks	*f*	*c.f.*
0–10	8	8
10–20	15	23
20–30	22	45
30–40	20	65
40–50	10	75
50–60	5	80
	N = 80	N = 80

Median. Median = size of $\frac{N}{2}$ item = $\frac{80}{2}$ = 40th item

Median lies in the class 20–30.

$$\text{Med} = L + \frac{N/2 - c.f.}{f} \times i$$

$$L = 20,\ N/2 = 40,\ c.f. = 23,\ f = 22,\ i = 10$$

$$\text{Med} = 20 + \frac{40 - 23}{22} = \times 10 = 20 + 7.73 = 27.73$$

Mode. By inspection mode lies in the class 20–30.

$$\text{Mo} = L + \frac{\Delta_1}{\Delta_1 + \Delta_2} \times i$$

$$L = 20,\ \Delta_1 = (22 - 15) = 7,$$

$$\Delta_2 = (22 - 20) = 2,\ i = 10$$

$$\text{Mo} = 20 + \frac{7}{7+2} \times 10 = 20 + 7.78 = 27.78.$$

Arithmetic Mean. Since we are asked to calculate arithmetic mean using median and mode we will apply the following formula:

Mode = 3 Median – 2 Mean

Substituting the values of mode and median

27.78 = 3 (27.73) – 2 Mean

$$-2\overline{X} + 83.19 = 27.78$$

$$-2\overline{X} = 27.78 - 83.19 = -55.41$$

$$\overline{X} = 27.705$$

Example 49:

A factory pays workers on piece rate basis and also a bonus to each worker on the basis of individual output in each quarter.

The rate of the bonus payable is as follows:

Output in units	***Bonus in Rs.***	***Output in units***	***Bonus in Rs.***
70–74	*40*	*90–94*	*70*
75–79	*45*	*95–99*	*80*
80–84	*50*	*100–104*	*100*
85–89	*60*		

The individual output of a batch of 50 workers is given below:

94	*83*	*78*	*76*	*88*	*86*	*93*	*80*	*91*	*82*
89	*97*	*92*	*84*	*92*	*80*	*85*	*83*	*98*	*103*
87	*88*	*88*	*81*	*95*	*86*	*99*	*81*	*87*	*90*
84	*97*	*80*	*75*	*93*	*101*	*82*	*82*	*89*	*72*
85	*83*	*75*	*72*	*83*	*98*	*77*	*87*	*71*	*80*

By suitable classification you are required to find:

(i) Average bonus per worker for the quarter.

(ii) Average output per worker.

Solution:

Frequency Distribution By Output & Bonus

Output	*Tallies*	*f*	*Bonus X*	*m.p. m*	*(m-87)/5 d*	*fd*	*fX*
70–74	III	3	40	72	–3	–9	120
75–79	IIII	5	45	77	–2	–10	225
80–84	IIII IIII IIII	15	50	82	–1	–15	750
85–89	IIII IIII II	12	60	87	0	0	720
90–94	IIII II	7	70	92	+ 1	+ 7	490
95–99	IIII I	6	80.	97	+ 2	+ 12	480
100–104	II	2	100	102	+ 3	+ 6	200
						Σfd = – 9	ΣfX = 2985

(i) Average bonus per worker for the quarter :

$$\overline{X} = \frac{\Sigma fX}{N} = \frac{2,985}{50} = 59.7$$

(ii) Total quartely bonus paid = 59.7 × 50 = Rs. 2,985

(iii) Average output per worker

$$\overline{X} = A + \frac{\Sigma fX}{N} \times i 87 - \frac{9}{50} \times 5 = 86.1 \text{ units.}$$

Example 50:

10 percent of the workers in a firm employing a total of 1,000 workers earn less than Rs. 5 per day 200 earn between Rs. 5 and 9.99 30 percent between 10 and 14.99 250 workers between 15 and 19.99 and the rest 20 and above. What is the median wage?

Solution:

First convert the given figure in a frequency distribution as follows:

Wages (Rs.)	*No. of workers f*	*c.f.*	
Below 5	1,000 × 10/100 = 100		100
5–9.99	200		300
10–14.99	1,000 × 30/100 = 300	1,000	600
15–19.99	250		850
20 and above	150		1,000

$$\text{Median} = \text{size of } \frac{N}{2} \text{th item} = \text{Size of } \frac{1{,}000}{2} = 500\text{th item}$$

Median lies in the class 10–14.99. But the real limit of this class is 9.995–14.995.

$$\text{Med.} = L + \frac{N/2 - \text{c.f.}}{f} \times i$$

$$L = 9.995,\ N/2 = 500,\ \text{c.f.} = 300,\ f = 300,\ i = 5$$

$$\text{Med.} = 9.995 + \frac{500 - 300}{300} \times 5 = 9.995 + 3.333 = \text{Rs. } 13.328.$$

Example 51:

From the following data compute the mean marks of all the students of 50 schools in a city.

Marks obtained	*No. of schools*	*Average No. of students in a school*
More than 35	*7*	*200*
30–35	*10*	*250*
25–30	*15*	*300*
20–25	*9*	*200*
15–20	*5*	*150*
Less than 15	*4*	*100*

Solution:

First rewrite the given data in ascending order and then calculate the mean.

Calculation of Mean Marks

Marks	*No. of schools*	*Average No. of students*	*Total No. of students (2 × 3)*	*m.p.m*	*(m-27.5)/5 d*	*fd*
(1)	*(2)*	*(3)*	*(f)*	*m*	*d*	*fd*
10–15	4	100	400	12.5	–3	–1,200
15–20	5	150	750	17.5	–2	–1,500
20–25	9	200	1,800	22.5	–1	–1,800
25–30	15	300	4,500	27.5	0	0
30–35	10	250	2,500	32.5	+1	+2,500
35–40	7	200	1,400	37.5	+2	+2,800
			N = 11,350			Σfd = +800

$$\overline{X} = A + \frac{\Sigma fX}{N} \times i = 27.5 + \frac{800}{11{,}350} \times 5 = 27.5 + 0.35 = 27.85$$

Example 52:

In a certain examination the average grade of all students in class A is 68.4 and students in class B is 71.2. If the average of both classes combined is 70, find the ratio of the number of students in class A to the number of students in class B.

Solution:

Let us assume that the number of students in class A was 'X' and in class B 'Y'.

We are given $\overline{X}_{12} = 70$, $\overline{X}_1 = 68.4$, $\overline{X}_2 = 71.2$

Substituting these values in the formula:

$$\overline{X} = \frac{N_1\overline{X}_1 + N_2\overline{X}_2}{N_1 + N_2} \quad 70 = \frac{68.4\,x + 71.2\,y}{x + y}$$

$$70\,(X + Y) = 68.4X + 71.2\,Y$$

$$70X - 68.4X + 70Y - 71.2Y = 0$$

$$\Rightarrow \quad 1.6X = 1.2Y$$

Suppose $X = 10$

$$1.2Y = 16 \text{ or } Y = \frac{16}{1.2} = \frac{40}{3}$$

Thus X and Y are in the ratio of $10 : \frac{40}{3}$ or 30 : 40.

Hence for every 3 students in class A there are 4 students in class B.

Note : We can assume X to be anything but the ratio would come out to be 30 : 40.

Example 53:

Find the class intervals if the arithmetic mean of the following distribution is 33 and assumed mean 35:

Step deviations	*–3*	*–2*	*–1*	*0*	*+1*	*+2*
Frequency	*5*	*10*	*25*	*30*	*20*	*10*

Solution:

Determination of Class Intervals

Step deviations d	***Frequency f***	***fd***
–3	5	–15
–2	10	–20
–1	25	–25
0	30	0
+1	20	+20
+2	10	+20
	N = 100	Σfd = –20

$$\overline{X} = A + \frac{\Sigma fX}{N} \times i$$

$$A = 35, \ \overline{X} = 33, \ N = 100, \ \Sigma fd = -20$$

Substituting the values $33 = 35 - \dfrac{20}{100} \times i$

$$33 - 35 = -\ 0.2\ i$$

$$0.2\ i = 2\ i = \frac{2}{0.2} = 10. \text{ Thus the class interval is 10}$$

Assumed mean lies in the mid-value of that class '0' as step deviation. The lower and upper limits of this class are:

$35 - \dfrac{10}{20} = 30$ and $35 + \dfrac{10}{2} = 40$, i.e., 30–40. The other classes will be:

0–10 10–20 20–30 30–40 40–50 50–60

Note. Since all the step deviations show equal gap, c = i in the formula for calculating arithmetic mean. As we are required to determine class interval we have substituted i in place of c.

Example 54:

The following is the age distribution of 2,000 persons working in a large textile mill:

Age group	*No. of persons*	*Age group*	*No. of persons*
15 but less than 20	*80*	*45 but less than 50*	*268*
20 but less than 25	*250*	*50 but less than 55*	*150*
25 but less than 30	*300*	*55 but less than 60*	*75*
30 but less than 35	*325*	*60 but less than 65*	*25*
35 but less than 40	*287*	*65 but less than 70*	*20*
40 but less than 45	*220*		

Because of the heavy losses the management decides to bring down the strength to 40% of the present number according to the following scheme:

(i) To retrench the first 10% from lower group.

(ii) To absorb the next 40% in other branches.

(iii) To mass 10% from the highest age group retire prematurely.

What will be the age limits of persons retained in the mill and of those transferred to other branches? Also calculate the average age of those retained.

Solution:

The number of persons to be retrenched from the lower group = $\frac{2,000 \times 10}{100}$ = 200 = 200. Eighty of these will be from 15–20 age group and the rest (200 – 80) = 120 from 20–25 age group.

The persons to be absorbed in other branches = $\frac{2,000 \times 40}{100}$ = 800. They belong to the following age groups:

Age group		**No. of persons**
20–25	(250 – 120)	130
25–30		300
30–35		325
35–40	(287 – 242)	45
		800

Those who are to retire are $\frac{2,000 \times 10}{100}$ = 200 in all and they belong to highest age group. Their age groups are:

Age group		*No. of persons*
65–70		20
60–65		25
55–60		75
50–55	(150 – 70)	80
		200

Hence the age limits of those who are retained in the mill are:

Age group	*No. of persons*
35–40	242
40–45	220
45–50	268
50–55	70
Total	800

Calculation of Average Age of Those Retained

Age group	*m.p. m*		*(m – 47.5)5 d*	*fd*
35–40	37.5	242	– 2	– 484
40–45	42.5	220	– 1	– 220
45–50	47.5	268	0	0
50–55	52.5	70	+ 1	+ 70
			N = 800	Σfd = – 634

$$\overline{X} = A + \frac{\Sigma fX}{N} \times i = 47.5 + \frac{634}{800} \times 5 = 47.5 - 3.96 = 43.54 \text{ years.}$$

Example 55:

Find the numbers whose arithmetic mean is 12.5 and geometric mean 10.

Solution:

Let the numbers of 'a' and 'b'

$$\text{G.M.} = \sqrt{a \times b} = 10 \quad \therefore \quad ab = (10)^2 = 100$$

$$\overline{X} = \frac{a+b}{2} = 12.5$$

$\therefore \quad a + b = 2 \times 12.5 = 25$

We know that $(a + b)^2 - (a - b)^2 = 4\,ab$.

Substituting the given values we have

$$(25)^2 - (a - b)^2 = 4 \times 100$$

$$625 - (a - b)^2 = 400$$

$$\Rightarrow \quad (a - b)^2 = 625 - 400 = 225$$

$$a - b = 225 = 15$$

$$a + b = 25 \quad \text{...(i)}$$

$$a - b = 15 \quad \text{...(ii)}$$

$$2a = 40 \text{ or } a = 20$$

Substituting the value of 'a' in equation (i).

$$b = 25 - 20 = 5$$

Hence the two numbers are 20 and 5.

Example 56:

A machinery depreciated by 40% in the first year 25 percent in the second and 10% p.a. during the next three years. What is the average rate of depreciation during the whole period?

Solution:

The cost of the machine will not affect of the rate of depreciation and hence it can be ignored. Average rate of depreciation would be obtained by applying geometric mean:

Year	*Diminishing value taking 100 as base X*	*log X*
1994	100 – 40 = 60	1,7782
1995	100 – 25 = 75	1,8751
1996	100 – 10 = 90	1,9542
1997	100 – 10 = 90	1,9542
1998	100 – 10 = 90	1,9542
		Σ log X = 9,5159

$$\text{G.M.} = \text{A.L.}\left(\frac{\Sigma \log X}{N}\right) = \left(\frac{9{,}5159}{5}\right) = \text{A.L.}1{,}9032 = 80.$$

Since the diminishing value is Rs. 80, the depreciation will be 100 – 80 = 20%.

Thus the average rate of depreciation charged during the whole period is 20 percent.

Example 57:

The median and mode of the following wage distribution are known to be Rs. 33.5 and 34 respectively. Three frequency values from the table are however missing. Find these missing values.

Wages in Rs.	*Frequencies*
0–10	*4*
10–20	*16*
20–30	*?*
30–40	*?*
40–50	*?*
50–60	*6*
60–70	*4*
	230

Solution:

Let the missing frequencies be:

20–30	x
30–40	y
40–50	230–30–x–y

Since median and mode are 33.5 and 34 they both lie in the class 30–40.

$$\text{Med.} = L + \frac{N/2 - \text{c.f.}}{f} \times i$$

Med. = 33.5, L = 30, N/2 = 115,

c.f. = 20 + x, f = y, i = 10

$$33.5 = 30 + \frac{115 - (20 + x)}{y} \times 10$$

$$3.5 = \frac{95 - x}{y} \times \text{ or } 3.5y = 950 - 10x$$

$$\Rightarrow \qquad 3.5y + 10x = 950 \qquad \ldots(i)$$

$$\text{Mode} = L + \frac{\Delta_1}{\Delta_1 + \Delta_2} \times i$$

Mode = 34 = 30;

$\Delta_1 = (y - x)$;

$\Delta_2 = [y - (230 - 30 - x - y)]$; i = 10

$$34 = 30 + \frac{y - x}{y - x + 2y - 200 + x} \times 10$$

$$= 30 + \frac{(y - x)}{3y - 200} \times 10$$

$$4 = \frac{10y - 10x}{3y - 200} \text{ or } 12y - 800 = 10y - 10x$$

$\Rightarrow$ $2y + 10x = 800$ ...(ii)

Multiplying Eqn. (i) by 10

$$3.5y + 10x = 950$$
$$2y + 10x = 800$$

$$1.5y = 150$$
$$y = 100$$

Substituting the value of y in Eqn. (i)

$$3.5 \times 100 + 10x = 950$$

$\Rightarrow$ $350 + 10x = 950$ or $x = 960$

Thus the missing frequencies are:

Class intervals	*Frequencies*
20–30	$x = 60$
30–40	$y = 100$
40–50	$200 - 60 - 100 = 40$

Example 58:

In a class of 50 students 10 have failed and their average of marks is 2.5. The total marks secured by the entire class were 281. Find the average marks of those who have passed.

Solution:

N = 50, No. of failures = 10, average marks of those who failed = 2.5.

Total marks secured by all 50 students = 281

Marks secured by 10 students who failed = $10 \times 2.5 = 25$

Marks secured by 50 students = 281

Out of 50 ten have failed.

Total marks of those who passed = $281 - 25 = 256$

Average marks of those who passed = $\frac{256}{400} = 6.4.$

Example 59:

An economy grows at the rate of 2% in the first year, 2.5% in the second year, 3% in the third year, 4% in the fourth year ... and 10% in the tenth year. What is the average rate of growth of the company?

Solution:

For finding the average rate of growth we shall have to calculate the geometric mean. Since the growth rate is 4% in the fourth year and 10% in the 10th year and in between the growth rates are not given we shall assume that the growth rates have been 5 6 7% etc. in the 5th, 6th, 7th year respectively.

Year	*Growth Rate*	*Value at the end of the year x*	*log x*
1st	2	102	2.0086
2nd	2.5	102.5	2.0107
3rd	3	103	2.0128
4th	4	104	2.0170
5th	5	105	2.0212
6th	6	106	2.0253
7th	7	107	2.0294
8th	8	108	2.0334
9th	9	109	2.0374
10th	10	110	2.0414
			Σ log X = 20.2372

$$\text{G.M.} = \text{AL}\left(\frac{\Sigma \log X}{N} = \text{AL}\left(\frac{20.2372}{10}\right)\right) = \text{AL}(2.022372) = 105.6$$

Average rate of growth = 105.6 – 100 = 5.6 percent.

Example 60:

Incomes of employees in an industrial concern are given below. The total income of the 10 employees in the class over Rs. 2,500 is Rs. 30,000. Compute the mean income. Every employee belonging to the top 25% of the earners is required to pay 5% of 'income to workers' relief fund. Estimate the contribution to this fund.

Income (Rs.)	*Frequency*	*Income (Rs.)*	*Frequency*
Below 500	*90*	*1500–2000*	*60*
500–1000	*150*	*2000–2500*	*70*
1000–1500	*100*	*2500 and cover*	*10*

Solution:

Compertation of Arithmetic Mean

Income (Rs.)	*M.P. m*	*f*	*fm*
0–500	250	90	22500
500–1000	750	150	112500
1000–1500	1250	100	125000
1500–2000	1750	80	140000
2000–2500	2250	70	157500
2500 and over	—	10	30000
		N = 500	Σ fm = 587500

$$\overline{X}\ \frac{\Sigma fm}{N} = \frac{587500}{500} = 1175$$

Note: If is given that the total income of 10 employees in the class over Rs. 2,500 is Rs. 30,000.

Number of employees belonging to the top 25% of the earners is

$$\frac{25}{100} \times 500 = 125$$

and the distribution of these top earners (as obvious from the above table) is as follows:

Distribution of Top 25% of Earners

Income (Rs.)	Frequency
2500 and over	10
2000–2500	70
1500–2000	45

45 persons in the last class i.e. 1500–2000 are to be taken with the highest income level starting from 2000 and below. Under the assumption that the frequencies are distributed uniformly over the entire class we interpolate this number as follow:

80 persons have income in the range 2000 – 1500 = Rs. 500

∴ 45 persons have income in the range

$$\frac{500}{60} \times 45 = 281.25 \text{ or } 281.$$

Since we are interested in the top 45 earners in the income group 1500 – 2600 their salaries will range from (2000 – 281) to 2000 i.e. 1719 to 2000.

The distribution of top 125 persons is as follows:

Income (Rs.)	*Mid-point m*	*f*	*Total Income fm*
2500 and over	—	10	30,000 (given)
2000–2500	2250	70	1,57,500
1719–2000	1859.5	45	83,677.5
		N = 125	2,71,177.5

Hence the total income of the top 25% of earners is Rs. 2,71,177.5.

5% contribution to the fund

0.05 of 2,71,177.5 = Rs. 93,598.

Example 61:

(a) A man travelled by cr for 3 days. He covered 480 km. each day. On the first day he drove for 10 hour on the second day he drove for 12 hours at 40 km. an hour and on last day he drove for 15 hours at 32 km. What was his average speed? (B. Com. Bombay Univ. Bangalore Univ. 1996)

Solution:

Since the distance travelled is constant i.e. 480 km. each day the appropriate erage is the Harmonic Mean.

$$\text{H.M.} = \frac{N}{\frac{1}{X_1} + \frac{1}{X_2} + \frac{1}{X_3}}$$

$$= \frac{3}{\frac{1}{48} + \frac{1}{40} + \frac{1}{32}} = \frac{\frac{3}{37}}{480} = \frac{3 \times 480}{37} = 38.92 \text{ km p.h}$$

Mr. Dushmanta of Bhubaneswar started for village which was at a distance of six km. travelled in his car at a speed of 40 km. per hour. After travelling for 4 km. the car topped running. He then travelled in a rickshaw at a speed of 10 km. per hour After travelling a distance of $1\frac{1}{2}$ km. he left rickshaw and covered the remaining distance on foot a speed of 4 km. per hour.

Find the average speed per hour of Mr. Dushmanta and verify the calculations.

Solution:

This problem can be solved with the help of Weighted harmonic Mean.

Speed X	*Distance W*	*W/X*
40	4.0	0.100
10	1.5	0.150
4	0.5	0.125
	$\Sigma W = 6$	$\Sigma(W/X) = 0.375$

$$\text{Average speed} = \frac{\sum W}{\sum (W/X)} \; \frac{6}{0.375} = 16 \text{ km. p.h.}$$

Example 62:

Calculate the median and mod (by grouping method) form the following data:

Central size:	*15*	*25*	*35*	*45*	*55*	*65*	*75*	*85*
Frequencies:	*5*	*9*	*13*	*21*	*20*	*15*	*8*	*3*

Solution:

Since we are given central values first we determine the lower and upper limits of the classes. The class interval is 10 and hence the first class would be 10-20 (because the mid-point is 15)

Calculation of Median and Mode

Class groups	*m.p.m*	*f*	*(m – 55)/10 d*	*fd*	*c.f.*
10-20	15	5	–4	–20	5
20-30	25	9	–3	–27	14
30-40	35	13	–2	–26	27
40-50	45	21	–1	–21	48
30-60	55	20	0	0	68
60-70	65	15	+1	+15	83
70-80	75	8	+2	+16	91
80-90	85	3	+3	+9	94
		N = 94		$\Sigma fd = 54$	

Calculation of Median: Med. = size of $\frac{N}{2}$ th item = $\frac{94}{2}$ = 47th item

Median lies in the class 40-50.

$$\text{Med.} = \frac{N/2 - \text{c.f.}}{f} \times i\ 40 + \frac{47 - 27}{21} \times 10 = 40 + 9.524 = 49.524$$

Class groups	*I*	*II*	*III*	*IV*	*V*	*VI*
10-20	5					
		14		27		
20-30	9		22		43	
30-40	13	34				
40-50	21		41			54
50-60	20			56		
		35			43	
60-70	15		23			
70-80	8					26
		11				
80-90	3					

Class in which Mode is Expected to lie

Col. No.	*40-50*	*50-60*	*60-70*
I	1		
II		1	1
III	1	1	
IV	1	1	1
V	1	1	
VI	1	1	1
	5	5	3

This series is bimodal and hence we shall determine the mode by the indirect method.

Mode = 3 Median – 2 Mean.

Calculation of Mean $\overline{X} = A + \frac{\Sigma fd}{N} \times i = 55 - \frac{54}{94} \times 10$

$$= 55 - 5.745 = 49.255$$

$$\text{Mode} = (49.524) - 2\,(49.255)$$
$$= 148.572 - 98.51 = 50.062$$

Example 63:

In a factory there are 100 skilled, 250 semi-skilled and 150 unskilled workers. It has be observed that on an average a unit length of a particular fabric is woven by a skilled worker in 3 hours by a semi-skilled worker in 4 hours and by an unskilled worker in 5 hours. After a training of 2 years the semiskilled workers are expected to become skilled and the training for weaving the unit length of fabric by an average worker?

Solution:

Average time per worker before training

$$\frac{(100 \times 3) + (250 \times 4) + (150 \times 5)}{100 + 250 + 150} = \frac{2050}{500} = 4.1 \text{ hrs.}$$

Now after training the composition of workers is as follows :

Skilled workers = 100 + 250 = 350

Semi-skilled workers = 150

Unskilled workers = Nil

Average time per worker after training is :

$$\frac{(350 \times 3) + (150 \times 4)}{350 + 150} = \frac{1050 + 600}{500}$$

$$= 3.3 \text{ hours.}$$

After 2 years one hour less would be required.

Note. An assumption has even been made that there has not been any turnover of workers.

Example 64:

Find the missing frequency from the following data :

Marks :	*0 – 10*	*10 – 20*	*20 – 30*	*30 – 40*	*40 – 50*	*50 – 60*
No. of students :	*5*	*15*	*20*	*–*	*20*	*10*

The arithmetic mean in 34 marks.

Solution:

Let the missing frequency be denoted by X.

Calculation of Missing Frequency

Marks	*f*	*m.p.*	*fm*
0-10	5	5	25
10-20	15	15	225
20-30	20	25	500
30-40	X	35	35*
40-50	20	45	900
50-60	10	55	550
		N = 70 + X	Σfm = 2200 + 35 X

$$\overline{X} = \frac{\Sigma fX}{N}, 34 = \frac{22000 + 35X}{70 + 35X}$$

$$34(70 \times X) = 2200 + 35X$$
$$2380 + 34X = 2200 + 35X$$
$$34X - 35X = 2200 - 2380$$

$-X = 180$ or $X = 180$ Hence the missing value is 180.

Note. By putting X = 180 we can calculate the arithmetic mean and verify the result. It comes out to be the same i.e. arithmetic mean is 34.

Example 65:

In a class of 50 students 10 have filed and their average of marks is 2.5. The total marks secured by the entire class were 28.1. Find the average marks of hose who have passed.

Solution:

N = 50 failed = 10

Mean marks of those who failed = 2.5

Total marks of 10 students who failed = 2.5 × 10 = 10 = 25

Total marks secured by entire class = 281

Total marks obtained by those who passed = 281 – 25 = 256

Average marks obtained by those who passed = $\frac{256}{40}$ = 6.4

Example 66:

Average rainfall of a city from Monday to Saturday is 0.3 inch. Due to heavy rainfall on Sunday the average rainfall for the week increased to 0.5 inch. What was the rainfall on Sunday?

Solution:

Average rainfall from Monday to Saturday = 0.3 (i.e. for six days).

Total rainfall for the entire week including Sunday = 0.5

∴ Total rainfall = 5 × 7 = 3.5

Hence rainfall on Sunday = 3.5 – 1.8 = 1.7.

Example 67:

From the following data calculate the missing value when mean is 115.86:

Wages (Rs.)	:	*110*	*112*	*113*	*117*	*(X)*	*125*
No. of workers	:	*25*	*17*	*13*	*15*	*14*	*08*

Solution:

Calculation of Missing Value

Wages (Rs.)	***f***	***fX***
110	25	2750
112	17	1904
113	13	1469
117	15	1755
X	14	14X
125	8	1000
128	6	768
130	2	260
	N = 100	ΣfX = 9906 = 14X

$$\overline{X} \frac{\Sigma fd}{N} \Rightarrow 115.86 = \frac{9906 + 14X}{100}$$

$$11586 = 9906 + 14X \quad 14X = 1680 \Rightarrow X = 120$$

Hence the missing value is 12.0

Example 68:

The rate of a certain commodity in the first week of January 1987 is 0.4 kg. per rupee : it is 0.6 kg per rupee in the third week. Therefore is it correct to say that the average price is 0.5 kg. per rupee? Verify.

Solution:

The answer given is not correct as it is based on arithmetic mean whereas in this base appropriate average is harmonic mean.

$$\text{H.M.} = \frac{N}{\frac{1}{a}+\frac{1}{b}+\frac{1}{c}}$$

$$= \frac{3}{\frac{1}{0.4}+\frac{1}{0.6}+\frac{1}{0.5}} = \frac{.3+.2+.24}{12} = \frac{3}{\frac{.74}{.12}} = \frac{3 \times .12}{.74} = 0.486$$

Hence the average price is 0.486 kg. rupee and not 0.5 kg. per rupee.

Example 69:

Find the missing frequency. If arithmetic mean is 28 of the data given below find the median of the series later.

Profits per shop :	*0 – 10*	*10 – 20*	*20 – 30*	*30 – 40*	*40 – 50*	*50 – 60*
No. of shops :	*12*	*18*	*27*	*–*	*17*	*6*

Solution:

Let the missing frequency by X.

Determination of Messing Frequency

Profits (Rs.)	*m.p.m*	*No. of shops f*	*(m – 35)/10 d*	*fd.*
0-10	5	12	–3	–36
10-20	15	18	–2	–36
20-30	25	27	–1	–27
30-40	35	X	0	0
40-50	45	17	+1	+17
50-60	55	6	+2	+12
		N = 80 + X		Σfd = – 70

$$\overline{X} = A + \frac{\Sigma fd}{N} \times i$$

A = 35 Σfd = – 70 N = 80 + Xi = 10 $\overline{X}$ = 28 (given).

$$28 = 35 \frac{70}{80 \times X} \times 10$$

28(80 + X) = 35(80 + X) – 700

2240 + 28X = 2800 + 35X – 700

28X – 35X = 2800 – 700 – 2240

–7x = – 140 or X = 20.

Hence the missing frequency is 20.

Calculation of Median

Profits (Rs.)	*No. of shops* *f*	*c.f*
0-10	12	12
10-20	18	30
20-30	27	57
30-40	20	77
40-50	17	94
50-60	6	100
		N = 100

Med. size of $\frac{N}{2}$th item = $\frac{100}{2}$ = 50th item

Median lies in the class 20-30.

$$\text{Med} = L + \frac{N/2 - c.f.}{f} \times i$$

$$L = 20 \quad N/2 = 50 \quad c.f. = 30 \quad f = 27 \quad i = 10$$

$$\text{Med.} = 20 \ \frac{50 - 30}{27} \times 10 = 20 + 7.407 = 27.\ 407.$$

Example 70:

Calculate Median and Arithmetic Mean from the following series:

05 men get less than Rs. 5
12 men get less than Rs. 10
22 men get less than Rs. 15
30 men get less than Rs. 20
36 men get less than Rs. 25
40 men get less than Rs. 30 .

Solution:

We are give cumulative frequencies. First find simple frequency and then compute arithmetic mean and median.

Income (Rs.)	*m.p.m*	*f*	*(m-12.5)/5d*	*fd*	*c.f.*
0-5	2.5	5	–2	-10	5
5-10	7.5	7	–1	–7	12
10-15	12.5	10	0	–0	22
15-20	17.5	8	+1	+8	30
20-25	22.5	6	+2	+12	36
25-30	27.5	4	+3	+12	40
			N = 40		Σfd = 15

$$\overline{X} = A + \frac{\Sigma fd}{N} \times 12.5 + \frac{15}{40} \times 5\ 12.5 + 1.875 = 14.375$$

$$\text{Med.} = \text{Size of } \frac{N}{2}\text{th item} = \frac{40}{2} = 20\text{th item}$$

Median lies in the class 10-15.

$$\text{Med} = L + \frac{N/2 - c.f.}{f} \times i$$

$$= 10 + \frac{20 - 12}{10} \times 5 = 10 + 4 = 14$$

Example 71:

Find the missing information in the following table:

	A	*B*	*C*	*Combined*
Number	*10*	*8*	–	*24*
Mean	*20*	–	*6*	*15*
geometric Mean	*10*	*7*	–	*8.397*

Solution:

Finding missing information:

Number: For c missing information shall be

$24 - (10 + 8) = 6$

Mean: Let x be the mean of B

Them $(20 \times 10) + (8 \times x) + (6 \times 6) = (15 \times 24)$

$$200 = 8x + 36 = 360$$

$$8x = 360 - 200 - 36 = 124$$

$$x = \frac{124}{8} = 15.5$$

Hence mean of B = 15.5

Geometric Mean: Let x be the geometric mean of C.

$$(10)^{10} \times (7)^8 \times x^6 = (8.397)^{24}$$

$$= 10 \log 10 + 8 \log 7 + 6 \log x = 24 \log 8.397$$

$$= 10 + (8 \times 0.8451) + 6 \log x = 24\ (.9241)$$

$$= 10 + 6.7608 + 6 \log x = 22.1784$$

$$6 \log x = 22.1784 + 10 - 6.7608 = 5.44176$$

$$\log x = 0.9029$$

$$x = \text{A.L. } .9029 = 7.997$$

Hence geometric mean of C is 7.997.

Example 72:

During a period of decline in stock market prices a stock sold at Rs. 50 per share on one day Rs. 40 on the next day and Rs 25 on the third.

(i) If an investor bought 100, 120 and 180 shares on the respective three days, find the average price paid per share.

(ii) If the investor bought Rs 1000 worth of shares on each of the three days find the average price paid per share.

Solution:

(i)

x	w	wx
50	100	5000
40	120	4800
25	180	4500
	$\Sigma w = 400$	$\Sigma wx = 14300$

$$\text{Average price paid per share} = \frac{\Sigma WX}{\Sigma W} = \frac{14300}{400} = 35.75$$

Hence average price per share = Rs. 35.75

(ii) The investor buys shares of Rs. 1,000 per day. We can calculate the number of shares bought on each day

Price	**Amount spent per day**	**No. of shares bought**
50	1000	20
40	1000	25
25	1000	40
		105

$$\text{Hence average price paid per share } \frac{105}{3} = \text{Rs. } 35.$$

Which Average to Use

We have explained above the methods of computing the various types of averages and also their distinctive features. At this point the reader has a right to ask "Which of these averages should I use"? or "When ought I to use one or the other of the averages described"? "or "Which of these is the best average to be used"

It must be clearly understood that no one average can be regarded as best for all circumstances. The following considerations influence the selection of an appropriate average:

(a) The type of data available. Are they badly skewed (avoid the mean), gappy around the middle (avoid the median) or unequal in class interval (avoid the mode)?

(b) would the average be sued for further computations?

(c) The typical value required in the particular problem. Within the framework of descriptive statistics the main requirement is to know what each average means and then select one that fulfils the purpose in hand. Is a composite average of all absolute or relative values needed (arithmetic mean or geometric mean) or is middle value wanted (median) or the most common value (mode)?

(d) The purpose which the average is designed to serve.

On occasions it may even be advisable to work out more than one average and present them although to be sure this procedure creates an added burden for the reader as well as for the statistician. But he added burden is preferable to the use of single average that may be an incomplete description. To use it also is like looking through a key hole the part of the room you can see cannot give a full idea of the whole room.

Median. The median is generally the best average in open-end grouped distributions especially where if plotted as a frequency curve one gets a J or reverse J curve; for example in case of price distribution or income distribution. In such cases very high or very low values would cause the mean to be higher to lower than the most "common" values. In such instances the median or middle value of the series may be more representative figure to use in describing the mass of data.

Mode. Generally speaking the need of mode lies in the fact that it can be used to describe quantitative data. The mode can be used in problems involving the expression of preferences where quantitative measurements are not possible. Thus the preferred type of package design among a number of alternative design would be the modal design. If we want to compare consumer preferences for different kinds of products or different kinds of advertising we can compare the modal preferences expressed by different groups of people but we cannot calculate the median or mean Mode is a particularly useful average for discrete series e.g., number of people wearing a given size of shoe or number of children per household etc. The mode is best suited where there is an outstandingly large frequency.

Geometric Mean. Geometric mean is useful in averaging ratio and percentages and in computing average rates of increase or decrease. It is particularly important in Economics and Business Statistics in index number construction.

Harmonic mean is useful in problems in which of a variable are compared with a constant quantity of another variable i.e., rates time distance covered within certain time and quantities purchased or sold per unit etc.

Harmonic Mean

In the following cases arithmetic mean should not be used:

(a) when there are very large and very small items arithmetic mean would be seriously misleading on account of undue influence from extreme items.

(b) The arithmetic mean should not be used to average ration and rates of change. In such cases the geometric mean is more suitable.

(c) When the distribution is unevenly spread concentration being smaller or large at irregular points.

(d) In distributions with open-end intervals.

(e) In highly skewed distributions.

Leaving aside the above specific cases where either median mode geometric mean or harmonic mean is more appropriate in other cases we should apply as a rule of thumb the arithmetic mean–the most popular and widely used average in practice.

It may also be pointed out that a complete description of distribution occasionally calls for two or more of these averages. It is true that presenting two or more averages creates an added burden for the investigator as well as for the consumer of statistics. However, the work this extra burden entails is fully justified if it presents a more complete description of the data that is possible from a single measure.

General Limitations of Average

(a) At times the average may give a very absurd result. For example if we are calculating size of a family we may get a value 4.8. But this is impossible as persons cannot be in fractions. However we should remember that it is an average value representing the entire group.

(b) An average may given us value that does not exist in the data. For example the arithmetic mean of 100, 300, 250, 50, 100 is = 160 a value that does not exist in the data.

(c) Measures of Central Tendency fall to give an idea about the formation of the series. Two or more series may have the same central value but may differ widely in composition. For example observe the following two series:

Series A	*Series B*
150	300
170	500
190	20
210	20
180	2
Total 900 $\overline{X}$ = 180	900 = 180

(d) Since an average is a single value representing a group of values it must be properly interpreted; otherwise there is every possibility of Jumping to wrong conclusions. This can best illustrated with the help of a story. A person had to cross the river from one bank to another. He was not aware of the depth of the river so he enquired of another man who told him that the average depth of water is 5' 4". The man was 5' 6" and the he though that he can very easily cross the river because at all time he would be above the level of water. So he started. In the beginning the level of water. So he reached the middle the water was 15 ft. deep and he lost his life. The man was drowned because he had a misconception that average depth means uniform depth throughout. But it is not so. An average represents a group of values and lies somewhere in between the tow extremes i.e. the largest and the smallest items of the series.

Example 77:

The price of a commodity doubles in a period of 4 years. What is the average percentage increase per year?

Solution:

Geometric mean would be more appropriate here. Applying the following formula:

$$P_n = P_0 \left(1 + \frac{r}{100}\right)^n$$

Here $P_n = 200 \; P_0 = 100 \text{ and } n = 4$

$$200 = 100 \left(1 + \frac{r}{100}\right)^n$$

$$\left(1 + \frac{r}{100}\right) = 2^{1/4}$$

Let $X = 1^{1/4}$

$$\text{Log } X = \frac{1}{4} \text{ Log } 2 \; \frac{1}{4} \; (0.301) = 0.07525$$

$$X = AL\ (0.07525) = 1.19$$

$$\left(1 + \frac{r}{100}\right) = 1.19 = 1.19 - 1 = 0.19 \text{ or } 19 \text{ per cent.}$$

$$\frac{r}{100} = 1.19 - 1 = 0.19 \text{ or } 19 \text{ per cent.}$$

FORMULAS

Individual Series	*Desecrate Series*	*Continuous Series*
Arithmetic Mean		
Direct Method:	Direct Method	Direct Method
$\overline{X} = \frac{\Sigma X}{N}$	$\overline{X} = \frac{\Sigma fX}{N}$	$\overline{X} = \frac{\Sigma fm}{N}$
Short-cut Method	Short-cut Method	Short-cut Method
$\overline{X} = A + \frac{\Sigma X}{N}$	$\overline{X} = A + \frac{\Sigma fX}{N}$	$\overline{X} = A + \frac{\Sigma fm}{N}$
Step Deviation Method	Stem Deviation Method	
$\overline{X} = A + \frac{\Sigma X}{N} \times i$	$\overline{X} = A + \frac{\Sigma fm}{N} \times i$	
Median		
Size of $\frac{N+1}{2}$ th item	Size of $\frac{N+1}{2}$ th item	Size of $\frac{N+1}{2}$ th item
		Med. $= L + \frac{N/2 - c.f.}{f} \times i$
Mode		
Either by inspection or	$M_0 = L + \frac{\Delta_1}{\Delta_1 + \Delta_2} \times i$	
The value that occurs largest number of times.	grouping method determine that value around which most of the fre-frequencies are concentrated.	$\Delta_1 = \lvert f_1 - f_2 \rvert$ $\Delta_1 = \lvert f_1 - f_2 \rvert$

Empirical Mode: Mode = 3 Median – 2 Mean

Geometric Mean		
G.M. $= AL\left(\frac{\Sigma \log X}{N}\right)$	G.M. $= AL\left(\frac{\Sigma f \log X}{N}\right)$	G.M. $= AL\left(\frac{\Sigma f \log m}{N}\right)$
Harmonic Mean		
H.M. $= AL\ \frac{N}{\Sigma(1/x)}$	H.M. $= AL\ \frac{N}{\Sigma(f/x)}$	H.M. $= \frac{N}{\Sigma(f/x)}$
Weighted Arithmetic Mean	Weighted Geometric Mean	Weighted Harmonic Mean

$$\overline{X}_w = \frac{\Sigma WX}{\Sigma W} \qquad G.M._w = AL\left[\frac{\Sigma(\log XW)}{\Sigma W}\right] \qquad H.M._w = \frac{\Sigma W}{\left(\frac{1}{a}\times W_1\right)+\left(\frac{1}{b}\times W_2\right)}$$

Empirical Mode: Mode = 3 Median – 2 Mean

Combined Arithmetic Mean	Weighted Geometric Mean
$\overline{X}_{12} = \frac{N_1 \overline{X}_1 + N_2 \overline{X}_2}{N_1 + N_2}$	$G.M._{12} = AL \left(\frac{N_1 \log GM_1 + N_1 \log GM_2}{N_1 + N_2}\right)$

EXERCISES

1. Define and explain mean, median and mode. Mention formula to find out mean, median and mode both for ungrouped and grouped data.
2. Calculate mean and median from the following ungrouped data:

   (i) 10, 8, 20, 22, 39 and 18 **[Ans.** $\overline{X}$ = 19.5, median = 19]

   (ii) 19, 21, 17, 16, 19, 21, 23 and 23. **[Ans.** $\overline{X}$ = 19.8, median = 20]

   (iii) 17, 19, 13, 17, 13, 11 and 21. **[Ans.** $\overline{X}$ = 15.8, median = 17]

   (iv) 15.8, 13. 5.2, 13.3, 17.8, 18.1 and 18.9. **[Ans.** X =16.05, median =15.8]

   (v) 1060, 1060, 1070, 1130, 1370, 1190, 1270 and 1170.

   (vi) No. of animals per cage

   3, 7, 8, 11, 13, 15, 16, 17, 18 and 20. **[Ans.** X= 12.8]

   (vii) Haemoglobin percent in g/100 ml

   6.0, 6.5, 7.5, 8.2, 8.5, 8.7, 8.8, 8 ,9, 9 and 9.5 **[Ans.** X = 8.16]
3. (a) The following table gives the marks obtained by 60 students in statistics in certain examination. Find mean and median from the data:

Examination Marks	*No. of students*
More than 70%	*7*
More than 60%	*18*
More than 50%	*40*
More than 40%	*40*
More than 30%	*63*
More than 20%	*65*

(b) An income tax assessed depreciated the machinery of his factory by 20 percent in each of the first two years and 40 per cent in the third year. How much average depreciation relief should be claimed from taxation department?
[Hint : G.M 27.32%]

(c) Compute the weighted mean of first natural numbers when weights are equal to the corresponding numbers.

4. The management of a college decides to give scholarship to the students who have scored marks 70 and above 70 in Business Statistics. The following are the marks scored by II B. Com. students:

71	73	74	85	86	88	91	94	96	99
74	74	76	93	91	94	96	98	88	94

The scholarship payable is given below:

Marks	Scholarship amount (Rs.)
70-75	100
75-80	200
80-85	300
85-90	400
90-95	500
95-100	600

Estimate the total scholarship payable and average scholarship

5. (a) Find the mean and mode from the following data:

% mark :	10-19	20-29	30-39	40-49	50-59	60-69	70-79
Students :	8	19	29	36	25	13	4

[$\overline{X}$ = 42.41. Mo. = 43.39]

(b) Draw ogive curves for the following data and hence find the median:

Age (years)	No. of persons	Age (years)	No. of persons
20-25	9	40-45	19
25-30	25	45-50	13
30-35	34	50-55	7
35-40	25	55-60	2

[35]

6. Tick the correct answer from the following

(a) The sum of the deviations of individual observations (i) mode, (ii) median, (iii) geometric mean, (iv) none of these.

(b) Which average is affected most by extreme observations (i) mode, (ii) median,
(iii) geometric mean,
(iv) arithmetic mean, (v) harmonic mean.

(c) In a moderately asymmetrical distribution:

(i) A.M. > G.M. > H.M. (ii) A.M > G.M. > H.M.

(iii) A.M. > G.M. > H.M. (iv) G.M > A.M. > H.M.

(v) H.M. > G.M. > A.M.

(d) Which of the following is the most unstable average: (i) mode, (ii) median, (iii) geometric mean, (iv) harmonic mean, (v) arithmetic mean.

(e) For dealing with qualitative data the best average is (i) geometric mean, (ii) median, (iii) harmonic mean. (iv) median, (v) mode.

(f) The positional measure of central tendency is (i) geometric mean, (ii) median, (iii) harmonic mean, (iv) arithmetic mean, (v) none of these.

(g) One of the methods of determining mode is (i) mode = 2 median –3 mean, (ii) mode = 2 median + 3 mean, (iii) mode = 2 median –2 mean. (iv) mode = 3 median + 2 mean, (v) none of these.

(h) The geometric mean of two numbers 8 and 18 shall be (i) 12. (ii) 13, (iii) 15, (iv) 11.08, (v) none of these.

(i) In a moderately skewed distribution the values of mean and median are 5 and 6 respectively. The value of mode in such a situation is approximately equal to:

(a) 8, (b) 11, (c) 16, (d) none of these

Ans. (a) (iv), (b) (iii), (c) (ii), (d) (i), (e) (iv), (f) (ii), (g) (iii), (h) (i).

7. Calculate mean, median and mode from the following data:

Marks more than :	0	20	40	60	80	100	120
No. of students :	80	76	50	28	18	9	3

[$\overline{X}$=50. Med.= 49.09, Mode=45]

8. (a) Calculate mean, median, mode from the following data of the heights in inches of a group of students:

61, 62, 63 61, 63, 64, 60, 65, 63, 64, 66, 64,

Now suppose that a group of students, whose heights are 60, 96, 59, 68,67 and 70 inches, is added to the original group, find the median and mode of the combined group.

[Mean = 67.79: Med. = 63.5: Mode = 64]

(b) From the following frequency distribution, find the class frequencies that are missing:

Intelligence Quotient	No. of students	Intelligence Quotient	No. of students
55-64	2	105-114	?
65-74	19	115-124	92
75-84	78	125-134	14
85-94	?	135-134	4
95-104	301		

You are given that the total frequency is 900 and the median 100.48. [18:304]

9. Find out the median and mode from the following table:

No. of days absent	No. of students	No. of days absent	No. of students
Less than 5	29	Less than 30	644
Less than 10	224	Less than 35	650
Less than 15	465	Less than 40	653
Less than 20	582	Less than 45	655
Less than 25	634		

[Med. = 12.75, Mo. = 11.35]

10. A laundry uses two different makes of washing machines. According to its past experience, the following results have been recorded:

Make of Machine	Median life	Mean Life
A	6,500 hours	6,000 hours
B	6,000 hours	6,500 hours

If both makes are of the same price, which make should the laundry purchase from now on? Give reasons.

11. A travelling salesman made five trips during the past two months, making the sales as tabulated below:

Trip	No. of days	Value of sales (Rs.)	Sales per day (Rs.)
1	4	500	125
2	5	380	76
3	8	576	72
4	3	108	63
5	4	260	65
	N = 24	1.824	374

(a) The sales manager criticised the salesman's performance as not very good since his mean sales are only Rs. 74.8 $\left(\frac{374}{5} = 74.8\right)$. The salesman replied that the sales manager was unfair in making such a statement, for his mean sales were as high as Rs. 76 $\left(\frac{1824}{24} = 76\right)$. What does each average mentioned here mean?

(b) Which mean seems to you appropriate in this case?

12. (a) What is statistical average? What are the desirable properties for an average to possess? Which of the average you know possess most of these properties?

(b) What is meant by measure of central tendency? What are the characteristics of a good measure of central tendency?

13. (a) Prove that the arithmetic mean of two positive numbers a and b is at least as large as their geometric mean.

(b) Find the missing figure:

Median = Mode + ? (Mean–Mode).

14. (a) What are the functions of an average?

(b) Explain the method of drawing a Lorenz curve with the help of an example.

15. (a) What characteristics a good average should possess? How far does the median meet these requirements? Explain the concept and uses of weighted average.

(b) "Every average has its own peculiar characteristics. It is difficult to say which average is the best". Comment briefly.

16. (a) Why is arithmetic mean considered to be the most suitable measure of central tendency?

(b) State the empirical relationship between mean, median, and mode.

17. (a) Discuss the merits and demerits of median as a measure of central tendency. Also calculate the two quartiles Q_1 and Q_3 from the following data:

(b) Describe the different Measure of central tendency of a frequency distribution, mentioning their merits and demerits.

(c) Briefly explain the role of Grouping Table and Analysis Table in calculation of Mode.

18. (a) What are the properties of a good average? Examine these properties with reference to the arithmetic mean and geometric mean.

(b) Point out the difference between the simple arithmetic mean and weighted arithmetic mean.

(c) A man travels 20 kms at 40 kms per hours, 10 kms at 60 kms per hours. What is his average speed?

19. (a) What do your understand by central tendency? Explain with the help of an example. What purpose does a measure of central tendency serve?

(b) Explain clearly the concept of central value taking a suitable example. Does central value always imply middle most value?

20. (a) What are the various Measure of location of distribution and for what purposes are they used?

(b) The purpose of an average is to represent a group of individual values in simple and concise manner so that the mind can get a quick understanding of general size of individuals in the group." Explain.

(c) State the various Measure of central tendency.

(d) "An average is a substitute for complex group of variables but its is not always safe to depend on the substitute all to the exclusion of individual members of the group Discuss.

21. (a) A motor car covered a distance of 50 miles four times. The first time at 50 m.p.h., the second at 20 m.p.h., the third at 40 m.p.h. and the fourth at 25 m.p.h. Calculate the average speed and explain the choice of the average.

(b) A man gets three annual raises in salary. At the end of the first year, he gets an increase of 4%, at the end of second. An increase of 6% on his salary as it was at the end of the year, and at the end of the third year, an increase of 9% on his salary as it was at the end of the third year. What is the average percentage increase?

(c) A machine is assumed to depreciate 40 per enc in value in the first year, 25% in the second year and 10% per annum for the next three years, each percentage being calculated on the diminishing value. What is the average percentage depreciation for five years? Ans. (a) 29.63, (b) 6.3%, (c) 20.02%

22. (a) Mr. A spends Rs. 100 for apples costing Rs. 10 per kilogram and another Rs. 100 for apples costing Rs. 8 kilogram. What is the average price per kilogram?

(b) A company expects the sale of its products to in crease by 50% in the following year. By what percentage should it increase the selling price in order to double its gross revenue?

(c) Three men take 12, 8, 6 hours respectively to husk an acre of corn. Determine the average number of hours to husk an acre.

(d) In a moderately skewed distribution arithmetic mean = 24.6 and the mode = 16.1. Find the value of the median and explain the reason for the method employed.

Ans. (a) 4.44. (c) $\overline{X}$ = 5.67, (d) 25.1

23. (a) The price of a commodity increased by 20% from 1997 to 1988, by 15% from 1998 to 1999 and by 10% from 1999 to 2000. Find the yearly average percentage increase from 1997 to 2000.

(b) Calculate geometric mean and harmonic mean of the following data:

125, 1462, 3, 7, 0.22, 0.08, 12.75, 0.5

(a) 14.9, (b) G.M. = 6.952 H.M.

24. Classify the following data of 50 marks obtained in statistics subject appropriately and compute the arithmetic and geometric means:

64,	26,	56,	85,	42,	38,	10,	55,	72,	65,
15,	47,	59,	62,	52,	54,	48,	62,	81,	31,
44,	54,	50,	52,	66,	38,	77,	88,	17,	58,
15,	25,	49,	51,	64,	68,	53,	50,	72,	58,
50,	61,	70,	80,	90,	05,	16,	51,	61,	52,

[Hint: Take classes 5-15, 15-25, etc.; $\overline{X}$ = 52.8]

25. (a) An investor purchased securities of a company investing a sum of Rs. 2,400 every month. If he bought these securities at Rs. 15 during the three months respectively what is the average price paid by him?

(b) Calculate the mean, median and mode for the following frequency distribution and verify the empirical relation connecting them:

Marks	:	0-20	20-40	40-60	60-80	80-100

Frequency :	3	17	27	20	9

[$\overline{X}$ = 54. Med = 53. Med. = 53.46, Mo. = 52]

26. (a) Draw a less than ogive from the following and locate the median:

Size :	10-20	20-30	30-40	40-50	50-60
Frequency :	20	60	100	150	75

[41.5]

(b) Calculate weighted arithmetic mean from the following data:

X :	1500	800	500	250	100
W :	10	20	70	100	150

[$\overline{X}_w$ = 302.86]

27. (a) From the information given below, find:
   (i) Which factory pays larger amount as daily wages?
   (ii) What is the average daily wage for the workers of two factories?

	Factory A	Factory B
No. of wage earners	250	200
Average daily wages	Rs. 12.0	Rs. 13.8

(b) A cyclist covers his first three kinds at an average speed of 8 km. Per hour, another two kms at 9 kms. per hour and the last two km. at 2 kms. per hour, Find the average speed for the entire journey.

(c) The average weight of a group of 25 boys was calculated to be 78.4 lb. It was later discovered that weight of one boy was misread as 69 lb. instead of the correct value 96 lb. Calculate the correct average.

(a) $\overline{X}_{12}$ = 12.8, (b) 4.3 km., (c) 79.5

28. Give a specific example of your own for each of the following cases:
   (a) The Indian is preferred to the arithmetic mean.
   (b) The geometric mean would be more satisfactory than the arithmetic mean.
   (c) The median would be preferred to the mode.
   (d) The mode would be preferred to the median.
   (e) The harmonic mean must be used instead of the arithmetic mean.
   (f) No average would be meaningful.

29. (a) Draw less than cumulative frequency curve for the following distribution. Read the median from the graph and verify your

result by the mathematical formula. Also obtain the limits of monthly income of central 50% of the employees:

Monthly Income (Rs.)	No. of Employees
Below 2000	3
2000-2200	7
2200-2400	25
2400-2600	30
2600-2800	24
2800-3000	9
3000 and above	2

[Med = 50, Q_1 = 2320, Q_3 = 2683.33, $Q_3 - Q_1$ = 363.33]

(b) Calculate mean and median for the following data:

Profits (Rs. Lakhs)	Frequency	Profits (Rs. Lakhs)	Frequency
10-20	4	10-60	124
10-30	16	10-70	137
10-40	56	10-80	146
10-50	97	10-90	150

[$\overline{X}$ = 46.33. Med. = 44.63]

(c) The mean age of a combined group of men and women is 25 years. If the mean age of the group of men is 26 and that of the group of women is 21. Find out the percentage of men and women in the group.

30. You are given below a distribution of income per month. Calculate the most suitable average giving reasons for your choice.

Income (Rs.)		Frequency
Less than 100	...	40
100-200	...	89
200-300	...	148
300-400	...	64
400 and above	...	39

[Since it is an open end distribution median would be more appropriate. Med. = 241. 22]

31. Calculate the Median from the following data, if mean value is 44:

Marks	No. of Students	Marks	No. of Students
70-80	10	30-40	12
60-70	10	20-30	7
50-60	20	10-20	8
40-50	?	0-10	5

(Missing value = 28, Med. = 50)

32. (a) What are averages? Calculate the mean and mode for the following frequency distribution:

Monthly wages (Rs.)	No. of workers	Monthly wages (Rs.)	No. of workers
Less than 200	78	600-800	42
200-400	165	800-1,000	12
400-600	93		

[$\overline{X}$ = 369.23. Mo = 309.43]

(b) Find out the quartiles and the mode for the following (take suitable class intervals):

Income (Rs.)	No. of Persons	Income (Rs.)	No. of Persons
Below 30	69	61-70	58
31-40	167	71-80	27
41-50	207	81 and above	10
51-60	65		

[Q_1 = 35.4, Q_2 = 43.66. Q_3 = 51.92, Mo = 42.7]

33. How would you account for the predominant choice of arithmetic mean of statistical data as a measure of central tendency? Under what circumstances would it be appropriate to use mean. mode and median? Discuss.

34. (a) Compare and contrast arithmetic mean. Geometric mean and harmonic mean. Which of them is least affected by extreme items?

(b) What are the desirable properties of a good average? Under what circumstances the use of the median harmonic and geometric means would give better results than the arithmetic mean? Explain with examples.

35. What do yōu mean by 'Central Tendency'? What are the desirable properties for an average? Which average possesses most of these properties?

36. (a) Discuss the merits and demerits of geometric mean. Explain its utility and algebraic characteristics.

    (b) Indicate briefly which of the properties of good measure of central tendency are possessed by any two of the following: Arithmetic mean, median, mode, geometric mean and harmonic mean.

37. (a) Explain what is meant by weighted average and discuss the effect of weighting.

    (b) Under what circumstances geometric mean and harmonic mean are suitable Measure for describing the central tendency of a frequency distribution?

    (c) What is an average? Explain the geometric mean and state its merits and limitations.

38. Construct a frequency table for the following data regarding annual profits (in thousands of rupees) of 50 firms, taking 25-34, 35-44, etc. a class intervals:

28	35	61	29	36	48	57	67	69	50
48	40	47	42	41	37	51	62	63	33
31	32	35	40	38	37	60	51	54	56
37	46	42	38	61	59	58	44	39	57
38	44	45	45	47	38	44	47	47	64

Construct a less then ogive and find:

(a) Number of firms having profit between 37.000 and Rs. 50,000

(b) Middle 50% group's range.

39. Calculate mean, median, media, mode for the following data:

Weekly Earnings (Rs) :	66-676768	68-69	69-70	70-71	71-22	
No. of Persons :	15	24	40	20	14	11

[$\overline{X}$ = 68.72. Med = 68.575, Mo = 68.44]

40. The following table shows the distribution of families according to their daily expenditure. The median and mode for distribution are Rs. 24.8 and Rs. 24, respectively. Find missing frequencies and A.M. ($\overline{X}$) of data.

Expenditure (Rs.) :	0-10	10-20	20-30	30-40	40-50
No. of Families :	14	–	27	–	15

41. An incomplete distribution is given as follows:

Variable :	0-10	10-20	20-30	30-40	40-50	50-60	60-70
Frequency :	10	20	?	40	?	25	15

You are given that the median value is 35. Using the median formula, fill up the missing frequencies.

[Hint : Refer to practical statistics $f_1 = 25$, $f_2 = 25$, $\overline{X} = 35.88$]

42. From a batch of 13 students, who had appeared for an examination. 4 students have failed. The marks of the successful candidates were 41, 57, 38, 61, 36, 35, 71, 50, 40, Calculate the median marks. [Med = 50]

43. The following table gives the weekly wages in rupees in a certain commercial organisation

Weekly wages (Rs.) :	30-32	32-34	34-36	36-38	38-40	40-42
Frequency :	2	9	25	30	49	62
Weekly wages (Rs.) :	42-44	44-46	46-48	48-50		
Frequency :	39	20	99	3		

Calculate from the above data:

(i) the median and the third quartile wages: and

(ii) the number of wage emmers receiving between Rs. 37 and Rs. 46 per week. Ans. (i) 41.74 : 46.35: (ii) 180

44. (a) Give a brief note of the Measure of central tendency together with their merits and demerits. Which is the best measure of central tendency and why?

(b) Under what circumstances would it be appropriate to use arithmetic mean, median and mode? Discuss.

45. (a) "The arithmetic mean is the best among all the averages." Given reasons to substantiate the statement. State the purpose of studying the other averages.

(b) Compare and contrast various Measure of central tendency.

46. (a) A book seller has 150 books of Economics and Accountancy. The average price on these books is Rs. per book. Average price of books on Economics is Rs. 43 and that of Accountancy is Rs. 35. Find out the number of books on Economics with the seller?

(b) Draw ogive by less than and more than methods for the following weekly income distribution:

Weekly Income (Rs.)	No. of employees	Weekly Income (Rs.)	No. of employees
Below 550	5	700-750	16
550-600	10	750-800	12
600-650	22	Above 800	15
650-700	30		

Read the value of median from the graph and verify your value from the formula of median. Also obtain the limits of weekly income of central 50 percent of the employees.

47. (a) A man travelled from one city to another. The distance between the two cities is 4 kms. He drives his car at 40 kms per hour. After travelling one km., the car stops running. He then travels on a tonga at 10 km, per hour. After travelling a distance of 1.5 km, he leaves and covers the remaining distance on foot at 4 km, per hour, Find the average speed per hour of that person.

(b) (i) Draw an ogive curve form the following data and measure the median value. Verify it by actual calculations.

Central Size :	5	15	25	35	45
Frequency :	5	11	21	16	10

(ii) The mean weight of 150 workers in a factory is 56 kgs. If the mean weight of men in the factory is 64 kgs. and that of the women is 48 kg., find the number of men and women is the factory.

48. Draw an ogive curve (less than type) for the following data and hence find: (i) median: (ii) inter-quartile range:

Daily Wages (Rs.)	No. of Workers	Daily Wages (Rs.)	No. of Workers
1–5	7	21–25	24
6–10	10	26–30	18
11–15	15	31–35	10
16–20	30	36–40	6

[Med = 20.2, (ii) 8.17]

49. (a) Define median. Indicate its merits and demerits.

(b) From the following data draw the Histogram and Ogive and determine the mode and the median graphically:

Marks :	0–6	6–12	12–18	18–24	24–30	30–36
No. of students:	4	8	15	20	12	6

(c) Calculate median and mode of the following series:

Size	:	6–10	11–15	16–20	21–25	26–30
Frequency	:	20	30	50	40	10

[Med. = 17.5. Mo. = 18.83]

50. (a) 'A' travelled some distance by cycle at a speed of 15 km. per hour. On return journey he travelled the same distance at a speed of 10 km per hour. What was his average speed per hour? [Hint : Calculate H.M. = 12]

(b) The average monthly wage of all workers in a factory is Rs. 444. If the average wages paid to male and female workers are Rs. 480 and Rs. 360 respectively, find the percentage of male and female works employed by the factory.

51. From the following data of the heights of 100 persons in a commercial concern, determine the modal height:

Height is inches :	58	60	61	62	63	64	65	66	68	70
No. of persons :	4	6	5	10	20	22	24	6	2	1

52. (a) The percentage of marks of 20 students, who were eligible for scholarships, are given below:

52	62	51	71	54	61	53	51	50	57
64	56	54	69	63	65	59	58	68	57

Calculate the monthly scholarship payable to the students as well as the average scholarship.

(b) Calculate mean, median and mode for the following frequency distribution:

Marks	No. of students	Marks	No. of students
0–20	21	51–60	24
21–30	19	61–70	18
31–40	60	71–80	15
41–50	42		

53. (a) The mean age of a group of 100 persons was found to be 32.02. Later, it was discovered that age 57 was misread as 27. Find correct mean.

(b) The goniometric mean of 10 observations was calculated as 28.6. It was later discovered that one of the observations was recorded as 23.4 instead of 32.4. Apply appropriate correction and calculate the correct geometric mean.

54. Calculate mean, median and mode from the following data:

Marks	No. of students	Marks	No. of students
10–20	4	10–60	124
10–30	16	10–70	137
10–40	56	10–80	146
10–50	95	10–90	150

[$\overline{X}$ = 46.33: Med. = 44.63: Mode = 40.67]

55. Compute Mean. Median and Mode from the following data:

Income (Rs.): More than	10	20	30	40	50	60	70	80
No. of persons:	72	67	59	50	36	21	9	3

[$\overline{X}$ = 39.03: Med. = 40: Mo = 42.5]

56. *(a) Calculate the weighted geometric mean of the following data:*

Weights (in lbs):	130	135	140	145	146	148	149	150
No. of Workers:	3	4	6	6	—	3	5	2	1

[142.5]

(b) Find out the missing frequencies for the class intervals using following data on 150 students:

Marks	No. of students	Marks	No. of students
0–10	4	50–60	25
10–20	8	60–70	40
20–30	11	70–80	?
30–40	15	80–90	2
40–50	?	90–100	1

Given that the mean marks are 65.

57. (a) Determine the healthier town from the following information:

Age group (years)	Town 'A' population	(Standard) Deaths	Town 'B' population	(Local) Deaths
0–15	15,000	300	20,000	500
15–50	20,000	400	52,000	1,040
50 & above	5,000	140	8,000	240

(b) Calculate the value of mean for the following data:

Mid point :	10	15	20	25	30	35	40
Frequency :	7	9	18	26	10	4	3

[$\overline{X}$ = 23.05]

58. (a) State, giving reasons, the average to be used in the following situations:
   (i) To determine the average size of the shoe sold in the shop.
   (ii) To determine the average wages in an industrial concern.
   (iii) To find the average beauty among a group of students in a class.
   (iv) To find the per capita income in different cities.
   (b) State the formula of median for grouped data with class intervals.
   (c) Under what circumstances is harmonic mean the most suitable?
   (d) What is an average? Under what circumstances would you use the geometric mean instead of the arithmetic average?
   (e) Explain the relationship between mean, median and mode in a symmetrical and moderately asymmetrical distribution.

59. (a) An aeroplane travels distances of d_1. d_2 and d_3 kms at speeds V_1, V_2 and V_3 kms per hour respectively. Show that the average speed is given by V_1 where:

$$\frac{d_1 + d_2 + d_3}{V} = \frac{d_1}{V_1} + \frac{d_2}{V_2} + \frac{d_3}{V_3}$$

   (b) Calculate Geometric mean:

   2 4, 8, 12, 16, 24.

60. (i) Determine the value of mode with mean and median whose values are 20 and 22 respectively.
   (ii) The arithmetic mean of two observations is 25 and their geometric mean is 15. Find their harmonic mean.

   [(i) 26 (ii)]
   (iii) A cyclist covers this first five km. at an average speed of 10 km. p.h., another 3 km. at 8 km. p.h. and the last 2 km. p.h. Find the average speed for the entire journey.

61. Tick whether the following statements are True or False:
   (i) Arithmetic mean is always the best measure of central tendency. T/F
   (ii) An average alone is sufficient to understand the basic characteristics of a frequency distribution. T/F
   (iii) The value of median and mode can be determined graphically. T/F
   (iv) The value of median is affected more by sampling fluctuations than the value of arithmetic mean. T/F

(v) Geometric mean is more appropriate for dealing with problems of rates and speed. T/F

(vi) Median is a computed measure of central tendency. T/F

(vii) The subtraction of a constant from each of a set of numbers to be averaged changes the average of the set by the value of the constant. T/F

(viii) The harmonic mean is useful when data are given is terms of rates. T/F

(ix) In a positively skewed distribution, the value of mode is greater than the mean. T/F

Ans. (i) F, (ii) F, (iii) T, (iv) T, (v) F, (vi) F, (vii) T, (viii) T, (ix) F.

62. An incomplete frequency distribution is given below:

Marks :	0–10	10–20	20–30	30–40	40–50	50–60	60–70	Total
Frequency :	4	16	–	–	–	6	3	230

Find the three missing frequencies of the table, given that median = 33.5 and mode = 34. Also calculate the mean using the empirical relation between mean. Median and mode.

63. (a) Obtain the value of median from the following data:

335, 384, 407, 672, 522, 777, 753, 2488, 1490, [67.2]

(b) Calculate the mean from the following frequency table:

Mid points	Frequency	Mid points	Frequency
1	2	6	155
2	60	7	79
3	101	8	40
4	152	9	1
5	205		

64. (a) Calculate mean, median and mode from the following frequency distribution of marks at a test in English.

Marks :	5	10	15	20	25	30	35	40	45	50
No. of students :	20	43	75	76	72	45	39	9	8	6

[$\overline{X}$ 22.16, Med, = 20, Mo, = 20]

(b) Geometric mean of 2 numbers is 15. If by mistake one figure is taken as 5, instead of 3, find correct geometric mean.

65. (a) A market with 330 firms has the following distribution of average number of laborers in different income groups:

Incommensurable (value) Rs.	:	450	750	1050	1350	1650
No. of firms	:	80	64	52	56	78
Average no. of workers	:	16	24	15	17	8

Find the median income of all the orders.

(b) Determine the Geometric mean for the following values:

X : 5 8 12 16

[9.363]

66. (a) How Would you account for the predominant choice of arithmetic mean as a measure of central tendency? Under what circumstances would it be appropriate to use mode of median?

(b) Ram buys 1 kg. oranges from each of the four places at the rate of 2kg 4kg., 5kg., 8 kg., per rupee respectively. Find out how many oranges per rupee he purchased on an average?

67. Fill is the blanks:

(i) Median is better suited for ... interval series.

(ii) In a moderately asymmetrical distribution, the distance between the ...... and the ...... is about ...... the distance between the ...... and the ......

(iii) Given mean 25, mode 24, the median would be......

(iv) For studying phenomena like intelligence and honesty...... is a better average to used, while for phenomena like the size of shoes or redeemed garments. The average w be used is...

(v) In a symmetrical distribution mean... median.... mode.

(vi) The geometric mean is the....root of the product of all the measurements.

(vii) The mode a distribution is the value that has the greatest of....

(viii) The harmonic mean is the ...... of the arithmetic mean of the values.

Ans. (i) positional, (ii) mean, median, 1/3 mean, mode, (iii) 24.67, (iv) median, mode, (v) is equal to, is equal to (vi) nth, (vii) concentration, frequencies. (viii) reciprocal of.

68. Compute N (no. of observation) when X (mean) = 5 and –X = 30.

[**Ans.** 6]

69. The following is frequency distribution of the diameter of 1,000 parts of a particular type produced by the ABC Co. Ltd. Find the mean diameter of these parts.

Diameter (in centimeters)	Number of parts	Diameter (in centimetres)	Number of parts
3.05 to 3.14	13	3.55 to 3.64	296
3.15 to 3.24	29	3.65 to 3.74	74
3.25 to 3.34	65	3.75 to 3.84	40
3.35 to 3.44	200	3.85 to 3.94	5
3.45 to 3.54	278		

[$\overline{X}$ = 3.5074]

70. The following data was obtained in a grass land community. Calculate mean, median and mode by ungrouped and grouped series.

Number of seeds per plant (Indigofera species):

39, 55, 35, 45, 49, 52, 48, 33, 48, 47, 50, 51, 53, 50. 55, 53, 54, 53, 50, 48, 49, 31, 33, 50, 55, 50, 51, 50, 53, 55, 52, 49, 51; 50, 50, 44, 51, 50, 58, 59, 57, 59, 60, 58, 51.

5

Hypothesis Testing

INTRODUCTION

The averages in themselves do not adequately describe a distribution. Averages locate the center of a distribution but tell us nothing about how the measurements are arranged in relation to the center. Whenever data are collected, whether for a sample or for a population, it is desirable to consider the dispersion in the data. *Variability or dispersion is defined as the extent of scatterness around a measure of central tendency.*

A measure of dispersion gives an idea of the extent to which the individual measures differ from the measure of central tendency. Let us take two illustrations. Suppose that we have two distributions on the height of plants each with the mean of 68 inches. In the first distribution the highest value is 72 inches and the lowest 62 inches. The second distribution has a high value of 107 inches and a low of 25 inches. The range of the first distribution is 11 and that of the second distribution is 83. Thus to give a better picture of a distribution we need both, a measure of central tendency and that of variability or dispersion.

As another example of the importance of variability, consider a nurse who takes a patient's temperature. She knows that the normal temperature is 98.6°F or in other words, the mean of a population of temperatures taken from healthy person's is 98.6°F. If a person's temperature is 98.9°F, she still considers it normal and does not conclude that the person is ill. From the experience she knows that the temperatures of healthy individuals are variable and 98.9°F can be obtained even in persons without any illness.

Variability in biological studies is very important as the living beings exhibit a great amount of variability. Some of the commonly used numerical measures of variability in a data set are

(1) Range
(2) Mean Deviation
(3) Variance
(4) Standard Deviation

(5) Coefficient of variation

It is clear from above that dispersion (also known as scatter, spread or variation) measures the extent to which the items vary from some central value. Since Measures of Dispersion (or Variation) give an average of the differences of various items from an average, they are also called averages of the *second order.*

An average is more meaningful when it is examined in the light of dispersion. For example, if the average wage of the workers of factory A is Rs. 3885 and that of factory B Rs. 3900. We cannot necessarily conclude that the workers of factory B are better of because in factory B there may be much greater dispersion in the distribution of wages.

The study of dispersion is of great significance in practice as could well be appreciated from the following example:

	Series A	**Series B**	**Series C**
	100	100	1
	100	105	489
	100	102	2
	100	103	3
	100	90	5
Total	500	500	500
$\overline{X}$	100	100	100

Since arithmetic mean is the same in all three series, one is likely to conclude that these series are alike in nature. But a close examination shall reveal that distributions differ widely form one another. In series A, each and every item is perfectly represented by the arithmetic mean or, in other words. None of the items of series A deviates from the arithmetic mean hence there is no dispersion. In series B, only one item is perfectly represented by the arithmetic mean and the other items vary but the variation is very small as compared to series C. In series C, not a single item is represented by the arithmetic mean and the items vary widely from one another. In series C dispersion is much greater compared to series B. Similarly, we may have two groups of laborers with the same mean salary and yet their distributions may differ widely. The mean salary may not be so important a characteristic as the variation of the items from the mean. To the student of social affairs, the mean income is not so vitally important as to know how that income is distributed. Are a large number receiving the mean income or are there a few with enormous incomes and millions with incomes far below the mean? The tree figures below represent frequency distributions with some of the characteristics we wish to emphasis here.

The two curves in diagram (a) represent two distributions with the same mean $\overline{X}$, but with different dispersions. The two curves in (b) represent two distributions with the same dispersion but with unequal means. $\overline{X}_1$ and $\overline{X}_2$. (c) represents two distributions with unequal dispersion.

The measures of central tendency are, therefore, insufficient. They must be supported and supplemented with other measures. In this chapter, we shall be especially concerned with the measures of variability, or spread or dispersion. A measure of variation or dispersion is one that measures the extent to which there are differences between in dividable observation and some central or average value. In measuring variation we shall be interested in the amount of the variation or its *decrepit* not in the *direction.* For example, a measure of 6 inches below the mean has just as much dispersion as a measure of six inches above the mean.

Significance of Measuring Variation

Measures of variation are needed for four basic purposes:

1. To determine the Reliability of an Average.
2. To serve as a basis for the control of the vaıiability.
3. To Compare Two or More Series with Regard to their Variability
4. To Facilitate the use of Other Statistical Measures.

A brief explanation of these points is given below:

(i) Measures of variation point out as to how far an average is representative of the mass. When dispersion is small. The average is a typical value in the sense that it closely represents the individual value and it is reliable in the sense that it is a good estimate of the average in the corresponding universe. On the other hand, when dispersion is large, the average is not so typical, and unless the sample is very large, the average may be quite unreliable.

(ii) Another purpose of measuring dispersion is to determine nature and cause of variation in order to control the variation itself. In matters of health variations in body temperature, pulse beat and blood pressure are the basic guides to diagnosis. Prescribed treatment is designed to control their variation. In industrial production efficient operation requires control of quality variation, the causes of which are sought through inspection is basic to the control of causes of specially important. In social sciences a special problem requiring the measurement of variability is the measurement of "inquality" of the distribution of income or wealth, etc.

(iii) Measures of Dispersion (or Variation) enable a comparison to be made of two or more series with regard to their variability. The study of variation may also be looked upon as a means of determining uniformity of consistency. A high degree of variation would mean little uniformity or consistency whereas a low degree of variation would mean great uniformity or consistency.

(iv) Many powerful analytical tools in statistics such as correlation analysis, the testing of hypothesis, analysis of variance, the statistical quality control, regression analysis are based on measures of variation of one kind or another.

Properties of A Good Measure of Variation

A good measure of dispersion should possess, as far as possible, the following properties:

(i) It should be simple to understand.

(ii) It should be easy to compute.

(iii) It should be rigidly defined.

(iv) It should be based on each and every item of the distribution.

(v) It should be amenable to further algebraic treatment.

(vi) It should be have sampling stability.

(vii) It should not be unduly affected by extreme items.

Methods of Studying Variation

The following are the important methods of studying variation:

1. The Range.
2. The Interquartile Range and the Quartile Deviation.
3. The Mean Deviation or Average Deviation.
4. The Standard Deviation, and
5. The Lorenz Curve.

Of these the first two, namely, the range and quartile deviations, are positional measures because they depend on the values at a particular position in the distribution. The other two, the average deviation and the standard deviation, are called calculation measures of deviation because all of the values are employed in their calculation and the last one is a graphic method.

Absolute and Relative Measures of Variation Measures of Dispersion (or Variation) may be either absolute or relative. Absolute Measures of Dispersion (or Variation) are expressed in the same statistical unit in which the original data are given such as rupees, kilograms, tonnes etc. These values may be used to compare the variations in two distributions provided the variables are expressed in the same units and of the same average size. In case the two sets of data are expressed in different units, however, such as quintals of sugar versus tonnes of sugarcane, or if the average size is very different such as manager's salary versus workers' salary, the absolute Measures of Dispersion (or Variation) are not comparable. In such cases measures of relative dispersion should be used.

A measure of relative dispersion is the ration of a measure of absolute dispersion to an appropriate average. It is sometimes called a coefficient of dispersion, because "coefficient" means a pure number that is in dependent of relative dispersion the average used as base should be the same one from which the absolute deviations were measured. This means that the arithmetic mean should be used with the standard deviation, and either the arithmetic mean or median with the mean deviation.

1. Range

Range is the simplest method of studying dispersion. It is defined as the difference between the value of the smallest item and the value of the largest item included in the distribution. Symbolically.

$$\text{Range} = L - S$$

where L = Largest item, and

S = Smallest item.

The relative measure corresponding to range, called the coefficient of range, is obtained by applying the following formula:

$$\text{Coefficient of Range} = \frac{L - S}{L + S}$$

If the averages of the two distributions are about the same, a comparison of the range indicates that the distribution with the smaller range has less dispersion, and the average of that distribution is more typical of the group.

Example 1:

Compute coefficient of quartile deviation from the following data:

Marks	*10*	*20*	*30*	*40*	*50*	*80*
No. of Students	*4*	*7*	*15*	*8*	*7*	*2*

Solution:

Calculation of coefficient of quartile deviation

Marks	Frequency	c.f.	Marks	Frequency	c.f.
10	4	4	40	8	34
20	7	11	50	7	41
30	15	26	60	2	43

$$Q_1 = \text{Size of } \frac{N+1}{4} \text{ th item} = \frac{43+1}{4} = 11\text{th item.}$$

Size of 11th item is 20. Thus $Q_1 = 20$

$$Q_3 = \text{Size of } 3\left(\frac{N+1}{4}\right) \text{ th item} = \frac{3 \times 44}{4} = 33\text{rd item.}$$

Size of 33rd item is 40. Thus $Q_3 = 40$

$$\text{Q.D.} = \frac{Q_3 - Q_1}{2} = \frac{40-20}{2} = 10$$

$$\text{Coefficient of Q.D.} = \frac{Q_3 - Q_1}{Q_3 + Q_1} = \frac{40-20}{40+20} = 0.333$$

$$Q_3 = \text{Size of } \frac{3N}{4} \text{th item } \frac{3 \times 200}{4} = 150\text{th item}$$

Q_3 lies in the class 38 – 40

$$Q_3 = L + \frac{3N/4 - \text{c.t.}}{f} \times i$$

$$L = 38,\ 3N/4 = 150,\ \text{c.f.} = 76,\ t = 99,\ i = 2$$

$$Q_3 = 38 + \frac{150 - 76}{99} \times 2 = 38 + 1.49 = 39.\ 49$$

$$\text{Q.D.} = \frac{39.49 - 36.16}{2} = 1.67$$

$$\text{Coefficient of Q.D} = \frac{Q_3 - Q_1}{Q_3 + Q_1} = \frac{39.49 - 36.16}{39.49 + 36.16} = \frac{3.33}{75.65} = 0.044$$

Merits and Limitations

Merits: In certain respects it is superior to range as a measure of dispersion.

(a) It is also useful in erratic or badly skewed distributions where the other Measures of Dispersion (or Variation) would be warped by extreme values. The quartile deviation is not affected by the presence of *extreme values.*

(b) It has a special utility in measuring in case of open end distributions or one in which the data may be ranked but measured quantitatively.

Limitations: Quartile deviation ignores 50% items, i.e., the first 25% and the last 25%. As the value of quartile deviation does not depend upon every item of the series it cannot be regarded as a good method of measuring dispersion.

(a) It is in fact not a measure of dispersion as it really does not show the scatter around an average but rather a distance on a scale, i.e., quartile deviation is not itself measured from an average, but it is a positional average. Consequently, some statisticians speak of quartile deviation as a measure of *partition* rather than a measure of dispersion. If we really desire to measure variation in the sense of showing the scatter round an average, we must include the deviation of each and every item from an average in the measurement.

(b) Its value is very much affected by sampling fluctuations.

(c) It is not capable of mathematical manipulation.

Because of the above limitations quartile deviation is not often useful for statistical inference

Percentile Range: Like semi-interquartile range, the percentile range is also used as a measure of dispersion. Percentile range of a set of data is defined as:

$$\text{Percentile Range} = P_{90} - P_{10}$$

where P_{90} and P_{10} are the 90th and 10th percentiles respectively. The semi-percentile range, i.e. $\left(\frac{P_{90} - P_{10}}{2}\right)$ can also be used, but is not commonly employed.

The Mean Elevation

The two methods of dispersion discussed above, namely, range and quartile deviation, are not Measures of Dispersion (or Variation) in the strict sense of the term because they do not show the scatterness around an average. However, to study the formation of a distribution we should take the deviations from an average. The two other measures namely, the average deviation and the standard deviation help us in achieving this goal.

The mean deviation is also known as the average deviation. It is the average difference between the items in a distribution and the median or mean of that series. Theoretically there is an advantage in taking the deviations from median because the *sum of deviations of items from median is minimum when signs are ignored.* However, in practice, the arithmetic mean is more frequently used in calculating the value of average deviation and this is the reason why it is more commonly called mean deviation. In any case the average used must be clearly stated in a given problem so that any possible confusion in meaning is avoided.

Computation of Mean Deviation-Individual Observations: If X_2, X_1, X_3, X_N are N given observations then the deviation about an average A is given by

$$\text{M.D.} = \frac{1}{n} \Sigma \mid X - A \mid \frac{1}{N} \Sigma \mid D \mid \quad \Rightarrow \quad \frac{\sum \mid D \mid}{N}$$

where | D | = X – A |. Read as mod (X – A) is the modulus value or absolute value of the deviation ignoring plus and minus signs.

Steps

(a) Obtain the total of these deviations, i.e., Σ | D |.

(b) Compute the median of the series.

(c) The deviations of items from median ignoring ± signs and denote these deviations by | D |.

(d) Divide the total obtained in step (iii) by the number of observations.

If a distribution is normal, the mean ± mean deviation is the range that will include 57.7 percent of the items in the series. If it is moderately skewed, then we may expect approximately 57.5 percent of the items to fall within this range. Hence, if average deviation is small, the distribution is highly compact or uniform, since more than half of the cases are concentrated within a small range around the mean.

Example 2:

The following are the prices of shares of AB Co. Ltd. from Monday to Saturday

Day	Price (Rs.)	Day	Price (Rs.)
Monday	200	Thursday	160
Tuesday	210	Friday	220
Wednesday	208	Saturday	250

Calculate range and its coefficient.

Solution:

Range = L – S

Here L = 250 and S = 160

Range = 250 – 160 = Rs. 90

$$\text{Coefficient of Range} = \frac{L - S}{L + S} = \frac{250 - 160}{250 + 160} = \frac{90}{160} = 0.22.$$

Continuous Series: There are two methods of determining the range from data grouped into a frequency distribution. The first method is to find the difference between the upper limit of the highest wage class and the lower limit of the lowest wage class. The other method is to subtract the midpoint of the lowest wage class from the mid-point of the highest wage class. In practice, both the methods are used.

Example 3:

Calculate the mean deviation and its coefficient of the two income groups of five and seven members given below:

I (Rs.): 4,000 4,200 4,400 4,600 4,800

II (Rs.): 3,000 4,000 4,200 4,400 4,600 4,800 5,800

Solution:

Calculation of mean deviation

	Group I	Group II	
	Deviation form median 4400 | D |		**Deviation from median 4400 | D |**
4,000	400	3,000	1,400
4,200	200	4,000	400
4,400	0	4,200	200
4,600	200	4,600	0
4,800	400	4,600	200
		4,800	400
		5,800	1,400
n = 5	Σ | D | = 1200	N = 7	Σ | D | = 4,000

Mean Deviation: Group I: M.D. $= \dfrac{\sum|D|}{N}$

$|D|$ = Deviation from median ignoring signs,

$$\text{Median} = \text{Size of } \frac{N+1}{2}\text{th item} = \frac{5+1}{2} = \text{3rd item}$$

Size of 3rd item is 4,400 M.D. $= \dfrac{1{,}200}{5} = 240$

This means that the average deviation of the individual incomes from the median income is Rs. 240.

Mean Deviation: Group II

$$\text{Mean} = \text{Size of } \frac{N+1}{2}\text{th item} = \frac{7+1}{2} = \text{4th item}$$

Size of 4th item is 4, 400

$$\Sigma\,|D| = 4{,}000, \quad N = 7.$$

$$\text{M.D.} = \frac{4{,}000}{7} = 571.43.$$

Note. If we were to compute coefficient of mean deviation we shall divide mean deviation by median. Thus for the first group:

$$\text{Coefficient of M.D.} = \frac{240}{4{,}400} = 0.\ 054$$

and for the second group

$$\text{Coefficient of M.D.} = \frac{571.43}{4{,}400} = 0.\ 130.$$

Calculation of Mean Deviation

Discrete Series : In discrete series the formula for calculating mean deviation is

$$\text{M.D.} = \frac{\sum f|D|}{N} \quad \text{(by the same logic as given before)}$$

$|D|$ denotes deviation from median ignoring signs.

Steps.

(a) Multiply these deviations by the respective frequencies and obtain the total $\sum f|D|$.

(b) Take the deviations of the items from median ignoring signs and denote them by $|D|$.

(c) Divide the total obtained in Step (ii) by the number of observations. This gives us the value of mean deviation.

(d) Calculate the median of the series.

Example 4:

Find out the value of quartile deviation and its coefficient from the following data:

Roll No.	*1*	*2*	*3*	*4*	*5*	*6*	*7*
Marks	*20*	*28*	*40*	*12*	*30*	*15*	*50*

Solution:

Calculation of Quartile Deviation

Marks arranged in ascending order: 12 15 20 28 30 40 50

$$Q_1 = \text{Size of } \frac{N+1}{4} \text{th item} = \text{Size of } \frac{7+1}{4} = \text{2nd item}$$

Size of 2nd item is 15. Thus $Q_1 = 15$

$$Q_3 = \text{Size of } 3\left(\frac{N+1}{4}\right) \text{th item} = \text{Size of } \left(\frac{3 \times 8}{4}\right) \text{th item} = \text{6th item}$$

Size of 6th item is 40. Thus $Q_3 = 40$

$$\text{Q.D.} = \frac{Q_3 - Q_1}{2} = \frac{40 - 15}{2} = 12.\ 5.$$

$$\text{Coefficient of Q.D.} = \frac{Q_3 - Q_1}{Q_3 + Q_1} = \frac{40 - 15}{40 + 15} = \frac{25}{55} = 0.455.$$

Example 5:

Calculate quartile deviation and the coefficient of quartile deviation from the following data:

Wages in Rupees per week	*less then 35*	*35-37*	*38-40*	*41-43*	*over 43*
Number of wage earners	*14*	*62*	*99*	*18*	*7*

Solution:

Calculation of Q.D. and its coefficient

Wages (Rs. per week)	**f**	**c.f.**
Less than 35	14	14
35–37	62	76
38–40	99	175
41–43	18	183
over 43	7	200

$$\text{Q.D.} = \frac{Q_3 - Q_1}{2}$$

$$Q_1 = \text{Size of } \frac{N}{4}\text{th item} = \frac{200}{4} = 50 \text{ th item}$$

Q_1 lies in the class 35–37.

$$Q_1 = L + \frac{N/4 - c.t.}{f} \times i$$

$$L = 35,\ N/4 = 50,\ c.f. = 14,\ f = 62,\ i = 2$$

$$Q_1 = 35 + \frac{50 - 14}{62} \times 2 = 35 + 1.16 = 36.16$$

The relative measure corresponding to the mean deviation, called the coefficient of mean deviation, is obtained by dividing mean deviation by the particular average used in computing mean deviation. Thus if mean deviation has been computed from median, the coefficient of mean deviation shall be obtained by dividing mean deviation by median.

$$\text{Coefficient of M.D.} = \frac{\text{M.D}}{\text{Median}}.$$

If mean has been used while calculating the value of mean deviation, in such a case coefficient of mean deviation shall be obtained by dividing mean deviation by the mean.

Example 6:

Calculate Coefficient of Range from the following data:

Marks	*No. of students*	*Marks*	*No. of students*
10–20	*8*	*40–50*	*8*
20–30	*10*	*50–60*	*4*
30–40	*12*		

Solution:

$$\text{Coefficient of Range} = \frac{L - S}{L + S} = \frac{60 - 10}{60 + 10} = \frac{50}{70} = 0.714$$

Merits and Limitations The merits and limitations of Range can be numerated here.

Imitations.

(a) Range cannot tell us anything about the character of the distribution within the two extreme observations. Observe the following three series:

Series A	46	6	46	46	46	46	46	46
Series B	6	10	6	6	46	46	46	46
Series C	6	6	15	25	30	32	40	46

In all the three series range is the same, i.e., (46–6) = 40. But it does not mean that the distributions are alike. The range takes no account of the form of the distribution within the range. Range is, therefore, unreliable as a guide to the dispersion of the value within a distribution.

(b) It is subject to fluctuations of considerable magnitude from sample to sample.

(c) Range is not based on each and every item of the distribution.

Merits.

(a) It takes minimum time to calculate the value of range. Hence, if one is interested in getting a quick rather than very accurate picture of variability one may compute range.

(b) Amongst all the methods of studying dispersion range is the simplest to understand and the easiest to compute.

Uses : Despite serious limitations range is useful in the following cases:

(i) *Quality Control.* The object of quality control is to keep a check on the quality of the product without 100% inspection. When statistical methods of quality control are used, control charts are prepared and in preparing these charts range plays a very important role. The idea basically is that if the range-the difference between the largest and smallest mass produced items-increases beyond a certain point. The production machinery should be examined to find out why the items produced have not followed their usual more consistent pattern.

(ii) *Fluctuations in the Share Prices.* Range is useful in studying the variations in the prices of stocks and shares and other commodities that are sensitive to price changes from one period to another. For example, by computing range we can get an idea about the range of variation of say, gold prices. If the minimum price for 10 gm. gold in the year 1999 was Rs. 4050 and the maximum price Rs. 4950 this at once tells us about the range of variation, i.e., Rs. 900 (4950 – 4050).

(iii) *Weather Forecasts.* The meteorological department does make use of the range in determining, say, the difference between the minimum temperature and the maximum temperature. This information is of great concern to the general public because they know as to within what limits the temperature is likely to vary on a particular day.

(iv) *Everyday Life.* The range is a most commonly used measure of dispersion in everyday life. Questions of the form "what is the minimum and maximum temperature on a particular day"? "What is the difference between the wages earned by workers of a particular factory"? How much one spends on petrol in his car/scooter in a month"–are all usually answered in the form of range. Answers to questions such as these are usually given in the form of 'Between such and such. Regardless of the crudity of expression the answer is still a range.

The Interquartile Range or The Quartile Deviation

The range as a measure of dispersion discussed above has certain limitations. It is based on two extreme items and it fails to take account of the scatter within the range. From this there is reason to believe that if the dispersion of the extreme items is discarded, the limited range thus established might be more instructive. For this purpose there has bee developed a measure called the *interquartile range,* the range which includes the middle 50 percent of the distribution. That is one quarter of the observations at the lower end, another quarter of the observations at the upper end of the distribution are excluded in computing the interquartile range. In other words, interquartile range represents the difference between the third quartile and the first quartile.

Symbolically.

Interquartile range = $Q_3 - Q_1$.

Very often the interquartile range is reduced to the form of the Semi-interquartile range or quartile deviation by dividing it by 2.

Symbolically.

$$\text{Quartile Deviation or Q.D} = \frac{Q_3 - Q_1}{2}.$$

Quartile deviation gives the average amount by which the two quartiles differ from the median. In symmetrical distribution the two quartiles (Q_1 and Q_3) are equidistant from the median, i.e., Med. – Q_1 = Q_3 – Med. and as such the difference can be taken as a measure of dispersion. The median ± Q.D. covers exactly 50 percent of the observations.

In reality, however, one seldom finds a series in business and economic data that is perfectly symmetrial. Nearly all distributions of social series are asymmetrical. In an asymmetrical distribution. Q_1 and Q_3 are not equidistant from the median. As a result an asymmetrical distribution includes only approximately 50 percent of observations.

When quartile deviation is very small, it describes high uniformity or small variation of the central 50% items and a high quartile deviation means that the variation of the central 50% items and a high quartile deviation means the variation among the central items is large.

Quartile deviation is an absolute measure of dispersion. The relative measure corresponding to this measure, called the coefficient of quartile deviation, is calculated as follows.

$$\text{Coefficient of Q.D.} = \frac{(Q_3 - Q_1)/2}{(Q_3 + Q_1)/2} = \frac{Q_3 - Q_1}{Q_3 + Q_1}$$

Coefficient of quartile deviation can be used to compare the degree of variation in different distributions.

Computation of Quartile Deviation : The process of computing quartile deviation is very simple, we have just to compute the values of the upper and lower quartiles. The following illustrations would clarify calculations.

Example 7:

X Ltd. is actively considering the following two mutually exclusive project for adoption.

Year	***Project X*** ***Cost profit (Rs. in Lakhs)***	***Project Y*** ***Cash profit (Rs. in Lakhs)***
1	*10*	*5*
2	*5*	*25*
3	*20*	*45*
4	*40*	*30*
5	*60*	*30*

Which is the most risky project (use coefficient of variation)

Solution:

For finding out which is more risky project out of X and Y, we compare coefficent of variation.

Calculation of Coefficient of variation

	Project X			**Project Y**	
X	$(X - \bar{X})$ **x**	**x²**	**Y**	$(Y - \bar{Y})$ **y**	**y²**
10	–17	289	5	–22	484
5	–22	484	25	–2	4
20	–7	49	45	+18	324
40	+13	169	30	+3	9
60	+33	1089	30	+3	9
$\Sigma x = 135$	$\Sigma x = 0$	$\Sigma x^2 = 2080$	$\Sigma y = 135$	$\Sigma y = 0$	$\Sigma y^2 = 830$

Project X

$$C.V. = \frac{\sigma}{X} \times 100$$

$$\bar{X} = \frac{\sum x}{N} = \frac{135}{7} = 27$$

$$\sigma = \sqrt{\frac{\sum x^2}{N}} = \sqrt{\frac{2080}{5}} = 20.4$$

$$C.V. = \frac{20.4}{27} \times 100 = 75.56$$

Project Y

$$C.V. = \frac{\sigma}{Y} \times 100$$

$$\bar{Y} = \frac{\sum y}{N} = \frac{135}{7} = 27$$

$$\sigma = \sqrt{\frac{\sum y^2}{N}} = \sqrt{\frac{830}{5}} = 12.88$$

$$C.V. = \frac{12.88}{27} \times 100 = 47.7$$

Since coefficient of variation is much more for project X hence it is a more risky project.

Which Measure of Dispersion to Use

Like measure of central value, in case of measures of variation also, the choice of a suitable measure depends on the following two factors:

(i) *The type of data available.* If they are few in number, or contain extreme values, avoid the standard deviation. If they are generally skewed, avoid the mean deviation as well. If they have gaps around the quartiles, the quartile deviation should be avoided. If there are open-end classes, the quartile measure of dispersion should be preferred.

(ii) *The purpose of investigation.* In an elementary treatment of statistical series in which a measure of variability is desired only for itself any of three measures, namely, range, quartile deviation and average deviation, would be acceptable. Probably the average deviation would be better. However, in usual practice, the measure of variability is employed in further statistical analysis. For such a purpose the standard deviation, by far, is the most popularly used. It is free from those defects from which other measures suffer. It leads itself to the analysis of variability in terms of normal curve of error. Practically all advanced statistical methods deal with variability and centre around the standard deviation. Hence, unless the circumstances warrant the use of any other measure, we should make use of standard deviation for measuring variability.

List of Formulae

Individual Observations	***Discrete & continuous Series***
Range Range = L = S Coeff. of Range = $\frac{L - S}{L + S}$	(same as on the left) But L i.e., largest value, will be upper limit of the highest class and S will be the lower limit of the lowest class
Quartile Devotion Q.D. = $\frac{Q_3 - Q_1}{2}$	(Same as on the left)
Coeff. of Q.D. = $\frac{Q_3 - Q_1}{Q_3 + Q_1}$ Mean Deviation M. D. = $\frac{\sum\lvert D\rvert}{N}$ Coeff. of M.D. = $\frac{M.D.}{Median}$ or $\frac{M.D.}{Mean}$ (if deviations are taken from mean)	 M.D. = $\frac{\sum f\lvert D\rvert}{N}$ Coeff. of M.D. = $\frac{M.D.}{Median}$ or $\frac{M.D.}{Mean}$ (if deviations are taken from mean)
Standard Deviation *Actual Mean Method*	*Actual Mean Method*

$$\sigma = \sqrt{\frac{\left(\sum(X - X)^2\right)}{N}}$$

$$\sigma = \sqrt{\frac{\sum(X - \overline{X})^2}{N}}$$

Assumed Mean Method

$$\sigma = \sqrt{\frac{\sum d^2}{N} - \left(\frac{\sum d}{N}\right)^2}$$

Assumed Mean Method

$$\sigma = \sqrt{\frac{\sum fd^2}{N} - \left(\frac{\sum fd}{N}\right)^2}$$

Step Deviation method

$$\sigma = \sqrt{\frac{\sum d^2}{N} - \left(\frac{\sum d}{N}\right)^2} \times i$$

Step Deviation method

$$\sigma = \sqrt{\frac{\sum fd^2}{N} - \left(\frac{\sum fd}{N}\right)^2} \times i$$

$$C.V. = \frac{\sigma}{X} \times 100$$

$$C.V. = \frac{\sigma}{X} \times 100$$

Combined Standard Deviation

$$\sigma_{12} = \sqrt{\frac{N_1\sigma_1^2 + N_2\sigma_2^2 + N_1d_1^2 + N_2\sigma_2^2}{N_1 + N_2}}$$

where $d_1 = |\overline{X}_1 - \overline{X}_{12}|$ and $d_2 = |\overline{X}_2 - \overline{X}_{12}|$

Example 8:

You are given below the daily wages paid to the workers in two factories X and Y:

Daily Wages	***No. of workers***	
	Factory X	***Factory Y***
12-13	*15*	*25*
13-14	*30*	*40*
14-15	*44*	*60*
15-16	*60*	*35*
16-17	*30*	*12*
17-18	*14*	*15*
18-19	*7*	*5*

Using appropriate measure answer the following?

(i) *Which factory pays higher average wage?*

(ii) *Which factory has more consistent wage structure.*

Solution:

For finding out which factory pays higher average wage, we have to compute the arithmetic means and for finding out which factory has more consistent wage structure, we have to compare coefficient of variation.

Calculation of $\overline{X}$ and C.V.

Wages	m.p	t	Factory X			Factory Y		
(Rs.)			(m – 15)/3	fd	fd²	f	fd	fd²
12-13	12.5	15	–3	–45	135	25	–75	225
13-14	13.5	30	–2	–60	120	40	–80	160
14-15	14.5	44	–1	–44	44	60	–60	60
15-16	15.5	60	0	0	0	35	0	0
16-17	16.5	30	+1	+30	30	12	+12	12
17-18	17.5	14	+2	+28	56	15	+30	60
18-19	18.5	7	+3	+21	63	5	+15	45
		N = 200		Σfd = – 70	Σfd² = 448		Σfd = –– 158	Σfd² = 562

Factory X factory Y

$$\overline{X} = A + \frac{\sum fd}{N} = 15.5 - \frac{70}{200} = 15.15$$

$$\overline{X} = A + \frac{\sum fd}{N} = 15.5 - \frac{158}{200} = 14.71$$

Since arithmetic mean is higher for factory X, hence factory X pays higher average wage.

$$\sigma = \sqrt{\frac{\sum fd^2}{N} - \left(\frac{\sum d}{N}\right)^2} \qquad \sigma = \sqrt{\frac{\sum fd^2}{N} - \left(\frac{\sum fd}{N}\right)^2}$$

$$= \sqrt{\frac{448}{200} - \left(\frac{-70}{200}\right)} \qquad = \sqrt{\frac{562}{200} - \left(\frac{-158}{200}\right)^2}$$

$$= \sqrt{2.24 - 1225} = 1.455 \qquad = \sqrt{2.81 - 624} = 1.479$$

$$\text{C.V.} = \frac{\sigma}{X} \times 100 = \frac{1.45}{15.15} = 9.60 \quad \text{C.V.} = \frac{\sigma}{X} \times 100 = \frac{0.479}{14.71} \times 100 = 10.5$$

Since coefficent of variation is less for factory X, hence factory X has more consistent wage.

Example 9:

The mean and the standard deviation of one sample are respectively 54.4 and 8; the mean and the standard deviation of another sample are 50.3 and

7 respectively. The size of the first sample is 50 and that of the second is 1000. Find the mean and standard deviation of the composite sample (size 150) combining the aforesaid to samples.

Solution:

We are given the following information:

$$\overline{X}_1 = 54.4,\ \sigma_1 = 8,\ n_1 = 50\ \overline{X}_2 = 50.3,\ \sigma_2 = 7,\ n_2 = 100$$

We have to find $\overline{X}_{12}$ and σ_{12} $\overline{X}_{12} = \dfrac{n_1\overline{X}_1 + n_2\overline{X}_2}{n_1 + n_2}$

$$= \frac{(50 \times 54.4) + (100 \times 50.3)}{50 + 100} = \frac{2720 + 5030}{150} = \frac{7750}{150} = 51.67$$

$$\sigma_{12} = \sqrt{\frac{n_1\sigma_1^2 + n_2\sigma_2^2 + n_1\sigma_1^2 + n_2\sigma_2^2}{n_1 + n_2}}$$

$$d_1 = |\overline{X}_1 - \overline{X}_{12}| = |54.4 - 51.67| = 2.73$$

$$d_2 = |\overline{X}_2 - \overline{X}_{12}| = |50.3 - 51.\ 67| = 1.37$$

$$\sigma_{12} = \sqrt{\frac{50(8)^2 + 100(7)^2 + 50(2.73)^2 + 100(1.37)^2}{150}}$$

$$= \sqrt{\frac{3200 + 4900 + 372.645 + 187.69}{150}} = \sqrt{\frac{8660.335}{150}} = 7.6$$

Example 10:

Two brands of tyres are tested with the following results:

Life (in '000 miles)	No. of tyres brand	
	X	Y
20-25	*1*	*0*
25-30	*22*	*24*
30-35	*64*	*76*
35-40	*10*	*0*
40-45	*3*	*0*

(a) *Which brand of tyres have greater average life?*

(b) *Compare the variability and state which brand of tyres would you use on your fleet of trucks?*

Solution:

In order to answer part :

(a) we have to compare the means and to answer part

(b) compare the coefficient of variation.

Calculation of Coefficient of Variation (brand X)

Life ('000 miles)	m.p. m	f	(m-32.5)/5 d	fd	fd²
20-25	22.5	1	–2	–2	4
25-30	27.5	22	–1	–22	22
30-35	32.5	64	0	0	0
35-40	37.5	10	+1	+10	10
40-45	42.5	3	+2	+6	12
		N = 100		Σfd = – 8	Σfd² = 48

$$\overline{X} = A + \frac{\sum fd}{N} \times i$$

A = 32.5, Σfd = –8,
N = 100, i = 5

$$\overline{X} = 32.5 - \frac{8}{100} \times 5 = 32.5 - 4 = 32.1$$

$$\sigma = \sqrt{\frac{\sum fd^2}{N} - \left(\frac{\sum fd}{N}\right)^2} \times i = \sqrt{\frac{48}{100} - \left(\frac{-8}{100}\right)^2} \times 5$$

$$= \sqrt{4. - 0064} \times 5 = 6274 \times 5 = 3.137$$

$$\text{C.V.} = \frac{\sigma}{X} \times 100 = \frac{3.137}{32.1} \times 100 = 9.773$$

Calculation of Coefficient of Variation (brand X)

Life ('000 miles)	m.p. m	f	(m-32.5)/5 d	fd	fd²
20-25	22.5	0	–2	–2	0
25-30	27.5	24	–1	–24	24
30-35	32.5	76	0	0	0
35-40	37.5	0	+1	0	0
40-45	42.5	0	+2	0	0
		N = 100		Σfd = – 24	Σfd² = 48

$$\overline{X} = A + \frac{\sum fd}{N} \times i$$

A = 32.5, Σfd= – 24, N = 100, i = 5

$$\overline{X} = 32.1 - \frac{24}{100} \times 5 = 32.5 - 12 = 31.3$$

$$\sigma = \sqrt{\frac{\sum fd^2}{N} - \left(\frac{\sum fd}{N}\right)^2} \times i \sqrt{\frac{24}{100} - \left(\frac{-24}{100}\right)^2} \times 5$$

$$= \sqrt{24 - 0576} \times 5 = 1824 \times 5 = 3.137$$

$$C.V. = \frac{\sigma}{X} \times 100 = \frac{0.912}{31.3} \times 100 = 2.914$$

(a) Since arithmetic mean is more for brand X of tyres, they have greater average life.

(b) Since coefficient of variation is less for brand Y of tyres, they are more consistent and should be preferred for use.

Example 11:

The mean and standard deviation of a set of 100 observations were worked out as 40 and 5 respectively by a computer which by a computer which by mistake took the value 50 in place of 40 one observation. Find the correct mean and variance.

Solution:

$$\overline{X} = \frac{\sum X}{N}, \; N\overline{X} \; \Sigma X, \; N = 100, \; \overline{X} = 40$$

$$\Sigma X = 100 \times 40 = 4{,}000$$

But this is not the correct ΣX because one item has been taken as 50 instead of 40.

$\therefore$ Correct $\Sigma X = 4000 - 50 + 40 = 3{,}990$

$$\text{Correct Mean} = \frac{3990}{100} = 39.90$$

Correct variance

$$\text{Variance} = \frac{\sum X^2}{N} - (\overline{X})^2$$

$$\text{Variance} = \sigma^2 = (5)^2 = 25, \; N = 100$$

$$25 = \frac{\sum X^2}{100} - (40)^2$$

$$2500 = \Sigma X^2 - 160000 \quad \Rightarrow \quad \Sigma X^2 = 160000 + 2500 = 162500$$

$$\text{Correct } \Sigma X^2 = 162500 - (50)^2 + (40)^2 = 162500 - 2500 + 1600 = 161600$$

$$\text{Correct Variance} = \frac{\text{Correct} \sum X^2}{N} - (\text{Correct } \overline{X})^2$$

$$= \frac{161600}{100} - (39.9)^2 = \frac{161600 - 159201}{100} = \frac{2399}{100} = 23.99$$

Thus, correct mean = 39.9 and correct Variance = 23.99

Example 12:

The mean of 5 observations is 4.4 and the variance is 8.24. If the three of the five observations are 1, 2 and 6, find the other two.

Solution:

$$\overline{X} = \frac{\sum X}{N}, \quad \Sigma X = N\overline{X}$$

Here N = 5, $\overline{X}$ = 4.4; ΣX = 5 × 4.4 = 22

Let the two missing items be x_1 and x_2

$\therefore 1 + 2 + 6 + x_1 + x_2 = 22$

$\Rightarrow \quad x_1 + x_2 = 22 - 9$

$\Rightarrow \quad x_1 + x_2 = 13$

$$\sigma^2 = \frac{\sum X^2}{N} - \overline{X}^2$$

$$\Rightarrow \quad 8.\,24 = \frac{\sum X^2}{5} - (4.4)^2$$

$41.2 = \Sigma X^2 - 19.36 \times 5$

$\Rightarrow \quad \Sigma X^2 = 96.80 + 41.20 = 138$

$\Sigma X^2 = x_1 + x_2^2 + 1^2 + 2^2 + 6^2 = x_1^2 + x_2^2 + 41$

$x_1^2 + x_2^2 = 138 - 41 = 97$

$(x_1 + x_2)^2 = x_1^2 + x_2^2 + 2x_1x_2$

$(13)^2 = 97 + 2x_1x_2 \Rightarrow x_1x_2 = 36$

$2x_1x_2 = 169 - 97 \Rightarrow x_1x_2 = 36$

$(x_1 - x_2)^2 = x_1^2 + x_2^2 - 2x_1x_2 = 97 - 2(36) = 25$ ...(i)

$x_1 - x_2 = 5$

Adding $\underline{x_1 + x_2} = 13$ ...(ii)

$2x_1 = 18$

$x_2 = 9$ ...(iii)

Putting the value of x_1 in equation (i)

$$9 + x_2 = 13 \quad \therefore \quad x_2 = 4$$

Thus the two missing values are $x_1 = 9$, $x_2 = 4$.

Tchebycheff's Theorem : As discussed earlier standard deviation is a most widely used measure of variation. It has certain mathematical same units as the variable, whereas the variable is expressed in squared units.

Deeper insight into the role and meaning of standard deviation can be gained through examination and use of a theorem developed by the Russian mathematician Tchebycheff. According to this theorem, given a group of N numbers, at least the proportion $1 - (1/K)^2$ of the N observations will lie within K standard deviations of the mean. The symbol K represents number of standard deviations. The theorem can be quantified for any desired value of K. Each different value of K produces a new interval with different minimum proportions of observations encompassed.

To illustrate let K = 1.2 and 3. When K = 1, this interval created is the mean ± 1 standard deviation, i.e., $\mu \pm \sigma$. For this value of K. The minimum proportion of observations contained in the interval is

$$1 + (1/K)^2 = 1 - 1/1 = 1 - 1 = 0.$$

When K = 2. Then interval is $\mu \pm 2\sigma$ and the minimum proportion of items in the interval is:

$$1 - (1/3)^2 = 1 - 1/9 = 8/9$$

As K increases, the fraction obtained above would continue to increase as can be seen from the following table:

Tchebycheff Proportions

K values	Ranges	Minimum Proportion of Observations			
1	$\mu \pm \sigma$				0
2	$\mu \pm 2\sigma$	3/4	or	75	percent
3	$\mu \pm 3\sigma$	8/9	or	88.89	percent
4	$\mu \pm 4\sigma$	15/16	or	93.75	percent
5	$\mu \pm 5\sigma$	24/25	or	96.00	percent
:	:		:		
K	$\mu \pm k\sigma$		$1 - (1/K)^2$		

Lorenz Curve

The lorenz Curve, devised by Max O. Lorenz, famous economic statistician, is a graphic method of studying dispersion. This curve was used by him for the first time to measure the distribution of wealth and in wages, turnover, etc. However, still the most common use of this curve is in the study of the degree of inequality in the distribution of income and wealth between countries or between different periods of time. It is a cumulative percentage curve in which the percentage of items is combined with the percentage of other things as wealth. profits, turnover, etc.

While drawing the Lorenz curve the following procedure is adopted:

(i) The size of items (variable values) and frequencies are both cumulated. Taking grand total for eachas 100, percentages are obtained for these various coumulative values.

(ii) On the X-axis start from 0 to 1000 and take the percent of comulative frequencies.

(iii) On the Y-axis start from 0 to 100 and take the percent of the cumulated values of the variable.

(iv) Draw a disgonal line joing O(0, 0) with the point P(100, 100) as shown in the diagram below. The line OP will make an angle of 45° with the Y-axis and is called the line of equal distribution. Any point on this diagonal shows that same percent on X as on Y.

(v) Plot the percentages of the cumulated values of the variable (Y) against percentages of the correseponding cumulated frequencies (X) for the given distribution and join these points with a smooth freehand curve. For any given distribution this will never cross the line of equal distribution OP. It will always lie below OP unless the distribution is uniform in which case it will coincide with OP. The greater the variability, the greater is the distance of the curve from OP.

In the above diagram OP is the line of equal distribution. The points lying on the curve OAP indicate a less degree of variability as compared to the points lying on the curve OBP. When the points lie on the curve OCP. variability is still greater. Thus a measure of variability of the distribution is provided by the distance of the curve of the cumulated percentages of the given distribution from the line of equal distribution.

Example 13:

A student obtained the mean and standard deviation of 100 observations as 40 and 5.1 respectively. It was later found that one observation was wrongly copied as 50, the correct figure being 40. Find the correct men and standard deviation.

Solution:

We are given $\overline{X} = 40, \sigma = 5.1, N = 100$

$$\overline{X} = \frac{\sum X}{N}$$

$$40 = \frac{\sum X}{100} \Rightarrow \sum X = 4{,}000$$

But correct $\sum X = \sum X$ – wrong items + correct items

$= 4{,}000 - 50 + 40 = 3990$

$$\therefore \text{Correct} = \overline{X} = \frac{\text{Correct} \sum X}{N} = \frac{3990}{100} = 39.9$$

Corect standard deviation

$$\sigma = \sqrt{\frac{\sum X}{N} - (\overline{X})^2} \Rightarrow 5.1 = \sqrt{\frac{\sum X^2}{100} - (40)^2}$$

Squaring $26.01 = \frac{\sum X^2}{100} - 1600$

$\Rightarrow 2601 = \sum X^2 - 1{,}60{,}000 \; \sum X^2 = 162601$

Correct $\sum X^2 = 162601 - (50)^2 + (40)^2 = 162601 - 2500 + 1600 = 161701$

$$\text{Correct } \sigma = \sqrt{\frac{\text{Correct} \sum X}{N} - (\text{Correct } \overline{X})^2} = \sqrt{\frac{161701}{100} - (39.9)^2}$$

$$= \sqrt{161701 - 1592.02} = \sqrt{25} = 5.$$

Example 14:

Find the interquartile range and the coefficient of quartile deviation from the following data:

Markes in Staistics: above	*0*	*10*	*20*	*30*	*40*	*50*	*60*	*70*	*80*
No. of Students:	*150*	*140*	*100*	*80*	*80*	*70*	*30*	*14*	*0*

Solution:

Calculation of Interoquartile Range and Coefficient of Q.D.

Maks	f	c.f
0.10	10	10
10-20	40	50
20-30	20	70
30-40	0	70
40-50	10	80
50-60	40	120
60-70	16	136
70-80	14	150

Interquartile Range = $Q_3 - Q_1$

$$Q_1 \text{ Size of } \frac{N}{4}\text{th item} = \frac{150}{4} = 37.5\text{th item}$$

Q_1 lies in the class 10-20

$$Q_1 = L + \frac{N/4 - c.f}{f} \times i$$

L = 10, N/4 = 375, c.f 10, f = 40, i = 10

$$Q_1 = 10 + \frac{37.5 - 10}{40} \times 10 = 10 + 6.875 = 16.875$$

$$Q_3 = \text{Size of } \frac{3N}{4}\text{th item} = \frac{3 \times 150}{4} = 112.5\text{th item}$$

Q_3 lies in the class 50-60

$$Q_3 = L + \frac{3N/4 = c.f.}{f} \times i$$

L = 50, 3N/4 = 112.5, c.f. = 80, f = 40, i = 10

$$Q_3 = 50 + \frac{112.5 - 80}{40} \times 10 = 50 + 8.125 = 58.125$$

$$Q_3 - Q_1 = 58.125 - 16.875 = 45.623$$

$$\text{Coeff. of Q.D.} = \frac{Q_3 - Q_1}{Q_3 + Q_1} = \frac{58.125 - 16.875}{58.125 + 16.875} = \frac{45.625}{75} = 0.608$$

Example 15:

From the following information, find the standard deviation of x and y variable:

$\Sigma x = 235, \quad \Sigma x = 250$

$\Sigma x^2 = 6750, \ \Sigma y^2 = 6840$

$N = 10$

Solution:

$$\text{S.D. of x} = \sqrt{\frac{\sum x^2}{n} - \left(\frac{\sum x}{n}\right)^2}$$

$\Sigma x^2 = 6750, \Sigma x = 235, \ n = 10$

$$\sigma_x = \sqrt{\frac{6750}{10} = \left(\frac{235}{10}\right)^2} = \sqrt{675 - 552.25} = 11.08$$

$$\text{S.D = of y} = \sqrt{\frac{\sum y^2}{n} - \left(\frac{\sum y}{n}\right)^2}$$

$\Sigma y^2 = 6840,$
$\Sigma y = 250,$
$n = 10$

$$\sigma_x = \sqrt{\frac{6840}{10} = \left(\frac{250}{10}\right)^2} = \sqrt{685 - 625} = 7.68$$

Normally x is used to denote deviations from mean i.e. $x = (X - \overline{X})$ and $\Sigma(x - \overline{X})$ is always zero. But in the given question $\Sigma x = 235$ which means deviation are not taken from actual mean and hence the formula for assumed mean is used.

Example 16(a):

(a) Calculate mean deviation from the following series:

X	*10*	*11*	*12*	*13*	*14*
f	*3*	*12*	*18*	*12*	*3*

Solution:

Calculation of Meandeviation

X	f	| D |	f | D |	c.f.
10	3	2	6	3
11	12	1	12	15
12	18	0	0	33
13	12	1	12	45
14	3	2	6	48
	N = 48		Σ f | D | = 36	

$$\text{M.D.} = \frac{\sum f\,|D|}{N}$$

$$\text{Median} = \text{Size of } \frac{N+1}{2}\text{th item} = \frac{48+1}{2} = 24.5\text{th item}$$

Size of 24.5th item is 12, hence Median = 12

$$\text{M.D.} = \frac{36}{48} = 0.75.$$

Example 16(b):

Calculate the mean deviation from the mean for the following data:

Size:	*2*	*4*	*6*	*8*	*10*	*12*	*14*	*16*
Frequency:	*2*	*2*	*4*	*5*	*3*	*2*	*1*	*1*

Solution:

Calculation of Mean Deviation from Mean

X	f	fX	| X – 8 |	f | D |
2	2	4	6	12
4	2	8	4	8
6	4	24	2	8
8	5	40	0	0
10	3	30	2	6
12	2	24	4	8
14	1	14	6	6
16	1	16	8	8
	N = 20	Σ f X = 160		Σ f | D | = 56

$$\overline{X} = \frac{\sum fX}{N} = \frac{160}{20} = 8$$

$$M.D = \frac{\sum f|D|}{N} = \frac{56}{20} = 2.8.$$

Calculation of Mean Deviation-Continuous Series

For calculating mean deviation in continuous series the procedure remains the same as discussed above. The only difference is that here we have to obtain the mid-point of the various classes and take deviations of these points from median. The formaula is same, i.e.,

$$M.D. = \frac{\sum f|D|}{N}.$$

Example 17:

(a) Find the median and mean deviation of the following data:

Size	*Frequency*	*Size*	*Frequency*
0-10	*7*	*40-50*	*16*
10-20	*12*	*50-60*	*14*
20-30	*18*	*60-70*	*8*
30-40	*25*		

Solution:

Calculation of Median and Mean Deviation

Size	f	c.f.	m.p. m	| m – 35.2 | | D |	f | D |
0-10	7	7	5	30.2	211.4
10-20	12	19	15	20.2	242.4
20-30	18	37	25	10.2	183.6
30-40	25	62	35	0.2	5.0
40-50	16	78	45	9.8	156.8
50-60	14	92	55	19.8	277.2
60-70	8	100	65	29.8	238.4
	N = 100				Σf | D | = 1314.8

$$\text{Med.} = \text{Size of } \frac{N}{2}\text{th item} = \frac{100}{2}$$

Median lies in the clas 30-40

$$\text{Med.} = L + \frac{N/2 - \text{c.f.}}{f} \times i$$

L = 30, N/2 = 50,
c.f. = 37,
f = 25, i = 10

$$\text{Med.} = 30 + \frac{50 - 37}{25} \times 10 + 30 + 5.2 = 35.2$$

$$\text{M.D.} = \frac{\sum f|D|}{N} = \frac{1314.8}{100} = 13.148$$

Example 18(a):

(b) Calculate the mean deviation and its coefficient from the following data:

Size	*Frequency*	*Class*	*Frequency*
0-10	*5*	*40-50*	*20*
10-20	*8*	*50-60*	*14*
20-30	*12*	*60-70*	*12*
30-40	*15*	*70-80*	*6*

Solution:

Since nothing is special we will calculate mean deviation from median:

Calculation of Mean Deviation

Size	frequency f	c.f.	m.p. m	| m – 43 | | D |	f | D |
0-10	5	5	5	38	190
10-20	8	13	15	28	224
20-30	12	25	25	18	216
30-40	15	40	35	8	120
40-50	20	60	45	2	40
50-60	14	74	55	12	168
60-70	12	86	65	22	264
70-80	6	92	75	32	192
	N = 92				$\Sigma f \mid D \mid = 1414$

$$\text{Med.} = \text{Size of } \frac{N}{2}\text{th item} = \frac{92}{2} = 46\text{th item}$$

Median lies in the calss 40-50.

$$\text{Med.} = L + \frac{N/2 - c.f.}{f} \times i$$

L = 40,
N/2 = 46,
c.f. = 40,
f = 20,
i = 10

$$\therefore \text{Med} = 40 = \frac{46 - 40}{20} \times 10 = 40 + 3 = 43$$

$$\text{M.D.} = \frac{\sum f\,|D|}{N} = \frac{1414}{92} = 15.37$$

$$\text{Coeff. of M.D.} = \frac{\text{M.D}}{\text{Median}} = \frac{15.37}{43}$$

$$= 0.357.$$

Example 18(b):

Blood serum cholesterol levels of 10 persons are as under:

240, 260, 290, 245, 255, 288, 272, 263, 277, 251.

Calculate standard deviation with the help of assumed mean.

Solution:

Calculation of Standard Deviation by the Assumed Mean Method

x	(X – 264)* d	d^2
240	–24	576
260	–4	16
290	+26	676
245	–19	361
255	–9	81
288	+24	576
272	+8	64
263	–1	1
277	+13	169
251	–13	169
$\Sigma X = 2641$	$\Sigma d = +1$	$\Sigma d^2 = 2689$

$$\sigma = \sqrt{\frac{\sum d^2}{N} - \left(\frac{\sum d}{N}\right)^2}$$

$$\Sigma d^2 = 2689, \ \Sigma d = +1, \ N = 10$$

$$\sigma = \sqrt{\frac{2689}{10} - \left(\frac{1}{10}\right)^2}$$

$$= \sqrt{268.9 - 0.01} = 16.398$$

Merits and Limitations

Merits. The outstanding advantage of the average deviation is the its relative simplicity. It is simple to understand and easy to compute. Any one familiar with the concept of the average can readily appreciate the meaning of the average deviation. If a situation requires a measure of dispersion that will be presented to the general public or any group not very familiar with in statistics, the average deviation is useful.

(a) Since deviations are taken from a central value, comparison about formation of different distributions can easily be made.

(b) Mean deviation is less affected by the value of extreme itreme items than the standard deviation.

(c) It is based on each and every iterm of the data. Consequently change in the value of any item would change the value of mean deviation.

Limitations : The greatest drawback of this method is that algebraic signs are ignored while taking the deviations of the items. For example if from twenty, fifty is deducted we write 30 and not –30. This is mathematically wrong and makes the method non-algebraic. If the signs of the deviations are not ignored the net sum of deviations will be zero if the reference point is the mean or approximately zero or the reference point is median.

(a) It is rarely used in sociological studies.

(b) It is not capable of further algebraic treatment.

(c) This method may not given us very accurate results. The reason is that mean deviation gives us best results when deviations are taken from median. But median is not a satisfactory measure when the degree of variability in series is very high. And if we compute mean deviation from mean that is also not desirable because the sume of the deviations from mean (ignoring signs) is greater than the sum of the deviations from median (ignoring signs). If mean deviation is computed from mode that is also not scientific because the value of mode cannot always be determined.

Because of these limitations its use is limited and it is overshadowed as a measure of variation by the superior standard deviation.

Usefulness : The serious drawbacks of the average deviation should not bind us to its preactical utility, because of its simplicity in meaning and computation. It is expecially effective in reports presented to the general public or to groups not familiar with statistical methods. This measure is useful for small samples with no elaborate analysis required. Incidentally. It may be mentioned that the National Bureau of Economic research has found. in its work on forecasting business cycle, that the average deviation is the most proactical measure of dispersion to use of this purpose.

THE STANDARD DEVIATION

The standard deviation concept was interoduced by Karl Pearson in 1823. It is by far the most important and widely used measure of studying dispersion. Its significance lies in the fact that it is free from those defects from which the earlier methods suffer and satisfies most of the properties of a good measure of dispersion. Standard deviation is also known as *root mean square deviation* for the reason that it is the square root of the mean of the squared deviation from the arithmetic mean. Standard deviation is denoted by the small Greek latter σ (read as sigma).

The standared deviation measures the absolute dispersion (or variability of distribution; the greater the amount of dispersion or variability), the greater the standared deviation, for the greater will be the magnitude of the deviations of the values from their mean. A small standared deviation means a high degree of uniformity of the observation as well as homogeneity of a series: a large standared deviation means just the opposite. Thus if we have two or more comparable series with identical or nearly identical means, it is the distribution with the smallest standard deviation that has the most represenctive means. Hence standared deviation is extremely useful in judgig the representativeness of the mean.

Difference Between Mean Deviation and Standared Deviation

Both these Measures of Dispersion (or Variation) are based on each and every item of the distribution. but they differ in the following respects:

(a) Mean deviation can be computed either from median or mean. The standard deviation, on the other hand, is always computed from the arithmetic mean because the sum of the squares of the deviation of items from arithmetic mean is the least.

(b) Algebraic signs ae ignored while calculating mean deviation whereas in the calculation of standared deviation signs are taken into account.

Calculation of Standared Deviation

Individual Observations In case of individual observations standard deviation may be computed by applying any of the following two methods:

1. By taking deviation fo the items from the actual mean.
2. By taking deviations of the items from an assumed mean.

Deviations taken from actual mean. When deviations are taken from actual mean the following formula is applied:

$$\sigma * = \sqrt{\frac{\sum x^2}{N}}$$

where. $x = (X - \overline{X})$.

Steps:

(a) Divide $\sum x^2$ by the total number of observations. i.e., N and extract the square-root. This gives us the value of standard deviation.

(b) Square these deviations and obtain the total $\sum x^2$.

(c) Take the deviations of the items from the mean, i.e., fidn $(X - \overline{X})$. Denote these deviations by x.

(d) Calculate the actual mean of the series, i.e., $\overline{X}$.

Deviations taken from Assumed Mean. When the actual mean is in fractions. say, it is 123.674 it would be too caumbersome to take deviations from it and then obtain squares of these deviations. In such a case either the mean may be approximated or else the deivations be taken from an assumed mean and the necessary adjustment made in the value of the standared deviation. The former method of approximation is less accurate and, therefore, invariably in such a case deviations are taken from assumed mean.

When deviations are taken from assumed mean the following formula is applied:

$$\sigma = \sqrt{\frac{\sum fd^2}{N} - \left(\frac{\sum d}{N}\right)^2}$$

Steps.

(a) substitute the values of Σd^2. Σd and N in the above formula.

(b) Square these deviations and obtain the total Σd^2.

(c) Take the deviation of the items from an assumed mean, i.e., obtain (X + A). Denote these deviations by d. Take the total of these deviations. i.e., obtain Σd.

Example 19:

The annual salaries of a group of employees are given in the following tabe:

Salaries (in Rs. 000)	*45*	*50*	*55*	*60*	*65*	*70*	*75*	*80*
Number or persons	*3*	*5*	*8*	*7*	*9*	*7*	*4*	*7*

Calculate the standard deviation of the saiaries.

Solution:

Calculation of Standard Deviation

Salaries X	**No. of persons f**	**(X – 60)/5 d**	**fd**	**fd²**
45	3	–3	–9	27
50	5	–2	–10	20
55	8	–1	–8	8
60	7	0	0	0
65	9	+1	+9	9
70	7	+2	+14	28
75	4	+3	+12	36
80	7	+4	+28	112
	N = 50		Σfd = 36	Σfd2 = 240

$$\sigma = \sqrt{\frac{\sum fd^2}{N} - \left(\frac{\sum fd}{N}\right)^2} \times i \; \Sigma fd^2 = 240,$$

$N = 50,$

$\Sigma fd = 36,$

$i = 5$

$$\sigma = \sqrt{\frac{240}{50} - \left(\frac{36}{50}\right)^2} \times 5 = \sqrt{4.8 - .5184} \times 5$$

$= 10.35$

Calculation of Standard Deviation-Continuous Series

In continuous series any of the methods discussed above for discrete frequency distribution can be used. However, in practice it is the step deviation method that is most used. The formula is

$$= \sqrt{\frac{\sum fd^2}{N} - \left(\frac{\sum fd}{N}\right)^2} \times i$$

$$d = \frac{(m - A)}{i} \quad \text{where } i = \text{class interval}$$

Steps.

(a) Square the deviations and multiply them with the respective frequencies of each class and obtain Σfd^2.

(b) Multiply the frequencies of each class with these deviations and obtain Σfd.

(c) Wherever possible take a common factor and denote this column by d.

(d) Take the deviations of these mid-points from an assumed mean and denote these deviations by d.

(e) Find the mid-points of various classes.

Thus the only difference in procedure in case of continuous series is to find mid-points of the various classes.

Example 20:

Calculate the standard deviation from the following observations:

240.12 240.13240.15240.12240.17

240.15 240.17240.16240.22240.21

Solution:

Calculation of Standard Deviation

X	(X – 240) d	d^2
240.12	+0.12	.0144
240.13	+0.13	.0169
240.15	+0.15	.0225
240.12	+0.12	.0144
240.17	+0.17	.0289
240.15	+0.15	.0225
240.17	+0.17	.0289
240.16	+0.16	.0256
240.22	+0.22	.0484
240.21	+0.21	.0441
N = 10	$\Sigma d = +1.60$	$\Sigma d^2 = 0.2666$

$$\sigma = \sqrt{\frac{\sum d^2}{N} - \left(\frac{\sum d}{N}\right)^2} = \sqrt{\frac{2666}{10} - \left(\frac{16}{20}\right)^2} = \sqrt{.02666 - .0256} = 0.035$$

Calculation of Standard Deviation-Discrete Series

For calculating standared deviation in discrete series any of the following methods may be applied:

1. Actual mean method
2. Assumed mean metho
3. Step deviation method

(a) *Actual Mean Method.* Whn this method is applied, deviations are taken from the actual mean. i.e., we find $(x - \overline{X})$ and denote these deviations by x. These deviations are then squared and multiplied by the respective frequencies. The following formula is applied:

$$\sigma = \sqrt{\frac{\sum fd^2}{N}} \text{ where } x = (X - \overline{X})$$

However, in paractice this method is rarely used because if the actual mean is in fractions the calculations take a tott of time.

(b) *Assumed Mean Method.* When this method is used, the following formula is applied:

$$\sigma = \sqrt{\frac{\sum fd^2}{N} - \left(\frac{\sum fd}{N}\right)^2} \quad \text{where } d = (X - A)$$

Steps.

(a) Multiply the squared deviations by the respective frequencies, and obtain the total Σfd^2.

(b) Obtain the squares of the deviations, i.e., calculate d^2.

(c) Multiply these deviations by the respective frequencies and obtain the total. Σfd.

(d) Take the deviations of the items from an assumed mean and denote these deviations by d.

Example 21:

Calculate mean and standard deviation of following frequency distribution of marks:

Marks	*No. of Students*	*Makes*	*No. of Students*
0.10	*5*	*40-50*	*50*
10-20	*12*	*50-60*	*37*
20-30	*30*	*60-37*	*21*
30-40	*45*		

Solution:

Calculation of Mean and Standard Deviation

Marks	**m.p.** **m**	**f**	**(m – 35)/10** **d**	**fd**	**fd²**
0-10	5	5	–3	–15	45
10-20	15	12	–2	–24	48
20-30	25	30	–1	–30	30
30-40	35	45	0	0	0
40-50	45	50	+1	+50	50
50-60	55	37	+2	+74	148
60-70	65	21	+3	+63	189
		N = 200		Σfd = 118	Σfd = 510

$$\overline{X} = A + \frac{\sum fd}{N} \times i = 35 + \frac{118}{200} \times 10 = 35 + 5.9 = 40.9$$

$$\sigma = \sqrt{\frac{\sum fd^2}{N} - \left(\frac{\sum fd}{N}\right)^2} \times 10 = \sqrt{\frac{510}{200} - \left(\frac{118}{200}\right)} \times 10$$

$$= \sqrt{2.55 - 3481} \times 10 = 1.4839 \times 10 = 14.839$$

(a) Substitute the values on the above formula.

Example 22:

Calculate the standared deviation from the data given below:

Size of item	*Frequency*	*Size of item*	*Frequency*
3.5	*3*	*7.5*	*85*
4.5	*7*	*8.5*	*32*
5.5	*22*	*9.5*	*8*
6.5	*60*		

Solution:

Calculation of Standard Deviation

X **Size of item**	**f**	**(X – 6.5)** **d**	**fd**	**fd²**
3.5	3	–3	–9	27
4.5	7	–2	–14	28
5.5	22	–1	–22	22
6.5	60	0	0	0
7.5	85	+1	+85	85
8.5	32	+2	+64	128
9.5	8	+3	+24	72
	N = 227		Σfd = 128	Σfd² = 362

$$\sigma = \sqrt{\frac{\sum fd^2}{N} - \left(\frac{\sum fd}{N}\right)^2}$$

$\Sigma fd^2 = 362$, $\Sigma fd = 128$, $N = 217$

$$\sigma = \sqrt{\frac{362}{217} - \left(\frac{128}{217}\right)^2}$$

$$= \sqrt{1.668 - .348} = 1.149$$

(c) *Step Deviation Method.* When this method is used we take deviations of midpoints from an assumed mean and divide these deviations by the with of lass interval, i.e., 'i'. In case class intervals are unequal, we divide the deviatons of midpoints by the lowest common factor and use 'c' instead of 'i' in the formula for calculating standard deviation. The formula for calculating standard deviation is:

$$\sigma = \sqrt{\frac{\sum fd^2}{N} - \left(\frac{\sum fd}{N}\right)^2} \times i$$

where, $d = \dfrac{(X - A)}{i}$ and i = class interval.

The use of the above formula simplifies calculations.

Example 23:

Find the standard deviation from the following data:

Age under:	*10*	*20*	*30*	*40*	*50*	*60*	*70*	*80*
No. of person dying:	*15*	*30*	*53*	*75*	*100*	*110*	*115*	*125*

Solution:

Calculation of Standard Deviation

Marks	**f** **m**	**m.p.** **m**	**(m – 35)/10** **d**	**fd**	**fd²**
0-10	15	5	–3	–45	135
10-20	15	15	–2	–30	60
20-30	23	25	–1	–23	23
30-40	22	35	0	0	0
40-50	25	45	+1	+25	25
50-60	10	55	+2	+20	40
60-70	5	65	+3	+15	45
70-80	10	75	+4	+40	160
	N = 125			$\Sigma fd = 2$	$\Sigma fd^2 = 488$

$$\sigma = \sqrt{\frac{\sum fd^2}{N} - \left(\frac{\sum fd}{N}\right)^2} \times i = \sqrt{\frac{448}{125} \times \left(\frac{2}{125}\right)^2} \times 10$$

$$= \sqrt{3.904 - 0003} \times 10 = 1.976 \times 10 = 19.76.$$

Example 24:

The tollowing are some of the particulars of the distribution of weight, of boys and giris in a class

	Boys	*Girls*
Number	*100*	*50*
Mean weight	*60 kg*	*45 kg*
Variance	*9*	*4*

(a) Find the standard deviation of the combined data

(b) Which of the two distributions is more variable?

Solution:

(a) Combined S.D. $\sigma_{12} = \sqrt{\dfrac{N_1\sigma_1^2 + N_2\sigma_2^2 + N_1\sigma_1^2 + N_2\sigma_2^2}{N_1 + N_2}}$

For finding combined standard deviation, we have to calculate combined mean.

$$\overline{X}_{12} = \frac{N_1\overline{X}_1 + N_2\overline{X}_2}{N_1 + N_2}$$

$$= \frac{100\ (60) + 50\ (45)}{100 + 50} = \frac{6000 + 2250}{150} = 55$$

$N_1 = 100,\ \sigma_1{}^2 = 9,\ N_2 = 50,\ \sigma_2{}^2 = 4,$

$d_1 = |\overline{X}_1 - \overline{X}_{12}| = 60 - 55 = 5$

$d_2 =- |\overline{X}_2 - \overline{X}_{12}| = |45 - 55| = 10.$

Substituting the values

$$\sigma_{12} = \sqrt{\frac{100\ (9) + 50\ (4) + 100\ (5)^2 + 50\ (10)^2}{150}}$$

$$= = \sqrt{\frac{900 + 200 + 2500 + 5000}{150}} = \sqrt{\frac{8600}{150}} = 7.57$$

(b) For finding which distribution is more variable compare the coefficient of variation of two distributions:

$$\text{C.V. (Boys)} = \frac{\sigma}{X} \times 100 = \frac{3}{60} \times 100 = 5.00$$

$$\text{C.V. (Girls)} = \frac{\sigma}{X} \times 100 = \frac{2}{45} \times 100 = 4.44$$

Since coefficint of variation is more for distribution of weight of boys hence this distribution shows greater variability.

Example 25:

Find the standard deviation of the following distribution:

Age:	*20-25*	*25-30*	*30-35*	*35-40*	*40-45*	*45-50*
No. of persons:	*170*	*110*	*30*	*45*	*40*	*35*

Take assumed average = 32.5.

Solution:

Calculation of Standard Deviation

Age	m.p. m	No. of persons f	(m-32.5)/5 d	fd	fd²
20-25	22.5	170	–2	–340	680
25-30	27.5	110	–1	–110	110
30-35	32.5	80	0	0	0
35-40	37.5	45	+1	+45	45
40-45	32.5	40	+2	+80	160
45-50	47.5	35	+3	+105	315
		N = 480		$\Sigma fd = -220$	$\Sigma fd^2 = 1310$

$$\sigma = \sqrt{\frac{\sum fd^2}{N} 0 \left(\frac{\sum fd}{N}\right)^2} \times i = \sqrt{\frac{1310}{480} - \left(\frac{-220}{480}\right)^2} \times 5$$

$$= \sqrt{2.729 - 21} \times 5 = \sqrt{2.519} \times 5 = 1.587 \times 5 = 7.936.$$

Methematical Properties of Standard Deviation : Standard deviation has some very important mathematical properties which considerably enhance its utility in statistical work.

Combined Standard Deviation : Just as it is possible to compute combined mean of two or more than two groups, similarly we can also compute combined standard deviation of two or more groups. Combined standared deviation is denoted by σ_{12} and is computed as follows:

$$\sigma_{12} = \sqrt{\frac{N_1\sigma_1^2 + N_2\sigma_2^2 + N_1\sigma_1^2 + N_2\sigma_2^2}{N_1 + N_2}}$$

σ_2 = combined standard deviation;

σ_1 = standard deviation of first group;

σ_2 = standard deviation of second group;

$d_1 = (\overline{X}_1 - \overline{X}_{12})$; $d_2 = (\overline{X}_2 - \overline{X}_{12})$

The above formula can be extended to find out the standard deviation of three or more groups. For example, combined standard deviation of three groups would be:

$$\sigma_{123} = \sqrt{\frac{N_1\sigma_1^2 + N_2\sigma_2^2 + N_3\sigma_3^2 + N_1d_1^2 + N_2d_2^2 + N_3d_3^2}{N_1 + N_2 + N_3}}$$

where $d_1 = (\overline{X}_1 - X_{123})$; $d_2 = (\overline{X}_2 - X_{123})$; $d_3 = (\overline{X}_3 - X_{123})$.

Example 26:

The following table gives the length of life of 400 radio tubes:

Langth of life (hours)	*No. of Radio tubes*	*Length of life(hours)*	*No. of Radio tubes*
1,000-1, 199	*12*	*2,000-2,199*	*55*
1,200-1,399	*30*	*2,200-2,399*	*36*
1,400-1599	*65*	*2,400-2,599*	*25*
1,600-1,799	*78*	*2,600-2,799*	*9*
1,800-1,999	*90*		

Calculate (i) the average length of life of a radio tube, (ii) the standard deviation of the length of a tube, and (iii) the percentage number of tubes where length of ife of a tube falls within $\overline{X} \pm 2\sigma$.

Solution:

Calculation of Mean and Standard Deviation

Length of life (hours)	**m.p. m**	**No. of Radio tubes f**	**(m – 1899.5)/200 d**	**fd**	**fd²**
1000-1199	1099.5	12	–4	–48	192
1200-1399	1299.5	30	–3	–90	270
1400-1599	1499.5	65	–2	–130	260
1600-1799	1699.5	78	–1	–78	78
1800-1999	1899.5	90	0	0	0
2000-2199	2099.5	55	+1	+55	55
2200-2399	2299.5	36	+2	+72	144
2400-2599	2499.5	25	+3	+75	225
2600-2799	2699.5	9	+4	+36	144
		N = 400	+1	$\Sigma fd = 108$	$\Sigma fd^2 = 1368$

(i) $\overline{X} = A + \frac{\sum fd}{M} \times i\ 1899.5 - \frac{108}{400} \times 200 = 1899.5 - 54 = 1845.5$

(ii) $\sigma = \sqrt{\frac{\sum fd^2}{N} - \left(\frac{\sum fd}{N}\right)^2} \times i = \sqrt{\frac{1368}{400} - \left(\frac{-108}{400}\right)^2} \times 200$

$= \sqrt{3.42 - 0.0\ 73} \times 200 = \sqrt{3.347} \times 200 = 1.829 \times 200 = 365.8$

(iii) $\overline{X} \pm 2\sigma = 1845.5 \pm 2\ (365.8) = 1113.9$ to 2577.1.

We have to determine the percentage of tubes lying between 1113.9 and 2577.1. For this we make an assumption that the litems are equally distributed within each class. In the class 1000-1199 there are 12 frequencies. At 1113.9 or 1114 there would be 6.84 frequencies $\left(\frac{12}{200} \times 114 = 6.84\right)$. Frequencies between 1114 and 1199 are 12-6.84 = 5.16. Between 2400 and 2599, there are 25 frequencies. At 2577.1 or 2577 there would be 22.1 frequencies $\left(\frac{25}{200} \times 117 = 22.1\right)$

Thus the total frequencies between 1113.9 and 2577.1 would be 5.16 + 30 + 65 + 78 + 90 + 55 + 36 + 22.1 = 381.26 or $\frac{381.26}{400} \times 100 = 95.32$ percent.

Coefficient of Variation : The standared deviation discussed above is an absolute measure of dispersion. The corresponding relative measure is known as the *coefficient of variation.* This measure developed by Karl Pearson is the most commonly used measure of relative variation. It is used in such problems where we want to compare the variability of two or more than two series. That series (or group) for which the coefficient of varition is greater is said to be more variable or conversely less consistent, less uniform, less stable or less homogeneous. On the other hand, the series for which coefficient of variation is less is said to be less variable or more consistent, more uniform, more stable or more homogeneous. Coefficient of varation is denoted by C.V. and is obtained as follows:

$$\text{Coefficient of variation or C.V.} = \frac{\sigma}{X} \times 100$$

It may be pointed out that although any measure of dispersion can be used in conjunction with any average in computing relative dispersion, statisticians, in fact, almost always use the standared deviation as the measure of dispersion and the arithmetic mean as the average. When the relative dispersion is stated in terms of the arithmetic mean and the standard deviation, the resulting percentage is known as the coefficient of variation or coefficient of variation or coefficient of variability.

Example 27:

Goals scored by two teams in a Football session were as follows:

No. of Goals Scored in a Football Match	***No. of Football Matches Played***	
	Team 'A'	***Team 'B'***
0	*15*	*20*
1	*10*	*10*
2	*07*	*05*
3	*05*	*04*
4	*03*	*02*
5	*02*	*01*
Total	*42*	*42*

Calculate coefficient of variation and state which team is more consistent.

Solution:

In order to find out which team is more consistent we shall have to compare the coefficient of variation.

X	**(X – 7)** **x**	**x^2**	**Y**	**(Y – 7)** **y**	**y^2**
15	+8	64	20	+13	169
10	+3	9	10	+3	9
7	0	0	5	–2	4
5	–2	4	4	–3	9
3	–4	16	2	–5	25
2	–5	25	1	–6	36
$\Sigma X = 42$	$\Sigma x = 0$	$\Sigma x^2 = 118$	$\Sigma Y = 42$	$\Sigma y = 0$	$\Sigma y^2 = 252$

Team A

$$C.V. = \frac{\sigma}{X} \times 100$$

$$\overline{X} = \frac{\sum X}{N} = \frac{42}{6} = 7$$

$$\sigma = \sqrt{\frac{\sum x^2}{N}} = \sqrt{\frac{118}{6}} = 4.43$$

$$C.V. = \frac{4.43}{7} \times 100 = 63.29$$

Team B

$$C.V. = \frac{\sigma}{Y} \times 100$$

$$\overline{Y} = \frac{\sum X}{N} = \frac{42}{6} = 7$$

$$\sigma = \sqrt{\frac{\sum y^2}{N}} = \sqrt{\frac{282}{6}} = 6.48$$

$$C.V. = \frac{6.48}{7} \times 100 = 92.57$$

Example 28:

The number of workers employed, the mean wage (in Rs.) per month and standared deviation (in Rs.) in each section of a factory are given below. Calculate the mean wages and standard deviation of all the workers taken together.

Section	***No. of workers employed***	***Mean wages in Rs.***	***Standard deviation in Rs.***
A	*50*	*1113*	*60*
B	*60*	*1120*	*70*
C	*90*	*1115*	*80*

Solution:

$$\overline{X}_{123} = \frac{N_1\overline{X}_1 + N_2\overline{X}_2 + N_3\overline{X}_3}{N_1 + N_2 + N_3}$$

$$= \frac{(50 \times 1113) + (60 \times 1120) + (90 \times 1115)}{50 + 60 + 90}$$

$$= \frac{55{,}650 + 67{,}200 + 1{,}00{,}350}{200}$$

$$= \frac{2{,}23{,}200}{200} = \text{Rs. } 1116.$$

Combined standard deviation of three series:

$$\sigma_{123} = \sqrt{\frac{N_1\sigma_1{}^2 + N_2\sigma_2{}^2 + N_3\sigma_3{}^2 + N_1\sigma_1{}^2 + N_2\sigma_2{}^2 + N_3\sigma_3{}^2}{N_1 + N_2 + N_3}}$$

$$d_1 = |\overline{X}_1 - \overline{X}_{123}| \text{ or } |1113 - 1116| = 3$$

$$d_2 = |\overline{X}_2 - \overline{X}_{123}| \text{ or } |1120 - 1116| = 4$$

$$d_3 = |\overline{X}_3 - \overline{X}_{123}| \text{ or } |1115 - 1116| = 1$$

$$\sigma_{123} = \sqrt{\frac{50(60)^2 + 60(70)^2 + 0(80)^2 + 50(3)^2 + 60(4)^2 + 90(1)^2}{50 + 60 + 90}}$$

$$= \sqrt{\frac{1{,}80{,}000 + 2{,}94{,}000 + 5{,}76{,}000 + 450 + 960 + 90}{200}}$$

$$= \sqrt{\frac{10{,}51{,}500}{200}} = \sqrt{5.257.5} = 72.51$$

Standard Deviation of n Natural Numbers. The standard deviation of the first n natural numbers can be obtained by the following formula:

$$\sigma = \sqrt{\frac{1}{12}(N^2 - 1)}$$

Thus the standard deviation of natural numbers 1 to 10 will be

$$\sigma = \sqrt{\frac{1}{12}(10^2 - 1)} = \sqrt{\frac{1}{12} \times 99} = \sqrt{8.25} = 2.87.$$

Note : The answer would be the same when direct method of ealculating standard deviation used. But this holds good only for natural numbers.

The Sum of the Squares of the Deviations of Items in the Series from their Arithmetic is Minimum. In order words, the sum of the squares of the deviations of items of any series from a value other than the arithmetic mean would always be greater. This is the why standard deviation is always computed from the arithmetic mean.

The *Standard Deviation Enables us to Determine, with a Great Deal of accuracy, where the Values of a Frequency Distribution are Located.* With the help of Tchebycheffs theorem given by mathematician P.L. Tchebycheff (1821-1894), no matter what the shape of the distribution is, at least 75 percent of the values will fall within ±2 standard deviations from the mean of the distribution, and at least 89 percent of the values will be within ±3 standard deviations from the mean. With the help of normal curve we can measure even with greater precision the number of items the fall within specific ranges.

For a symmetrical distribution, the following relationships hold good:

Mean ±1 σ covers 68.27% of the items.

Mean ±2 σ covers 95.45% of the items.

Mean ±3 σ covers 99.73% of the items.

This can be illustrated by the following diagram:

Relation Between Measures of Dispersion (or Variation) : In a normal distribution there is a fixed relationship between the three most commonly used Measures of Dispersion (or Variation). The quartile deviation is smallest, the mean deviation next and the standard deviation is largest, in the following proportions :

$$\text{Q.D.} = \frac{2}{3}\sigma \text{ or } \sigma = \frac{3}{2}\text{Q.D.}$$

and $$\text{M.D.} = \frac{4}{5}\sigma \text{ or } \sigma = \frac{5}{4}\text{ M.D.}$$

These relationships can be easily memorized because of the sequence 2. 3. 4. 5. The same properties tend to hold true for many distributions that are *quite normal.* They are useful in eastimating one measure of dispersion when by natural numbers we mean only positive integers, e.g., 1, 2, 3, 4, 5, ... n. Another is known or in checking roughly the accuracy of a calculated value. If the computed σ differs very widely from its value estimated from Q.D. or M.D. either an error has been made or the distribution differes considerably from normal.

Another comparison may be made of the proportion of items that are typically included within the range of one Q.D., M.D. or S.D. measured both above and below the mean. In a normal distribution.

$\overline{X} \pm$ Q.D. includes 50 per cent of the items.

$\overline{X} \pm$ M.D. includes 57.31 per cent of the items

$\overline{X} \pm \sigma$ includes 68.27 per cent or about two-thirds of items.

Example 29:

From the prices of shares of X and Y below find out which is more stable in value:

X	*35*	*54*	*52*	*53*	*56*	*58*	*52*	*50*	*51*	*49*
Y	*108*	*107*	*105*	*105*	*106*	*107*	*104*	*103*	*104*	*101*

Solution:

In order to find out which shares are more stable, we have to compare coefficient of variations.

Calculation of Coefficient of Variation

X	**(X-$\overline{X}$)** x	**x^2**	**Y**	**(Y-$\overline{Y}$)** y	**y^2**
35	–16	25	108	+3	9
54	+3	9	107	+2	4
52	+1	1	105	0	0
53	+2	4	105	0	0
56	+5	25	106	+1	1
58	+7	49	107	+2	4
52	+1	1	104	–1	1
50	–1	1	103	–2	4
51	0	0	104	–1	1
49	–2	4	101	–4	16
ΣX = 510	Σx = 0	Σx^2 = 350	ΣY = 1050	Σy = 0	Σy^2 = 40

Coefficient of Variation X:

$$C.V. = \frac{\sigma}{X} \times 100$$

$$\overline{x} = \frac{\Sigma x}{N} = \frac{510}{10} = 51$$

$$\sigma = \sqrt{\frac{\Sigma x^2}{N}} = \sqrt{\frac{350}{10}} = 5.916$$

$$C.V. = \frac{5.916}{51} \times 100 = 11.6$$

Coefficient of Variation Y:

$$C.V. = \frac{\sigma}{Y} \times 100$$

$$\overline{Y} = \frac{\Sigma Y}{N} = \frac{1050}{10} = 105$$

$$\sigma = \sqrt{\frac{\Sigma y^2}{N}} = \sqrt{\frac{40}{10}} = 2$$

$$C.F. = \frac{2}{105} \times 100 = 1.905$$

Since coefficient of variation is much less in case of shares Y, hence they are more stable in value.

Example 30:

The following table shows that monthly expenditures of 80 students of a University on morning breakfast:

Expenditure (in Rs.)	***No. of Students***	***Expenditure (in Rs.)***	***No. of Students***
78 – 82	*2*	*53 – 57*	*13*
73 – 77	*6*	*48 – 52*	*9*
68 – 72	*7*	*43 – 47*	*7*
63 – 67	*12*	*38 – 42*	*4*
58 – 62	*18*	*33 – 37*	*2*

Calculate arithmetic mean, standard devaition and coefficient of variation of the above data.

Solution:

Calculation of $\overline{X}$ S.D. and C.V.

Expenditure (Rs.)	m.p. m	f	(m – 60)/5 d	fd	fd²
78 – 82	80	2	+4	+8	32
73 – 77	75	6	+3	+18	54
68 – 72	70	7	+2	+14	28
63 – 67	65	12	+1	+12	12
58 – 62	60	18	0	0	0
53 – 57	55	13	–1	–13	13
48 – 52	50	9	–2	–18	36
43 – 47	45	7	–3	–21	63
38 – 42	40	4	–4	–16	64
33 – 37	35	2	–5	–10	50
		N = 80		Σfd = – 26	Σ fd² = 352

$$\text{Mean}: \overline{X} = A + \frac{\Sigma fd}{N} \times i = 60 - \frac{26}{80} \times 5 = 60 - 1.625 = 58.375$$

$$\text{S.D.}: \sigma = \sqrt{\frac{\Sigma fd^2}{N} - \left(\frac{\Sigma fd^2}{N}\right)^2} \times i = \sqrt{\frac{252}{80} - \left(\frac{-26}{80}\right)^2} \times 5$$

$$= \sqrt{4.4 - .106 \times 5 = 2.072 \times 5 = 10.36}$$

$$\text{C.V.} = \frac{\sigma}{\overline{X}} \times 100 = \frac{10.36}{58.375} \times 100 = 17.75\%.$$

Example 31:

An analysis of the monthly wages paid to workers in two firms A and B, belonging to the same industry, gives the following result :

	Firm A	***Firm B***
Number of wage earners	550	650
Average monthly wages	Rs. 1450	Rs. 1400
Standard deviation of the distribution of wages	Rs. $\sqrt{10{,}000}$	Rs. $\sqrt{19{,}600}$

Answer the following questions with proper justifications :

(a) Which firm, or B, pays out larger amount as weekly wages?

(b) In which firm, A or B, is there greater variability in individual wages?

(c) What are the measures of :

(i) average weekly wages and

(ii) standard deviation of individual wages of all workers in the two firms taken together?

Solution:

In order to find out which firm A or B pays larger amount of weekly wages, we compare the total wages bill of firm A and firm B.

Firm A Total wage bill = 550 × 1450 = Rs. 7,97,500

Firm B Total wage bill = 650 × 1400 = Rs. 9,10,000

(b) To determine the firm in which there is greater variability in individual wages, we shall compare the coefficient of variation.

$$\text{Firm } A \quad \text{C.V.} = \frac{\sigma}{\overline{X}} \times \frac{\sqrt{10{,}000}}{1450} \times 100 = 6.89 \text{ per cent}$$

$$\text{Firm } B \quad \text{C.V.} = \frac{\sigma}{\overline{X}} \times 100 = \frac{\sqrt{19{,}600}}{1400} \times 100 = 10 \text{ per cent.}$$

Since coefficient of variation is more in case of firm B, hence there is greater variation in the distribution of wages of firm B.

(c) Combined Mean and Standard Deviation

$$\overline{X}_{12} = \frac{N_1 \overline{X}_1 + N_2 \overline{X}_2}{N_1 + N_2}$$

$$= \frac{(550 \times 1450) + (650 \times 1400)}{550 + 650} = \frac{7{,}97{,}500 + 9{,}10{,}000}{1200} = \text{Rs. } 1422.9$$

$$\sigma_{12} = \sqrt{\frac{N_1\sigma_1^2 + N_2\sigma_2^2 + N_1 d_1^2 + N_2 d_2^2}{N_1 + N_2}}$$

$$d_1 = \left|\overline{X}_1 - \overline{X}_{12}\right| = \left|1450 - 1422.92\right| = 27.08$$

$$d_2 = \left|\overline{X}_2 - \overline{X}_{12}\right| = \left|1400 - 1422.92\right| = -22.92$$

$$\sigma_{12} = \sqrt{\frac{(550 \times 10{,}000) + (650 \times 19{,}600) + 550\,(27.08)^2 + 650\,(-22.92)^2}{1200}}$$

$$= \sqrt{\frac{5{,}00{,}000 + 1{,}27{,}40{,}000 + 4{,}03{,}329.52 + 3{,}41{,}462.16}{1200}}$$

$$= \sqrt{\frac{1{,}89{,}89{,}791.68}{1200}} = 39.78$$

Variance : The term variance was used to describe the square of the standard deviation by R.A. Fisher in 1913. The concept of variance is highly important in advanced work where it is possible to split the total into several parts, each attributable to one of the factors causing variation in their original series. Variance is defined as follows :

$$=\sqrt{\frac{18984791.68}{1200}} = \sqrt{15820.66} = 125.78$$

$$\text{Variance} = \frac{\Sigma\left(X-\overline{X}\right)^2}{N}$$

For details please refer to chapter on 'Analysis of Variance'.

Thus, variance is nothing but the square of the standard deviation

i.e. $\text{Variance} = \sigma^2 \Rightarrow \sigma = \sqrt{\text{Variance}}$

In a frequency distribution where deviations are taken from assumed mean variance may directly be computed as follows:

$$\text{Variance} = \left\{\frac{\Sigma\, fd^2}{N} - \left(\frac{\Sigma\, fd}{N}\right)^2\right\} \times i^2$$

when $d = \frac{(X-A)}{i}$ and i = common factor.

Variance and Standard Deviation Compared

Both the variance and the standard deviation are measures of variability in a population. These two measures are closely related as is clear from the formula : Variance = σ^2. Variance is the average squared deviation from the arithmetic mean and standard deviation is the square root of the variance. In a subsequent chapter the significance of variance analysis will be discussed at length. The smaller the value of σ^2 the lesser the variability or greater the uniformity in the population.

Example 32:

The following table gives the marks obtained by a group of 80 students an examination. Calculate the variance.

Marks obtained	*No. of students*	*Marks obtained*	*No. of students*
10 – 14	2	34 – 38	10
14 – 18	4	38 – 42	8
18 – 22	4	42 – 46	4
22 – 26	8	46 – 50	6
26 – 30	12	50 – 54	2
30 – 34	16	54 – 58	4

Solution:

Calculation of Variance

Marks	m.p. m	f	(m – 32)/4 d	fd	fd²
10 – 14	12	2	–5	–10	50
14 – 18	16	4	–4	–16	64
18 – 22	20	4	–3	–12	36
22 – 26	24	8	–2	–16	32
26 – 30	28	12	–1	–12	12
30 – 34	32	16	0	0	0
34 – 38	36	10	+1	+10	10
38 – 42	40	8	+2	+16	32
42 – 46	44	4	+3	+12	36
36 – 50	48	6	+4	+24	96
50 – 54	52	2	+5	+10	50
54 – 58	56	4	+6	+24	144
		N = 80		Σ fd = 30	Σ fd² = 562

For details please refer to chapter on 'Theoretical Distributions'.

$$\text{Variance} = \left[\frac{\Sigma\ fd^2}{N} - \left(\frac{\Sigma\ fd}{N}\right)^2\right] \times i$$

$\Sigma\ fd^2 = 562$, $\Sigma\ fd = 30$, $N = 80$, $i = 4$

Substituting the values

$$\text{Variance} = \left\{\frac{562}{80} - \left(\frac{30}{80}\right)^2\right\} \times 4^2 = (7.025 - 0.141) \times 16$$

$$= 6.884 \times 16 = 110.44.$$

Merits and Limitations

Merits

The standard deviation is the best measure of variation because of its mathematical characteristics. It is based on every item of the distribution. Also it is amenable to algebraic treatment and is less affected by fluctuations of sampling than most other Measures of Dispersion (or Variation).

(a) It is possible to calculate the combined standard deviation of two or more groups. This not possible with any other measure.

(b) For comparing the variability of two or more distributions coefficient of variation is considered to be most appropriate and this is based on mean and standard deviation.

(c) Standard deviation is most prominently used in further statistical work. For example, in computing skewness, correlation, etc. use is made of standard deviation. It is keynote in sampling and provides a unit of measurement for the normal distribution.*

Limitations : As compared to other measures it is difficult to compute. However, this does not reduce the importance of this measure because of high degree of accuracy of results it gives.

(a) It gives more weight to extreme items and less to those which are near the mean. It is because of the fact that the squares of the deviations which are big in size would be proportionately greater than the squares of those deviations which are comparatively small. The deviations 2 and 8 are in the ratio of 1 : 4 but their squares, i.e. 4 and 64, would be in the ratio of 1 : 16.

Correcting Incorrect Values of Mean and Standard Deviation

Mistakes in calculations are always possible. Sometimes it so happens that while calculating mean and standard deviation we unconsciously copy out wrong items. For example, an item 21 may be copied as 12. Similarly, one item 127 may be taken as only 27. In such cases if the entire calculations are done again, it would become too tedeious a task. By adopting a very simple procedure we can correct the incorrect values of mean and standard deviation. For obtaining correct mean we find out correct ΣX by deducting from the original ΣX the wrong item and adding to it the correct item. Similarly for calculating correct standard deviation we obtain the value of correct ΣX^2. The following illustrations shall clarify the calculations.

Example 33:

The number of employees, wages per employee and the variance of the wages per employee for two factories is given below:

	Factory A	*Factory B*
Number of employees	*100*	*150*
Average per employee per month (Rs.)	*3200*	*2800*
Variance of the wages per employee per month (Rs.)	*625*	*729*

(a) In which factory is there greater variation in the distribution of wages per employee?

(b) Suppose in factory B, the wages of an employee were wrongly noted as Rs. 3050 instead of Rs. 3650, what would be the correct variance for factory B?

Solution:

(a) For finding out in which factory there is greater variation we compare the coefficient of variation.

Factory A | *Factory B*

C.V. = $\frac{\sigma}{\overline{X}} \times 100$ C.V. = $\frac{\sigma}{\overline{X}} \times 100$

$\sigma = \sqrt{625} = 25, \overline{X} = 3200$ · $\sigma = \sqrt{729} = 27, \overline{X} = 2800$

C.V. = $\frac{25}{3200} \times 100 = 0.781$ C.V. = $\frac{27}{2800} \times 100 = 0.964$

Since coefficient of variation is more in factory B, there is greater variation in the distribution of wages per employee.

(b) **Correct variance :** For finding out correct variance, we have to find out the correct mean.

$$\overline{X} = \frac{\Sigma X}{N}, \; N\overline{X} = \Sigma X = 150 \times 2800 = 420000$$

Correct $\Sigma X = 420000 - 3050 + 3650 = 420600$

$$\text{Correct } \overline{X} = \frac{420600}{150} = 2804$$

$$\text{Variance} = \frac{\Sigma X^2}{N} - \left(\overline{X}\right)^2$$

$$729 = \frac{\Sigma X^2}{150} - (2800)^2$$

$109350 = \Sigma x^2 - 11760000$

$\Sigma x^2 = 1176109350$

Correct $\Sigma x^2 = 1176109350 - (3050)^2 + (3650)^2$

$= 1176109350 - 9302500 = 1180129350$

$$\text{Correct variance} = \frac{\text{Correct } \Sigma X^2}{N} - \left(\text{Correct } \overline{X}\right)^2$$

$$= \frac{1180129350}{150} - (2804)^2$$

$= 7867529 - 7862416 = 5113$

Example 34:

The first of two sub-groups has 10 items with mean 15 and standard deviation 3. If the whole group has 250 items with mean 15.6 and standard deviation $\sqrt{13.44}$*, find the standard deviation of the second sub-group.*

Solution:

$N_1 = 100,$

$\overline{X}_1 = 15, \sigma_1 = 3$

$$N_1 + N_2 = 250,$$
$$\overline{X}_2 = 15,$$
$$\sigma_{13} = \sqrt{13.44}$$

Since $N_1 + N_2 = (250 - 100) = 150,$

$$\Rightarrow \quad 15.6 = \frac{100 \times 15 + 150\overline{X}_2}{250}$$
$$3900 = 1500 + 150\overline{X}_2$$
$$150\overline{X}_2 = 2{,}400$$
$$\Rightarrow \quad \overline{X}_2 = 16$$
$$\sigma_{12}^2 = \frac{N_1\sigma_1^2 + N_2\sigma_2^2 + N_1 d_1^2 + N_2 d_2^2}{N_1 + N_2}$$
$$d_1 = \left|\overline{X}_2 - \overline{X}_{12}\right| = |15 - 15.6| = -0.6$$
$$d_2 = \left|\overline{X}_2 - \overline{X}_{12}\right| = |16 - 15.6| = +0.4$$
$$13.44 = \frac{100\,(3)^2 + 150\sigma_2^2 + 100\,(0.6)^2 + 150\,(0.4)^2}{250}$$
$$13.44 = \frac{900 + 150\sigma_2^2 + 36 + 24}{250}$$
$$1540\,\sigma_2^2 = 3{,}360 - 960 = 2{,}400$$
$$\Rightarrow \sigma_2^2 = 16$$
$$\Rightarrow \sigma_2 = 4$$

The standard deviation of the second sub-group is 4.

Example 35:

Prepare a frequency table with each class interval of 10 kg. and first class interval as 40-50. Also find out the coefficient of variation:

72	74	40	60	82	115	41	61	65	83
53	110	46	84	50	67	78	79	56	65
68	69	104	80	79	79	52	73	59	81
66	49	77	90	84	76	42	64	64	70
72	50	79	52	103	96	51	86	78	94

Solution:

Class groups	Tally	Frequency f	m.p. m	(m – 75)/10 d	fd	fd²
40-50		4	45	–3	–12	36
50-60		8	55	–2	–16	32
60-70		9	65	–1	–9	9
70-80		16	75	0	0	0
80-90		6	85	+1	+6	6
90-100		3	95	+2	+6	12
100-110		2	105	+3	+6	18
1100-120		2	115	+4	+8	32
		N = 50			Σfd = – 11	Σfd² = 145

$$C.V. = \frac{\sigma}{X} \times 100$$

$$\overline{X} = A + \frac{\sum fd}{N} \times i = 75 - \frac{11}{50} \times 10 = 75 - 2.2 = 72.8$$

$$\sigma = \sqrt{\frac{\sum fd^2}{N} - \left(\frac{\sum fd}{N}\right)^2} \times i = \sqrt{\frac{145}{50} - \left(\frac{11}{50}\right)^2} \times 10$$

$$= \sqrt{2.9 - 0.0484} \times 10 = 1.6887 \times 10 = 16.887$$

$$C.V = \frac{16.887}{72.8} \times 100 = 23.2 \text{ percent.}$$

Example 36:

A company paid bonus to its employees as under:

Monthly salary (Rs.)	*Bondus paid (Rs.)*	*Monthly Salary (Rs.)*	*Bonus paid (Rs.)*
100-120	*500*	*180-200*	*900*
120-140	*600*	*200-220*	*1000*
140-160	*700*	*220 and over*	*1100*
160-180	*800*		

The actual salaries of the employees were as given below:

Re. : 205, 190, 195, 218, 187, 168, 250, 168, 190, 168, 170, 175, 178, 175, 150, 125, 148, 168, 156, 145, 125, 110, 162, 130, 150, 184.

Restate the data in the form of a frequency distribution and find out:

(i) *The total bonus paid.*

(ii) *The average salary paid per empoyee.*

(iii) *The standard deviation of the distribution.*

Solution:

For finding out the total bonus paid first we will clasify the data and the data and then determine the desired amount.

Monthly Salary (Rs.)	Tally Bars	f	Bonus X	fx
100-120	|	1	500	500
120-140	|||	3	600	1800
140-160	|||	5	700	3500
160-180	|||| ||||	9	800	7200
180-200	||||	5	900	4500
200-220	||	3	1000	2000
220 and over	|	1	1100	1100
		Total = 26		$\Sigma fX = 20600$

(i) Total Bonus paid = Rs. 20,600

Calculation of Average Salary Paid and the Standard Deviation

Salarly	m	(m – 170)/20 d	f	fd	fd^2
100-120	110	–3	1	–3	9
120-140	130	–2	3	–6	12
140-160	150	–1	5	–5	5
160-180	170	0	9	0	0
180-200	190	+1	5	+5	5
200-220	210	+2	2	+4	8
220-240	230	+3	1	+3	9
			N = 26	$\Sigma fd = -2$	$\Sigma fd^2 = 48$

$$\overline{X} = A + \frac{\sum fd}{N} \times i$$

$A = 170, \quad \Sigma fd = -2, \quad N = 26, \quad i = 20$

$$\overline{X} = 170 - \frac{2}{26} \times 20 = 170 - 1.536 - 1.538 = 168.462 \text{ or } 168.46$$

$$\sigma = \sqrt{\frac{\sum fd^2}{N} - \left(\frac{\sum fd}{N}\right)^2} \times i = \sqrt{\frac{48}{26} - \left(\frac{-2}{26}\right)^2} \times 20$$

$$= \sqrt{1.846 - 0059} \times 20 = 1.3565 \times 20 = 27.13$$

Example 37:

Mean and standard deviation of the foolowing continuous series are 31 and 15.9 respectively. The distribution after taking step deviations is as follows:

d	*–3*	*–2*	*–1*	*0*	*1*	*2*	*3*
f	*10*	*15*	*25*	*25*	*10*	*10*	*5*

Determine the actual class intervals.

Solution:

In order to ascertain the calss groups we need two values-the values-the class interval and the assumed mean. From the formula for finding out standard deviation we can determine the class interval and form the formula for calculating mean we can determine the assumed mean.

Computation for Determining Class Groups

d	f	fd	fd²
–3	10	–30	90
–2	15	–30	60
–1	25	–25	25
0	25	0	0
+1	10	+10	10
+2	10	+20	40
+3	5	+15	45
	N = 100	Σfd = – 40	Σfd² = 270

$$\sigma = \sqrt{\frac{\sum fd^2}{N} - \left(\frac{\sum fd}{N}\right)^2} \times i$$

$\Sigma fd^2 = 270$, $\Sigma fd = -40$, $N = 100$ and $\sigma = 15.9$

$$15.9 = \sqrt{\frac{270}{100} - \left(\frac{-40}{100}\right)^2} \times i$$

$$\Rightarrow \quad \sqrt{2.7 - 0.16} \times i = 1.59 \times i$$

$$i = \frac{15.9}{1.59} = 10$$

$$\overline{X} = A + \frac{\sum fd}{N} \times i$$

$$\overline{X} = 31, \ \Sigma fd = -40, \ i = 10$$

$$31 = A - \frac{40}{100} \times 10$$

$$\Rightarrow \quad A - A = 31$$

$$\Rightarrow \quad A = 31 + 4 = 35.$$

Hence, assumed mean from which deviations have been taken =- 35 and class intervals is 10. The lower and upper limits of this class would be 30 and 40. This class will correspond to zero (0) in the question given. The class preceding to this would be 20-30 and the class succeeding to this 40-50 and likewise we get other classes. Thus actual class group will be as follows:

Class group	Frequency	Class group	Frequency
0-10	10	40-50	10
10-20	15	50-60	10
20-30	25	60-70	5
30-40	25		

Example 38:

The mean and standard deviation of normal distribution are 60 and 5 respecitvely. Find the inter-quartile range and the mean deviation of the distribution:

Solution:

Given $\overline{X} = 60,$

$\sigma = 5.$

We have to find out the inter-quartile range and the mean deviation.

$$\text{M.D.} = \frac{4}{5}, \sigma = \frac{4}{5} \times 5 = 4$$

$$\text{Q.D.} = \frac{2}{3}, \sigma = \frac{8}{3} \times 5 = \frac{10}{3}$$

$$\frac{Q_3 - Q_1}{2} = \frac{10}{3} \Rightarrow Q_3 - Q_1 = \frac{20}{3} = 6.67$$

Hence, inter-quartile range = 6.67.

Example 39:

The sharehoders Research Centre of India has conducted recently a rsearch-study on price behaviour of three leading industrial shares, A, B and C for the period 1979 to 1985, the results of which are published as follows in its Quarterly Journal:

Share	*Average Price*	*Standard Deviation*	*Current selling Price*
A	*18.2*	*5.4*	*36.00*
B	*22.5*	*4.5*	*34.75*
C	*24.0*	*6.0*	*39.00*

The above figures are given is Rs.

(a) *Which share, in your opinion, appear to be more stable in value?*

(b) *If your are the holder of all the three shares, which one would you like to dispose of at present, and why?*

Solution:

(a) For finding out which share is more stable in value, we have to compare the coefficient of variation of shares A, B and C.

Share A: $\text{C.V.} = \frac{\sigma}{X} \times 100$

$\overline{X} = 18.0,$

$\sigma = 5.4$

$\text{C.V.} = \frac{5.40}{18} \times 100 = 30$

Share B: $\overline{X} = 22.5, \quad \sigma = 4.5$

$\text{C.V.} = \frac{4.0}{22.50} \times 100 = 20$

Share C: $\overline{X} = 24.0, \quad \sigma = 6.0$

$\text{C.V.} = \frac{6.00}{24.00} \times 100 = 25$

Since coefficient of variation is less for share B, hence share B is more stable in value.

(b) I would like to dispose of shares A, since their coefficient of variation is highest, meaning thereby that there is maximum variation in prices.

Example 40:

Calculate the quartile deviation for the data given below:

Daily wages (Rs.)	*No. of wage eamers*
35-36	*14*
36-37	*20*
37-38	*42*
38-39	*54*
40-41	*45*
42-42	*21*
42-43	*8*

Solution:

Calculation of Quartile Deviation

Daily wages (Rs.)	**f**	**c.f.**
35-36	14	14
36-37	20	34
37-38	42	76
38-39	54	130
40-41	45	175
41-42	21	196
42-43	8	204

$$Q.D. = \frac{Q_3 - Q_1}{2}$$

$$Q_1 = \text{Size of } \frac{N}{4}\text{th item} = \frac{204}{4} = 5\text{th item}$$

Q_1 lies in the class 37-38

$$Q_1 = L + \frac{N/4 - c.f.}{f} \times i$$

$L = 37$, $N/4 = 51$, $c.f. = 34$, $f = 42$, $i = 1$

$$Q_1 = 37 + \frac{51 - 34}{42} = 37 + 0.405 = 37.405$$

$$Q_3 = \text{Size of } \frac{3N}{4}\text{th item} = \frac{3 \times 204}{4} = 153\text{rd item}$$

Q_3 lies in the class 40-41

$$Q_3 = L + \frac{3N/4 - c.f.}{f} \times i$$

$L = 40$, $3N/4 = 153$, $c.f. = 130$, $f = 45$, $i = 1$

$$Q_3 = 40 + \frac{153 - 130}{45} = 40 + 0.511 = 40.511$$

$$\text{Q.D.} = \frac{40.511 - 37.405}{2} = \frac{3.106}{2} = 1.553$$

Example 41:

An algebra test was given to 400 high school children of whom 150 were and 250 girls. The results were as follows:

$n_1 = 150$	$n_2 = 250$
$\overline{X}_1 = 72$	$\overline{X}_2 = 73$
$s_1 = 7.0$	$s_2 = 6.4$

Find the mean and the standard deviation of combined group.

Solution:

$$\overline{X}_{12} = \frac{n_1\overline{X}_1 + n_2\overline{X}_2}{n_1 + n_2}$$

$$n_1 = 150,\ \overline{X}_1 = 72,$$

$$n_2 = 250,\ \overline{X}_2 = 73$$

$$\therefore\ \overline{X}_{12} = \frac{(150 \times 72) + (250 \times 73)}{150 + 250} = \frac{10800 + 18250}{400} = \frac{29050}{400}$$

$$= 72.625$$

$$\sigma_{12} = \sqrt{\frac{n_1\sigma_1^2 + n_2\sigma_2^2 + n_1 d_1^2 + n_2 d_2^2}{n_1 + n_2}}$$

$$\sigma_1 = 7,\ \sigma_2 = 6.4,$$

$$d_1 = \left|\overline{X}_1 - \overline{X}_{12}\right| = |72 - 72.625| = 0.6.25$$

$$d_2 = \left|\overline{X}_2 - \overline{X}_{12}\right| = 73 - 72.625 = 0.375$$

$$\sigma_{12} = \sqrt{\frac{150(7)^2 + 250(6.4)^2 + 150(.625)^2 + 250(.375)^2}{400}}$$

$$= \sqrt{\frac{7350 + 10240 + 58.594 + 35.135}{400}} = \sqrt{\frac{17683.7}{400}} = 6.65$$

Thus the combined mean was Rs. 72.63 and standard eviation 6.65.

Example 42:

The mean weight of 150 students is 60 kg. The mean weight of boys is 70 kg with a standard deviation of 10 kg. For the girls, the mean weight is 55 kg and the standard deviation is 15 kg. Find the number of boys and the combined standard deviation.

Solution:

$$\overline{X}_{12} = \frac{N_1\overline{X}_1 + N_2\overline{X}_2}{N_1 + N_2}$$

Given $\overline{X}_{12} = 60, \overline{X}_1 = 70, \overline{X}_2 = 55, N_1 + N_2 = 150$

We have to determine the no. of boys

Hence N_2 will be the no. of girls

$N_2 = (150 - N_1)$

Putting the values

$$60 = \frac{N_1\ 70 + (150 - N_1)\ 55}{150}$$

$$9000 = 70N_1 + 8250 = 55N_1$$

$\Rightarrow$ $15N_1 = 9000 - 8250 = 750$

$\Rightarrow$ $N_1 = 750/15 = 50$

Hence $N_1 = 150 - 50 = 100$

Thus the no. of boys and girls is 50 and 100 respectively.

Combined Standard Deviation

$$\sigma_{12} = \sqrt{\frac{N_1\sigma_1^2 + N_2\sigma_2^2 + N_1d_1^2 + N_2d_2^2}{N_1 + N_2}}$$

$N_1 = 50, \sigma_1 = 10, N_2 = 100, \sigma_2 = 15$

$d_1 = \left|\overline{X}_1 - \overline{X}_{12}\right| = 60 - 70 = 10,\ d_2 = \left|\overline{X}_2 - \overline{X}_{12}\right| = |55 - 60| = 5$

$$\sigma_{12} = \sqrt{\frac{50(10)^2 + 100(15)^2 + 50(10)^2 + 100(5)^2}{50 + 100}}$$

$$\sigma_{12} = \sqrt{\frac{5000 + 22500 + 5000 + 2500}{150}} = \sqrt{\frac{35,000}{150}} = 15.28.$$

Example 43:

Calculate coefficient of quartile deviation and coefficient of variation from the following data:

Maks	***N. of Students***
Below 20	*8*
" 40	*20*
" 60	*50*
" 80	*70*
" 100	*80*

Solution:

Calculation of Coeff. of Q.D. and Coeff. of Variation

Marks	m.p. m	No. of Students	(m – 50)/20 d	fd	fd²	c.f
0-20	10	8	–2	–16	32	8
20-40	30	12	–1	–12	12	20
40-60	50	30	0	0	0	50
60-80	70	20	+1	+20	20	70
80-100	90	10	+2	+20	40	90
		N = 80		Σfd = 12	Σfd² = 104	

$$\text{C.V.} = \frac{\sigma}{X} \times 100$$

$$\overline{X} = A \frac{\sum fd}{N} \times i = 50 + \frac{12}{80} \times 20 = 50 + 3 = 53$$

$$\sigma = \sqrt{\frac{\sum fd^2}{N} - \left(\frac{\sum fd}{N}\right)^2} \times i = \sqrt{\frac{104}{80} - \left(\frac{12}{80}\right)^2} \times 20$$

$$= \sqrt{1.3 - 0.225} \times 20 = 1.1303 \times 20 = 22.606$$

$$\text{C.V} = \frac{22.606}{53} \times 100 = 42.65 \text{ percent}$$

$$\text{Coeff. of Q.D.} = \frac{Q_3 - Q_1}{Q_3 + Q_1}$$

$$Q_1 = \text{Size of } \frac{N}{4}\text{th item} = \frac{80}{4} = 20\text{th item}$$

Q_1 lies in the class 20-40

$$Q_1 = L + \frac{N/4 - \text{c.f.}}{f} \times i = 20 + \frac{20 - 8}{12} \times 20 = 20 + 20 = 40$$

$$Q_3 = \text{Size of } \frac{3N}{4}\text{th item} = \frac{3(8)}{4} = 60\text{th item}$$

Q_3 lies in the class 60-80

$$Q_3 = L + \frac{3N/4 - \text{c.f.}}{f} \times i = 60 + \frac{60 - 50}{20} \times 20 = 60 + 10 = 70$$

$$\text{Coeff. of Q.D.} = \frac{70 - 40}{70 + 40} = \frac{30}{110} = 0.273.$$

Example 44:

Find the standard deviation, and coefficient of variation from the following data:

Wages	*No. of workers*	*Wages*	*No. of workers*
Up to Rs. 10	*12*	*Up to Rs. 50*	*157*
" " " 20	*30*	*" " " 60*	*202*
" " " 30	*65*	*" " " 70*	*222*
" " " 40	*107*	*" " " 80*	*230*

Solution:

Calculation of Coefficient of Variation

Wages (Rs.)	**m.p. m**	**No of workers**	**(m – 35)/10 d**	**fd**	**fd^2**
0-10	5	12	–3	–36	108
10-20	15	18	–2	–36	72
20-30	25	35	–1	–35	35
30-40	35	42	0	0	0
40-50	45	50	+1	+50	50
50-60	55	45	+2	+90	180
60-70	65	20	–3	+60	180
70-80	75	8	+4	–32	128
		N = 230		$\Sigma fd = + 125$	$\Sigma fd^2 = 753$

$$\text{C.V.} = \frac{\sigma}{\text{X}} \times 100$$

$$\sigma = \sqrt{\frac{\sum fd^2}{N} - \left(\frac{\sum fd}{N}\right)^2} \times i = \sqrt{\frac{753}{230} - \left(\frac{125}{230}\right)^2} \times 10$$

$$= \sqrt{3.274 + 0.295} \times 10 \sqrt{2.979} \times 10 = 1.726 \times 10 = 17.26$$

$$\overline{X} = A + \frac{\sum fd}{N} \times i = 35 + \frac{125}{230} \times 10 = 35 + 5.43 = 40.\ 43$$

$$\text{C.V.} = \frac{17.26}{40.43} \times 100 = 42.69 \text{ percent.}$$

Example 45:

For a group containing 100 observations, the arithmetic mean and standard deviation are 8 and $\sqrt{10.5}$ *. For 50 observations selected from these*

100 observations the mean and the standard deviation are 10 and 2 respectively. Find the arithmetic mean and the standard deviation of the other half.

Solution:

$$N_1 + N_2 = 100, \quad \overline{X}_{12} = 8 \quad \sigma_{12} = \sqrt{10.5}$$

$$N_1 = 50, \quad \overline{X}_1 = 10, \sigma_1 = 8.$$

We have to find the mean and standard deviation of the other helf.

$$\overline{X}_{12} = \frac{N_1\overline{X}_1 + N_2\overline{X}_2}{N_1 + N_2}; 8 = \frac{50 \times 10 + 50\overline{X}_2}{100}$$

$$800 = 500 + 50\overline{X}_2 \Rightarrow 50\overline{X}_2 = 300 \Rightarrow \overline{X}_2 = 6$$

$$\sigma_2^2 = \frac{N_1\sigma_1^2 + N_2\sigma_2^2 + N_1 d_1^2 + N_2 d_2^2}{N_1 + N_2}$$

$$d_1 = \left|\overline{X}_1 - \overline{X}_{12}\right|. = 10 - 8 = 2$$

$$d_2 = \left|\overline{X}_2 - \overline{X}_{12}\right| = 6 - 8 = -2$$

$$10.5 = \frac{50(2)^2 + 50\sigma_2^2 + 50(2)^2 + 50(-2)^2}{100}$$

$$10.5 \times 100 = 200 + 50\sigma_2^2 + 200 + 200$$

$$50\sigma_2^2 = 1050 - 600 = 450$$

$$\sigma_2^2 = \frac{450}{50} = 9 \Rightarrow \sigma_2 = 3.$$

Hence the mean of the remaining 50 items is 8 and the standard deviation 3.

Example 46:

In a small town, a survey was conducted in respect of profits made by retail shops. The following resuits were obtained:

Profite or Loss in '000 Rs.	*No. of shops*	*Profit or Loss in '000 Rs.*	*No. of shops*
–4 to –3	*4*	*1 to 2*	*56*
–3 to –2	*10*	*2 to 3*	*40*
–2 to –1	*22*	*3 to 4*	*24*
–1 to – 0	*28*	*4 to 5*	*18*
–0 to –1	*38*	*5 to 6*	*10*

Calculate

(i) The average profit made by a retail shop.

(ii) Total profit by all shops.

(iii) The coefficient of variation of earnings.

Solution:

Calculation of Mean and Standard Deviation

Wages (Rs.) Loss in '000 rs.	m.p. m	No of Shops f	(m – 1.5) d	fd	fd^2
–4 to –3	–3.5	4	–5	–20	100
–3 to –2	–2.5	10	–4	–40	160
–2 to –1	–1.5	22	–3	–66	198
–1 to –0	–0.5	28	–2	–56	112
0 to 1	+0.5	38	–1	–38	38
1 to 2	+1.5	56	0	0	0
2 to 3	+2.5	40	+1	+40	40
3 to 4	+3.5	24	+2	+48	96
4 to 5	+4.5	18	+3	–54	162
5 to 6	+5.5	10	+4	+40	160
		N = 250		$\Sigma fd = -38$	$\Sigma fd^2 = 1{,}066$

$$\overline{X} = A + \frac{\sum fd}{A} = 1.5 - \frac{38}{250} = 1.5 = 1348 \Rightarrow \text{Rs. } 1.348$$

$$\sigma = \sqrt{\frac{\sum fd^2}{A} - \left(\frac{\sum fd}{A}\right)^2} = \sqrt{\frac{1066}{250} - \left(\frac{-38}{250}\right)^2}$$

$$= \sqrt{4.2824 - 0.023} = 2.059 \Rightarrow \text{Rs. } 2{,}059$$

$$\text{C.V} = \frac{\sigma}{X} \times 100 = \frac{2059}{1348} \times 100 = 152.74\%$$

Total profit, i.e., $\Sigma X = N\overline{X} = 250 \times 1.318 = \text{Rs. } 3.37{,}00$

Example 47:

The mean and standared deviation of 200 items are found to be 60 and 20 respectively. If at the time of calculations, two items were wrongly taken as 3 and 67 instead of 13 and 17, find the correct mean and standard deviation. What is the correct coefficient of variation?

Solution:

We are given $\overline{X} = 60$, $\sigma = 20$, $N = 200$

$$\overline{X} = \frac{\sum X}{N} \Rightarrow 60 = \frac{\sum X}{200} \Rightarrow \Sigma X = 12{,}000$$

But correct $\sum X = \sum X$ – Wrong items + correct items

$= 12000 - 3 - 67 + 17 = 11960$

$$\therefore \text{Correct Mean} = \frac{\text{Correct} \sum X}{N} = \frac{11.960}{200} = 59.8$$

Correct Standard Deviation

$$\sigma = \sqrt{\frac{\sum X^2}{N} - (\overline{X})^2} \Rightarrow 10 = \sqrt{\frac{\sum X^2}{200} - (60)^2}$$

$$\text{Squaring } 400 = \frac{\sum X^2}{N} - 3{,}600$$

$80{,}000 = \sum X^2 = 3600 \times 200$

$\sum X^2 = 80{,}000 + 7{,}20{,}000 = 8{,}00{,}000$

Correct $\sum X^2$ = Incorrect $\sum X^2$ – Wrong Items square + Correct items squares

Correct $\sum X^2 = 8{,}00{,}000 - (3)^2 - (67)^2 + (13)^2 + (17)^2$

$= 8{,}00{,}000 - 9 - 4{,}489 + 169 + 289 = 7{,}95{,}960$

$$\text{Correct } \sigma = \sqrt{\text{Correct} \frac{\sum X^2}{N} - (\text{Correct } \overline{X})^2} = \sqrt{\frac{795960}{200} - (59.8)^2}$$

$$= \sqrt{3979.8 - 3576.04} = \sqrt{403.76} = 20.094.$$

Example 48:

A sample of 35 values has mean 80 and standard deviation 4. A second sample of 65 values has mean 70 and standard deviation 5. Find the standard deviation of the combined sample of 100 values.

Solution:

Combined Standard Deviation of two series is given by the formula

$$\sigma_{12} = \sqrt{\frac{N_1\sigma_1^{\ 2} + N_2\sigma_2^{\ 2} + N_1 d_1^{\ 2} + N_2\sigma_2^{\ 2}}{N_1 + N_2}}$$

$N_1 = 35,\ \overline{X}_1 = 80,\ \sigma_1 = 4,\ \sigma_2 = 5,\ N_2 = 65,\ \overline{X}_2 = 70,\ \sigma_2 = 5$

$$\overline{X}_{12} = \frac{N_1\overline{X}_2 + N_2\overline{X}_2}{N_1 + N_2}$$

$$= \frac{(35 \times 80) + (65 \times 70)}{35 + 65} = \frac{2800 + 4550}{100} = \frac{7350}{100} = 73.5$$

$d_1 = |\overline{X}_1 - \overline{X}_{12}| = |80 - 72.5| = 6.5$

$d_2 = |\overline{X}_2 - \overline{X}_{12}| = |70 - 73.5| = 3.5$

$$\sigma_{12} = \sqrt{\frac{35(4)^2 + 65(5)^2 + 35(6.5)^2 + 65(3.5)^2}{100}}$$

$$= \sqrt{\frac{560 + 1625 + 1478.75 + 796.25}{100}} = \sqrt{\frac{4460}{100}} = 6.68$$

Example 49:

Calculate deviation and its coefficient of the following distribution of 'collar' measurements:

Mid-value (inches)	*12.5*	*13.0*	*13.5*	*14.0*	*14.5*	*15.0*	*15.5*	*16.0*	*16.5*
No. of students	*4*	*19*	*30*	*63*	*66*	*29*	*18*	*1*	*1*

Solution:

Since we are given the midvalues we first determine the lower and upper limits. Since the difference between first and second midvalue is 0.5, deduct half of it, ie., 25 from first value i.e., lower limit in 12.25 and add 25 to given first value, i.e., it becomes 12.75 etc.

Calculation of Q.D. and its Coefficient

Limits	**m.p. m**	**f**	**c.f**
12.25–12.75	12.5	4	4
12.75–13.25	13.0	19	23
13.25–13.75	13.5	30	53
13.75–14.25	14.0	63	116
14.25–14.75	14.5	66	182
14.75–15.25	15.0	29	211
15.25–15.25	15.5	18	119
15.75–16.25	16.0	1	130
16.25–16.75	16.5	1	231

$$\text{Q.D.} = \frac{Q_3 - Q_1}{2}$$

$$Q_1 = \text{Size of } \frac{N}{4}\text{th item} = \frac{231}{4} = 57.75\text{th item}$$

Q_1 lies in the calss 13.75 – 14.25

$$Q_1 = L + \frac{N/4 - \text{c.t.}}{f} \times i$$

L = 13.75,

N/4 = 57.75,

c.f. = 53,

f = 63, i = 05

$$Q_1 = 13.75 + \frac{57.75 - 53}{63} \times 0.5 = 13.75 + 0.04 = 13.79$$

$$Q_3 = \text{Size of } \frac{3N}{4}\text{th item} = \frac{3 \times 231}{4} = 173.25\text{th item}$$

Q_3 lies in the class 14.25 – 14.75

$$Q_3 = L + \frac{3N/4 - c.f.}{f} \times i$$

L = 14.25,

3N/4 = 173.25,

c.f. = 116,

f = 66, i = 0.5

$$Q_3 = 14.25 + \frac{173.25 - 116}{66} \times 0.5$$

$$= 14.25 + 0.433 = 14.683$$

$$Q.D = \frac{Q_3 - Q_1}{2} = \frac{14.683 - 13.79}{2} = 0.447$$

$$\text{Coeff. of Q.D.} = \frac{Q_3 - Q_1}{Q_3 + Q_1}$$

$$= \frac{14.683 - 13.79}{14.683 + 13.79} = \frac{0.893}{28.473} = 0.031$$

Example 50:

A consignment of 180 articies is classified according to the size of the article as under. Find the standard deviation and its coefficient:

Measurement	*No. of articles*	*Measurement*	*No. of articles*
More than 80	*5*	*More than 30*	*150*
More than 70	*14*	*More than 20*	*170*
More than 60	*34*	*More than 10*	*176*
More than 50	*65*	*More than 0*	*180*
More than 40	*110*	*More than 90*	*0*

Solution:

This is a cumulative frequency distribution. First convert it to a simple frequency distribution in an ascending order.

Measurement	f	m.p. m	(m – 45)/10 d	fd	fd^2
0-10	4	5	–4	–16	64
10-20	6	15	–3	–18	54
20-30	20	25	–2	–40	80
30-40	40	35	–1	–40	40
40-50	45	45	0	0	0
50-60	31	55	+1	+31	31
60-70	20	65	+2	+40	80
70-80	9	75	+3	+27	81
80-90	5	85	+4	+20	80
	N = 180			$\Sigma fd = 4$	$\Sigma fd^2 = 510$

$$\sigma = \sqrt{\frac{\sum fd^2}{N} - \left(\frac{\sum fd}{N}\right)^2} \times i = \sqrt{\frac{510}{180} = \left(\frac{4}{180}\right)^2} \times 10$$

$$= \sqrt{2.833 - 0.0005} \times 10 = 1.683 \times 10 = 16.83$$

Coeff. of standard deviation. It can be obtained by dividing standard deviation by mean.

$$\overline{X} = A + \frac{\sum fd}{N} \times i = 45 + \frac{4}{180} \times 10$$

$$= 45 + 222 = 45.222$$

$$\text{Coeff. of S.D.} = \frac{\sigma}{X} = \frac{16.83}{45.222} = 0.372$$

Example 51:

(a) The mean and S.D of a set of 100 observations were worked out as 40 and 5 respectively. But by mistake a value 50 was taken in place of 40 for one observation, Re-calclulate the correct mean and S.D.

Solution:

$$\overline{X} = 40, \sigma = 5, N = 100$$

$$\overline{X} = \frac{\sum X}{N} \Rightarrow 40 = \frac{\sum X}{100}$$

$$\Rightarrow \Sigma X = 40 \times 100 = 4000$$

$$\text{Correct } \Sigma X = 4000 - 50 + 40 = 3990$$

$$\text{Correct } \overline{X} = \frac{3990}{100} = 39.9$$

$$\overline{X} = \sqrt{\frac{\sum X^2}{N} - (\overline{X})^2} \Rightarrow \sigma^2 = \frac{\sum X^2}{N} - (\overline{X})^2$$

$$2.5 = \overline{X} = \frac{\sum X^2}{100} - (40)^2$$

$$2500 = \sum X^2 - 160000$$

$$\Rightarrow \quad \sum X^2 = 162500$$

$$\text{Correct } \sum X^2 = 162500 - (50)^2 + (40)^2$$

$$= 1625500 - 2500 + 1600 = 161600$$

$$\text{Correct } \sigma = \sqrt{\frac{\text{Correct} \sum X^2}{100} - (\text{Correct } \overline{X})^2}$$

$$= \sqrt{\frac{161600}{100} - (39.9)^2} = \sqrt{1616 - 1592.01} = 4.9$$

(b) coefficients of variation of two series are 75% and 90% and their standard deviations 15 to 18 respectively. Find their mean.

Solution:

Series A

$$\text{C.V.} = \frac{\sigma}{X} \times 100$$

$$75 = \frac{15}{X} \times 100$$

$$75\,\overline{X} = 1500$$

$$\Rightarrow \quad \overline{X} = 20$$

Series B

$$\text{C.V.} = \frac{\sigma}{X} \times 100$$

$$90 \times \frac{18}{X} \times 100$$

$$90\,\overline{X} = 1800$$

$$\Rightarrow \quad \overline{X} = 20$$

The mean of the two series is the same, i.e., 20.

Example 52:

The scores of two batsmen A and B in ten innings during a certain season are:

A:	*32*	*28*	*47*	*63*	*71*	*39*	*10*	*60*	*96*	*14*
B:	*19*	*31*	*48*	*53*	*67*	*90*	*10*	*62*	*40*	*80*

Find (using coefficient of variation(which of the two batsman, A or B, is more consistent in scoring.

Solution:

In order to find out which of the two batsman is more consistent, we have to compare the coefficient of variation.

X	$(X - \overline{X})$ x	x^2	Y	$(Y - \overline{Y})$ y	y^2
32	–14	196	19	–31	961
28	–18	324	31	–19	361
47	+1	1	48	–2	4
63	+17	289	53	+3	9
71	+25	625	67	+17	289
39	–7	49	90	+40	1600
10	–36	1296	10	–40	1600
60	+14	196	62	+12	144
96	+50	2500	40	–10	100
14	–32	1024	80	+30	900
$\Sigma X = 460$	$\Sigma X = 0$	$\Sigma X^2 = 6500$	$\Sigma Y = 500$	$\Sigma y = 0$	$\Sigma y^2 = 5968$

Batsman A

$$\overline{X} = \frac{\sum X}{N} = \frac{460}{10} = 46$$

$$\sigma = \sqrt{\frac{\sum x^2}{N}} = \sqrt{\frac{6500}{10}} = 25.495$$

$$\text{C.V.} = \frac{25.49}{46} \times 100 = 55.41$$

Batsman B

$$\overline{Y} = \frac{\sum Y}{N} = \frac{500}{10} = 50$$

$$\sigma = \sqrt{\frac{\sum y^2}{N}} = \sqrt{\frac{5968}{10}} = 24.43$$

$$\text{C.V.} = \frac{24.43}{50} \times 100 = 48.86$$

Since coefficient of variation is less in case of batsman B, hence batsman B is more consisent.

Standard deviation of two organisations taken together = Rs. 12.637.

For finding out which orgaination is more equitable in regard to wages, we have to compare the coefficients of variation.

Organisation C

$$\text{C.V.} = \frac{\sigma}{X} \times 100$$

$$\sigma = \sqrt{\text{Variance}} = \sqrt{100} = 10$$

$$\overline{X} = 60$$

$$\therefore \text{C.V.} = \frac{10}{60} \times 100 = 16.67$$

Organisation D

$$\text{C.V.} = \frac{\sigma}{X} \times 100$$

$$\sigma = \sqrt{\text{Variance}} = \sqrt{144} = 12$$

$$\overline{X} = 48$$

$$\therefore \text{C.V.} = \frac{12}{48} \times 100 = 25.$$

Since coefficient of variation in organisation C is less hence it is more equitable in regard to wages

Example 53:

An analysis of monthly wages of workers of two organisations C and D yielded the follwing results:

	Organisation	
	C	*D*
No. of workers	*50*	*60*
Average monthly wages	*Rs. 60*	*Rs. 48*
Variance	*100*	*144*

Obtain the average monthly wage and the standard deviation of wages of all workers in the two organisations taken together. Which organisation is more equitable in regard to wages?

Solution:

$$\overline{X}_{12} = \frac{N_1\overline{X}_1 + N_2\overline{X}_2}{N_1 + N_2}$$

$N_1 = 50,\ \overline{X}_1 = 60,\ N_2 = 60,\ \overline{X}_2 = 48$

$$\therefore \overline{X}_{12} = \frac{(50 \times 60) + (60 \times 48)}{50 + 60} = \frac{3000 + 2880}{100} = \frac{5880}{110} = \text{Rs. } 53.45$$

$$\sigma_{12} = \sqrt{\frac{N_1\sigma_1{}^2 + N_2\sigma_2{}^2 + N_1 d_1{}^2 + N_2 d_2{}^2}{N_1 + N_2}}$$

$N_1 = 50,\ \sigma_{12} = 100,\ N_2 = 60,\ \sigma_1{}^2 = 144$

$d_1 = (\overline{X}_1 - \overline{X}_{12}) = (60 - 53.45) = 6.55$

$d_2 = (\overline{X}_2 - \overline{X}_{12}) = (48 - 53.45) = -5.45$

$$\sigma_{12} = \frac{\sqrt{50 \times 100 + 60 \times 144 + 50\,(6.55)^2 + 60(-5.45)^2}}{50 + 60}$$

$$\sigma_{12} = \frac{\sqrt{5000 + 8640 + 2145.125 + 1782.15}}{110}$$

$$= \sqrt{\frac{17567.275}{100}} = \sqrt{159.7025} = 12.637$$

Example 54:

Compete the table showing the frequencies with which words of differant number of the letters occur in the passage given below (omiting punctuation marks) treating as the variable the number of letters in each word, and calculate the coefficient of variation of the distribution. "Statistics are like proposals of marriage—they shold be, they rarely are, studied and considered, very deliberately upon their alround menits."

Solution:

Here the variable (X) is the number of letters in each word. In the given passage there are words with number of letters ranging from 1 to 12. Hence the variable X would take values from 1 to 11. The frequency distribution can easily be formed by using tally bars.

No. of Letters in a word X	Tally Bars	f	(X – 6) d	fd	fd^2
1	–	0	–5	0	0
2	||	2	–4	–8	32
3	||||	5	–3	–15	45
4	||||	4	–2	–8	16
5	||	2	–1	–2	2
6	|||	3	0	0	0
7	|	1	+1	+1	1
8	|	1	+2	+2	4
9	|	1	+3	+3	9
10	||	2	+4	+8	32
11	–	0	+5	0	0
12	|	1	+6	+6	36
		N = 22		$\Sigma fd = -13$	$\Sigma fd^2 = 177$

$$C.V. = \frac{\sigma}{X} \times 100$$

$$\overline{X} = A + \frac{\sum fd}{N} = 6 - \frac{13}{22} = 5.40$$

$$\sigma = \sqrt{\frac{\sum fd^2}{N} - \left(\frac{\sum fd}{N}\right)^2} = \sqrt{\frac{177}{22} - \left(\frac{-13}{22}\right)^2} = \sqrt{8.045 - 0.349} = 2.774$$

$$C.V. = \frac{2.779}{5.409} \times 100 = 51.28 \text{ percent.}$$

Example 55:

Find the mean and standard deviation of first n natural numbers.

Solution:

The first n natural numbers are 1, 2, 3, 4,..., n.

$$\sum X = 1 + 2 + 3 + 4 + ... + n = \frac{n(n+1)}{2}$$

$$\sum X^2 = (1)^2 + (2)^2 + (3)^2 + ... + n^2$$

$$= \frac{n(n+1)(2n+1)}{6}$$

$$\overline{X} = \frac{\sum X}{n} = \frac{n(n+1)}{2n} = \frac{n+1}{2} \quad \sigma^2 = \frac{\sum X^2}{n} - \left(\frac{\sum X}{n}\right)^2$$

$$= \frac{n(n+1)(2n+1)}{6n} - \left\{\frac{n(n+1)}{2n}\right\}^2$$

$$= \frac{(n(n+1)(2n+1)}{6} - \frac{(n+1)^2}{4}$$

$$= \frac{2(n+1)(2n+1) - 3(n+1)^2}{12}$$

$$= \frac{(n+1)(4n + - 3n - 3)}{12}$$

$$= \frac{(n+1)(n-1)}{12} = \frac{n^2 - 1}{12} \Rightarrow \sigma = \sqrt{\frac{n^2 - 1}{12}}$$

Thus the mean of n natural numbers is $\frac{n+1}{2}$ and standard deviation $= \sqrt{\frac{n^2 - 1}{2}}$

Example 56:

The table below gives ths weight measurements of 200 castings:

Weight in Kg.	*No. of Catings*	*Weight in Kg.*	*No. of Castings*
81-90	*2*	*141-150*	*37*
91-100	*5*	*151-160*	*29*
101-110	*13*	*161-170*	*11*
11-120	*20*	*171-180*	*3*
121-130	*30*	*181-190*	*1*
131-140	*49*		

Calculate arithmetic median, mode and standard deviation.

Solution:

Calculation of $\overline{X}$, Med, Mode and S.D.

Weight (Kgs.)	m.p. m	No. of castomgs	(m – 125.3)/10 d	fd	fd^2	c.f.
81-90	85.5	2	–4	–8	32	2
91-100	95.5	5	–3	–15	45	7
101-110	105.5	13	–2	–26	52	20
111-120	115.5	20	–1	–20	20	40
121-130	125.5	30	0	0	0	70
131-140	135.5	49	+1	+49	49	119
141-150	145.5	37	+2	+74	148	156
151-160	155.5	29	+3	+87	261	185
161-170	165.5	11	+4	+44	176	196
171-180	175.5	3	+5	+15	75	199
181-190	185.5	1	+6	+6	36	200
		N = 200		$\Sigma fd = 206$	$\Sigma fd = 894$	

$$\overline{X} = A + \frac{\sum fd}{N} \times i = 125.5 + \frac{206}{200} \times 10 = 124.5 + 10.3 = 135.8$$

$$\text{Med. Size of } \frac{N}{2}\text{th item} = \frac{200}{2} = 100\text{th item}$$

Median less in the class 131-140, But the real limits of this class is 130.5 – 140.5

$$\text{Med.} = L \ \frac{N/2 - c.f.}{f} \times i = 130.5 + \frac{100 - 70}{49} \times 10$$

$$= 130.5 + 6.12 = 136.62$$

Mode. By inspection mode lies in the class 131-140. The real limits of this class is 130.5 – 140.5.

$$\text{Mode} = L + \frac{\Delta_1}{\Delta_1 + \Delta_2} \times i$$

$$L = 13.5,\ \Delta_1 = (49 - 20) = 19,\ \Delta_2 = (49 - 37) = 12,\ i = 10$$

$$\therefore \text{Mode} = 130 + \frac{19}{19 + 20} \times 10 = 130.5 + 6.13 = 136.63$$

$$\sigma = \sqrt{\frac{\sum fd^2}{N} - \left(\frac{\sum fd}{N}\right)^2} \times 10 \sqrt{\frac{894}{200} - \left(\frac{206}{200}\right)^2} \times 10$$

$$\sqrt{4.471.061} \times 10 = 1.843 \times 10 = 18.46.$$

Example 57:

For two groups of observations the following results were available:

Group I	*Group II*
$\Sigma(X-5) = 8$	$\Sigma(X-5) = -10$
$\Sigma(X-5)^2 = 40$	$\Sigma(X-8)^2 = 70$
$N_1 = 20$	$N_2 = 25$

Find the mean and the standard deviation of the 45 observations obtained by combining the two groups.

Solution:

Group I	**Group II**
$\Sigma(X-5) = 8$	$\Sigma(X-5) = -10$
$\Sigma X - \Sigma 5 = 8$	$\Sigma X - \Sigma 8 = -10$
$\Sigma X - 5 \times 20 = 8$	$\Sigma X - 8 \times 25 = -10$
$\Sigma X = 8 + 100 = 108$	$\Sigma X = -10 + 200 = 190$
$\Sigma(X-5)^2 = 40$	$\Sigma(X-8)^2 = 70$
$\Sigma(X^2 - 10X + 25) = 40$	$\Sigma(X^2 - 16X + 64) = 70$
$\Sigma X^2 - 10\,\Sigma X + \Sigma 25 = 40$	$\Sigma X^2 - 16\,\Sigma X + \Sigma 64 = 70$
$\Sigma X^2 - 10 \times 108 + 25 \times 20 = 40$	$\Sigma X^2 - 16 \times 190 + 65 \times 25 = 70$
$\Sigma X^2 - 40 + 1080 - 500 = 620$	$\Sigma X^2 - 70 + 3040 - 1600 = 1510$

For 45 observations in the combined group we have:

$\Sigma X = 108 + 190 = 298$

$\Sigma X^2 = 620 + 1510 = 2130$

Mean of the two groups $= \dfrac{298}{45} = 6.222$

Standard deviation of the two groups

$$\sigma = \sqrt{\frac{\Sigma X^2}{N} - \left(\frac{\Sigma X}{N}\right)^2} = \sqrt{\frac{2130}{45} - \left(\frac{298}{45}\right)^2} = \sqrt{47.333 - 42.854}$$

$= 1.865.$

Example 58:

if the values of the mean and standard deviation of the following frequency distribution (obtained by step deviation method) are 135.3 and 9.6 respectively, determine the actual class-intervals:

d	*–4*	*–3*	*–2*	*–1*	*0*	*+1*	*+2*	*+3*	*Total*
t	*2*	*5*	*8*	*18*	*22*	*13*	*8*	*4*	*80*

Solution:

We will first determine the value of i then A and finally find actual class-interval.

d	f	fd	fd^2
–4	2	–8	32
–3	5	–15	45
–2	8	–16	32
–1	18	–18	18
0	22	0	0
+1	13	+13	13
+2	8	+16	32
+3	4	+12	36
	N = 80	$\Sigma fd = -16$	$\Sigma fd^2 = 208$

$$\sigma = \sqrt{\frac{\sum fd^2}{N} - \left(\frac{\sum fd}{N}\right)^2} \times i \Rightarrow 9.6 = \sqrt{\frac{208}{80} - \left(\frac{-16}{80}\right)^2} \times i$$

$$= \sqrt{2.6 - 0.04} \times i = 9.6$$

$$\Rightarrow 1.6i = 9.6 \Rightarrow i = 6$$

$$\overline{X} = A + \frac{\sum fd}{N} \times i$$

$$\Rightarrow 135.3 + 1.2 = 136.5$$

The mid-points coresponding to various step deviations shall be

d	–4	–3	–2	–1	0	1	2	3
m.p.	112.5	118.5	124.5	130.5	136.5	142.5	148.5	154.5

The various clase limits shall be obtained as follows:

$$\text{m.p.} \pm \frac{1}{2} \quad \Rightarrow \quad \text{m.p.} = \frac{6}{2}$$

The actual class limits shall be

Class limits	Frequency	Class limits	Frequency
109.5-115.5	2	133.5-139.5	22
115.5-121.5	5	139.5-145.5	13
121.5-127.5	8	145.5-151.5	8
127.7-133.5	18	151.5-157.5	4

EXERCISES

1. What do you mean by dispersion or deviation or scatter or variability? How many types of variability are there ? Describe them in brief.
2. What do you mean by measures of variability ? Name measures of variability of individual observations.
3. Define Range. Find out Range of following series:

   (i) 60, 72, 81, 5, 70, 72, 78, 66, 55, 58, 90.

   (ii) 11, 11.5, 18.9, 17.2, 14, 18, 16, 16.2, 16.2, 13.2, 22.4.
4. Prices of a particular item in 10 years in two cities are given below, which city has more stable prices?

City A :	55	54	52	53	56	58	52	50
City B :	108	107	105	105	106	107	104	103

   [C.V. (A) = 4.99, C.V. (B) = 1.97]

5. Calculate the coefficient of variation of the following two series and show which series is more variable:

Weight in kg. :	0–10	10–20	20–30	30–40	40–50	50–60	60–70
Class A: f:	1	2	9	8	5	4	1
Class B: f:	1	3	7	8	7	3	1

   [C.V. = 39 for both the series]

6. (a) Following data represent life of two models of refrigerators A and B:

Life (No. of years) Model A	Refrigerator Model B
0–25	2
2–416	7
4–613	12
6–87	19
8–105	9
10–124	1

   Find the average life of each model. What model has greater uniformity? Also obtain mode for both models.

   [Model A, C.V. =54,92, Model B. C.v. = 36.2]

   (b) For a group of 50 male workers, the mean and standard deviation for wages are Rs. 63 and Rs. 9 respectively. For a group of 40 female workers, these are Rs. 54 and Rs. 6 respectively. Find the combined mean and standard deviation.

[$\overline{X}_{12} = 59, \sigma_{12} = 9$]

7. Following are the records of two players regarding their performance in cricket matches:

Score of Player A:	48	52	55	60	65	45	63	70
Score of Player :	33	35	80	70	100	15	11	25

(a) Which player has scored more on an average?

(b) Which player is more consistent in his performance.

8. Claculate mean deviation from the median for the following data:

Age (years)	4-6	6-8	8-10	10-12	12-14	14-16	16-18
No. of Students:	30	90	120	150	80	60	20

9. (a) The following table gives the number of finished articles turned out daily by the different number of workers in the factory. Find the mean value and the standard deviation of output of finished articles daily.

No. of Articles:	18	19	20	21	22	23	24	25	26	27
No. of Workers:	3	7	11	14	18	17	13	8	5	4

(b) Explain, with a suitable example, the term 'Variation'. Mention some common measures of variation and describe the one which you think to be most important of them.

10. (a) The following are the scores of Manoj and Rajeev for 8 innings:

Manoj :	12	115	76	42	7	19	49	80
Rajeev :	47	12	76	73	24	51	63	54

Who of the two is a nor consistent batsman?

[C.V. (M) = 71.14, C.V. (R) = 41.96, Rajeev]

(b) Calculate the mean deviation from mean from the following series and find its coefficient:

Marks :	0–10	10–20	20–30	30–40	40–50
No. of Students :	5	8	15	16	6

[M.D. = 9.44, Coeff. of M.D. = 0.35

11. Find the mode, median, lower quartile (Q_1) and upper quartile (Q_3) and Coeff. of Q.D. from the following data:

Wages :	0–10	10–20	20–30	30–40	40–50
No. of workers :	22	38	46	35	20

$$\begin{bmatrix} \text{Mode} = 24.21\text{: Med.} = 24.46, \\ Q_1 = 14.803\text{: } Q_3 = 24.21 \\ \text{Coeff. of Q.D} = 0.396 \end{bmatrix}$$

12. Life of Allwyn and Godrej refrigerators that are currently popular were observed to be the following in a survey:

Life of refrigerators (in months)	No. of refrigerators Allwyn	Godrej
0–20	10	8
20–40	28	20
40–60	24	32
60–80	10	28
80–100	8	12

(a) Compute coefficients of variations for the life of Allwyn and Godrej refrigerators.

(b) Which model do you prefer? Give reasons.

13. Fill in the blanks:

(i) When mean is 79 and variance is 64. C.V.......

(ii) If in a series the coefficient of variation is 20 and mean 40. The standard deviation shall be.......

(iii) If $Q_1 = 30$ and $Q_2 = 50$, the coefficient of quartile deviation shall be.......

(iv) If the coefficient of variation of distribution is 50 and its standard deviation 20, the arithmetic mean shall be.....

(v) Quartile deviation is ...... of standard deviation.

[(i) 10.26, (ii) 8, (iii) 0.21, (iv) 40, (v) 0.6745]

14. Indicate when the following statements are True or False:

(i) Range is the best measure of dispersion. T/F

(ii) Absolute measure of dispersion can be used for the purpose of comparison. T/F

(iii) Quartile deviation is more suitable in case of open end distributions. T/F

(iv) Standard deviation can be calculated from any average. T/F

(v) There is no difference between variance and coefficient of variation. T/F

(vi) Mean deviation is least when deviations are taken from median. T/F

(vii) Lorenz curve was used for the first time for measuring the distribution of profits. T/F

(viii) Average deviation is most widely used in statistical work. T/F

(ix) In a moderately asymmetrical distribution, QD < MD < SD. T/F

(x) The variance is equal to square of standard deviation. T/F

Ans. (i) F, (ii) T, (iii) T, (iv) F, (v) F, (vi) T, (vii) F, (viii) F, (ix) T, (x) F

15. A purchasing agent obtained samples of 60 watt bulbs from two companies. He had the samples tested in his own laboratory for lengths of life with the following results:

Length of Life (in hours)	Samples From Co. A	Co.B
1700 and under 1900	10	3
1900 and under 2100	16	40
2100 and under 2300	20	12
2300 and under 2500	8	3
2500 and under 2700	6	2

(i) Which co's, bulbs do you think are better terms of average life?

(ii) If prices of both the co's, are same which co's, bulbs would you buy and why?

[C.V. (A) 11%. C.V. (B) 7.67% (i) A (ii) B]

16. (a) What do your understand by 'Dispersion'? What purpose does a measure of dispersion serve?

(b) Define mean deviation. How does it differ from standard deviation?

17. (a) In what way measures of variation supplement measures of central tendency? Explain.

(b) What is coefficient of variation? What purpose does it serve? Also distinguish between 'variance' and 'coefficient of variation'.

18. (a) "Measures of Dispersion (or Variation) and central tendency are complementary to each other in highlighting the characteristics of a frequency distribution." Explain the statement giving suitable examples.

(b) Discuss various Measures of Dispersion (or Variation). Explain each with the help os one illustration.

19. (a) What do you mean by dispersion? Sate one property of standard deviation which you consider to be the most important.

(b) Explain with suitable example the term variation. What purpose does a measure of variation serve? Comment on some of the well-known insures of variation along with their respective merits and demerits.

(c) Give the uses of standard deviation.

(d) Define dispersion. Discuss briefly the absolute and relative Measures of Dispersion (or Variation).

20. (a) List the various methods of measuring dispersion.

(b) Describe the various methods of measuring variation along with their respective merits and demerits.

(c) Calculate the standard deviation from the marks obtained by 5 students.

Marks out of 25:	8	12	13	15	22

(d) Explain with suitable example the term 'variation'. Mention some common measures of variation and describe the one which you think is widely used.

(e) What is 'dispersion'? What are the various Measures of Dispersion (or Variation)? Explain them fully.

(f) Fill is the blanks:

If C.V. of distribution is 50, SD = 20 the $\overline{X}$ shall be....

[$\overline{X}$ = 40]

21. (a) Which measure of dispersion would be most useful for:

(i) a social worker,

(ii) an actuary.

(iii) A public relations man for industry, and

(iv) a spokesman for organised labour?

(b) What is coefficient of dispersion? What purpose does it serve?

22. (a) What is a measure of dispersion? Discuss four important measures of spread indicating their uses.

(b) Prove that the standard deviation is independent of the change of origin but not of scale.

23. (a) Explain the concept of dispersion in statistical analysis. Describe its various measures and discuss their merits and demerits.

(b) Write down the formula for mean deviation and standard deviation in a discrete series.

24. During the 10 weeks of a session, the marks scored by two candidates, Jayanth and Vasanth, taking the computer programs course are given below:

Jayanth	:	58	59	60	54	65	66	52	75	69	62
Vasanth	:	87	89	78	71	73	84	65	66	56	46

(a) Who the better score. Jayanth or Vasanth?

(b) Who is more consistent?

[Jayanth: $\overline{X} = 61$: C.V. = 11.89: Vasanth, $\overline{X} = 71,5$, C.V. = 18.29]

25. From some financial statistics, it is found that the monthly average of electricity charges was Rs. 2.460 and S.D. Rs. 120. The monthly average of direct wages was Rs. 42,000 and S.D. Rs. 1,200. State, which is more variable?

[3.88, 2,86, Electricity charges]

26. Based on the frequency distribution given below, compute the following statistical measures to characterise the distribution:

(i) Coefficient of variation

(ii) Inter-quartile range,

(iii) Modal value.

Annual tax paid (Rs. Thousand)	No. of Managers	Annual tax paid (Rs. Thousand)	No. of managers
5–10	18	25–30	20
10–15	30	30–35	12
15–20	46	35–40	6
20–25	28		

[C.V. = 40.16 (ii) 10.97, (iii) 17.35]

27. (a) What Measures of Dispersion (or Variation) will you choose for:

(i) the movement of prices in stock market,

(ii) reporting on the rainfall in a certain region, and

(iii) a frequency distribution with an open interval either at the beginning or at the end.

(b) Correct the following statements:

(i) the mean deviation of a frequency distribution about an origin is minimum when the origin is the mean.

(ii) the average deviation of a distribution is greater that standard deviation.

28. Calculate S.D. and C.V. for the following frequency distribution:

Class	Frequency	Class	Frequency
4–8	11	24–28	9
8–12	13	28–32	17
12–16	16	32–36	6
16–20	14	36–40	4
20–24	14		

[σ = 9.287. C.V. = 46.62]

29. A factory produces two types of electric lamps. A and B. In an experiment relating to their life, the following results were obtained:

Length of life (in hrs.)	No. of Lamps A	No. of Lamps B
500–700	5	4
700–900	11	30
900–1,100	26	12
1,100–1,300	10	8
1,300–15,00	8	6
Total	60	60

Compare the variability of the two varieties using coefficient of variations.

[C.V. (A) = 21.6: C.V. (B) = 23.4]

30. From the following figures, determine the percentage of case which lie outside the mean at distance $\overline{X} \pm 1\sigma$, $\overline{X} \pm 2\sigma$, $\overline{X} \pm 3\sigma$.

115	117	121	125	116	120	118	117	119	116
122	124	123	118	120	118	126	127	122	123

[$\overline{X} \pm \sigma$ = 15%, $\overline{X} \pm 2\sigma$ = 85% and X$\pm$ 3σ = nil]

31. Find the coefficient of variation from the following:
   (a) $\Sigma X = 250$, N = 10 s = 8
   (b) Mean = 40, Variance = 25.
   (c) $\Sigma dx^2 = 272$, $\Sigma dx = 25$, N = 100.
   Assumed mean = 4.

32. (a) The index number of prices of Cotton and Coal shares in 1988 were as under:

Month	Jan.	Feb.	March	April	May	June	July	Aug.	Sept.	Oct.
Cotton	188	178	173	164	172	184	184	185	211	217
Coal	131	130	130	129	129	120	127	127	130	137

Which of the two shares you consider more variable in price?

(b) The variable x takes only two values x_1 and x_2 with frequencies f_1 and f_2 respectively. If S be the standard deviation of p, show that:

$$S^2 = f_1 f_2 \left(\frac{x_1 - x_2}{f_1 + f_2} \right)^2$$

33. An analysis of the monthly wages paid to workers in two firms A and B, belonging to the same industry, gave the following results:

		Firm A	Firm B
No. of workers	:	160	150
Average wage	:	560	575
Variance of wage distribution	:	400	625

Find out:

(a) Which firing pays larger amount as monthly wages?

(b) In which firm is there greater variability in individual in individual wages?

(a) Total Wage bill (A) = 89,600, (B) = 86.250. [(A)]

(b) C.V. (A) = 357, C.V. (B) = 4.35 [(B)]

34. Following are the data for marks obtained by students in a paper. The top 20% students will qualify for a prize. What is the lower limit of marks, above which the student will get the prize?

Marks	No. of students	Marks	No. of students
0–10	5	50–60	10
10–20	7	60–70	4
20–30	8	70–80	4
30–40	10	80–90	2
40–50	10		

35. (a) The following table gives the height of students in a class. Find out the quartile deviation.

Height (inches) :	50–53	53–56	56–59	59–62	62–65	65–68
No. of students :	2	7	24	27	13	3

[Q.D. = 2.21]

(b) Calculate the standard deviation for the following data:

X:	50	60	70	80	90	100	110	120
f:	14	40	54	46	26	12	6	2

36. A student obtained the mean and standard deviation of 100 observations as 40 and 5.1 respectability. It was later found that one observation was wrongly coupled as 50, the correct figure being 40. Find the correct mean and standard devotion.

    [$\overline{X}$ = 39.9: σ = 5]

37. Tick the correct answer:

    I. Coefficient of quartile deviation is calculated by the formula:

    (i) $\frac{Q_2 + Q_1}{4}$, (ii) $\frac{Q_3 + Q_1}{2}$, (iii) $\frac{Q_3 - Q_1}{Q_3 + Q_1}$, (iv) $\frac{Q_2 +}{Q_3 -}$

    (v) None of these

    II. Mean ± 3σ covers: (i) 93.37% items, (ii) 90%, (iii) 99.73%, (v) all the items.

    III. Quartile deviation is : (i) 4/5 σ, (ii) 3/2 σ, (iii) 2/3 σ, (iv) 4/5 σ, (v) None of these.

    IV. Coefficient of varination is calculated by the formula:

    (i) $\frac{\overline{X}}{s} \times 100$ (ii) $\frac{\overline{X}}{s}$, (iii) $\frac{s}{\overline{X}} \times 100$,(iv) $\frac{s}{\overline{X}} \times 10($

    (v) None of these.

    V. The measure of variation that is least affected by extreme observations is: (i) Range, (ii) Mean deviation, (iii) Standard deviation, (iv) Quartile deviation, (v) All of these,

    [I. (iii), II (v), III, (i), IV, (iv), V, (ii)]

38. Find the actual class group from the information given below:

d :	-3	-2	-1	0	1	2	3	4
f :	15	15	23	22	25	10	5	10

You are told $\overline{X}$ of the distribution is 36.16 and σ = 19.76.

[A = 35, i = 10: Class groups 0–10, 10–20, ... 70–80]

39. Define dispersion as a concept. State the measures in use for it. Calculate one of these for the following data:

Infection Score (%) :	0–10	10–20	20–30	30–40	40–50
No. of affe-etude fields :	2	3	8	15	26
Infection Score (%) :	50–60	60–70	70–80	80–90	90–100
No. of affe-eted fields :	30	6	5	4	1

40. A survey of taxi drivers in Mumbai gave the following information:

Kilometres driven per month	Number of drivers	Kilometres driven per month	Number of drivers
2,000–2,999	5	6,000–6,999	16
3,000–3,999	12	7,000–7,999	9
4,000–4,999	18	8,000–8,999	3
5,000–5,999	38		

Find (i) the range which contains the middle 50% of the drivers, and (ii) the coefficient of variation. [(ii) C.V. = 54.19]

41. (a) The mean and standard deviation of series of seventeen items are 25 and 5 respectively. While calculating these measures, a measurement 53 was wrongly read as 35, Correct this error and find out the correct standard deviation.

[Correct: $\overline{X}$ = 2606, σ = 8]

(b) The daily temperature recorded in a city in Russia in a year is given below:

Temperature 'C	No. of days	Temperature 'C	No. of days
-40 to -30	10	0 to 10	65
-30 to -20	28	10 to 20	180
-20 to -10	20	20 to 30	10
-10 to 0	42		

Calculate the mean, standard deviation and coefficient of variation.

[$\overline{X}$ = 4.3, s = 14.5]

42. For two firms A and B belonging to same industry, the following details are available

	Firm A	*Firm B*
Number of Employees :	100	200
Average wage per month :	Rs. 240	Rs. 170
Standard deviation of the	Rs. 6	Rs. 8

Find: (i) Which firm pays out larger amount as monthly wages? (ii) Find average monthly wage and the standard deviation of the wages of all employees in both the firms.

[(i) B, (ii) B, (iii) $\overline{X}_{12}$ = 193.33]

43. (a) The mean of two samples of size 50 and 100 respectively are 54.1 and 50.3 and standard deviations are 8 and 7. Find the mean and the standard deviation of the sample of size 150 obtained by combining the two samples.
    (b) Two samples of size 40 to 50 have the same mean 53, but different standard deviation of the combined sample of size 90.
    (c) For a group of 200 candidates, the mean and standard deviations were found the be 40 and 15. Later on, it was discovered that the score 43 was missed as 53. Find the correct mean and standard deviation after correction.
    (d) Mean and standard deviation of 200 items are found to be 60 and 20. If at the time of calculations two items are wrongly taken as 3 and 67 instead of 13 and 17, find the correct mean and standard deviation.

    [(a) $\overline{X}_{12} = 51.57, \sigma = 7.56$: (b) $\sigma_{12} = 14$: (c) 39.95: 14.97: 14.97; (d) 59.8, 10.09]

44. Comment briefly on the following statements:
    (a) The median is the point about which the sum of the squared deviations is minimum.
    (b) A computer found that the standard deviation of a set of 40 observations, whose values ranged between 116 and 136, is 22,
    (c) The range is the mean perfect measure of variability because it includes all the measurements.
    (d) After the settlement of dispute, the average weekly wage in a factory had increased from Rs. 8 to 12 and the standard deviation had increased from 1 to 15. After settlement, the wages has become higher and more uniform.

45. Find the mean, median and standard deviation of the weight of bullets in guns as given in the following table:

Variable	Frequency	Variable	Frequency
210–215	8	230–235	14
215–220	13	235–240	10
220–225	16	240–245	7
225–230	29	245–250	3

[$\overline{X}$ = 227.55; Med, = 227.24; σ = 8.73]

46. The profits (in Rs. lakhs) earned by 100 companies during 1997-98 are shown below:

Profits	No. of Companies	Profits	No. of Companies
20–30	4	60–70	15
30–40	8	70–80	10
40–50	18	80–90	8
50–60	29	90–100	7

Compute (a) Mean, (b) Median, (c) Standard deviation.

[$\overline{X}$ = 59.1: Med = 56.67: σ = 17.56]

47. Calculate the mean and standard deviation of the following distribution:

Age (Years) :	25–30	30–35	35–40	40–35	45–50	50–55
No of workers :	70	51	47	31	29	22

[$\overline{X}$ = 36.78: σ = 8.19]

48. (a) Your are supplied the following data about height of boys and girls in a college:

	Number	Average Height	Variance
Boys	72	68"	9"
Girls	38	61"	4"

You are required to find out (i) the combined mean and S.D. of heights of boys and girls taken together, and show (ii) whose height is more variable?

{ $\overline{X}_{12}$ = 65.58, σ_{12} = 3.62]

(b) Find mean deviation and its coefficient of the following series:

Size :	10	11	12	13	14
Frequency :	5	12	18	12	3

[M.D. = 08, Coeff. of M.D. = 0.03]

49. (a) Prove that the some of the squared deviations is least when taken from the mean.

    (b) Dispersion is known as the average of the second order. Give reasons.

50. Examine the different methods of measuring relationship between two variables. Point out the usefulness of each method taking suitable examples.

51. (a) "Averages and Measures of Dispersion (or Variation) are useful in understanding a frequency distribution". Educate the statement giving illustrations

    (b) What is a Lorenz Curve? How it is useful in measuring become inequalities between two regions?

52. (a) What are measures of disporting? Examine the qualities of standard deviation as a measure of dispersion.
    (b) Why is standard deviation considered to be the most reliable measure of dispersion?
53. (a) Critically examine the different methods of measuring vacation.
    (b) Describe the relative Measures of Dispersion (or Variation).
54. Distinguish between mean deviation and standard deviation Which is considered better and why?
55. (a) What are the properties of a good measure of variation?
    (b) Mention some common Measures of Dispersion (or Variation) and explain the one which you think to be most important of them.
56. (a) Determine the standard deviation for the following frequency distribution:

Wages per day :	2–4	4–6	6–8	8–10	10–12	12–14	Total
Number of workers :	9	12	18	;6	8	7	72

[σ = 2.93]

(b) Calculate the quartile deviation for the following frequency distribution:

X :	60	62	64	66	68	70	72
Frequency :	12	16	18	20	15	13	9

[Q.D. = 3]

57. (a) From the following information, find the standard deviation of x and y variables:
    $\Sigma x = 235, \Sigma y = 250, \Sigma x^2 = 6750, \Sigma y^2 = 6840, N = 10$
    [7.68]
    (b) Find the coefficient of variation if variance is 16, number of items is 20 and sum 7 of the items is 160.
    (c) From the following data, calculate (i) the coefficient of rənge, (ii) inter-quartile range, (iii) percentile range.

Marks :	5–9	10–14	15–19	20–24	25–29	30–34	35–39
No. Students:	1	3	8	5	4	2	2

58. Quartile deviations ? Mention the formula of Quartile deviation (Q) and calculate it with the help of following data:

Class-interval	*Frequency*
30-39	6
40-49	8 .

50-59	9
60-69	12
70-79	7
80-89	4
90-99	2

59. (i) What do you mean by deviation ? Calculate mean deviation of following data both in ungrouped and grouped condition.

Water % in the body of 15 fishes of a species is

62, 62, 63, 64, 67, 68, 71, 72. 71, 69, 68, 64, 62, 66, 68.

[**Ans.** 2.97]

(ii) 60, 72, 81, 5, 70, 72, 78, 66, 55, 58, 90. [**Ans.** 13.60]

(iii) 11, 11.5, 18.9, 17.2, 14, 18, 16, 16.2, 13.2, 22.4. [**Ans.** 2.77]

60. Standard deviation ? Why standard deviation is popular than mean deviatio in Biological statistical analysis ?

61. Calculate standard deviation and variance in the following set of data:

No. of flowers/plant (X)	*Frequency (f)*
1-3	6
4-6	14
7-9	9

62. For the data on the number of red flowers on a plant 2, 9, 10, 5, 15, 19 calculate (a) range (b) sample variance, (c) standard deviation. (**Ans.** (a) 17, (b) 39.20, (c) 6.26)

63. State whether the following statement is true or false: The variance in a sample remains unchanged if the value of each observation was decreased by 5.

64. Calculate the mean deviation for the following data on the number of seeds per pod of a plant. 2, 8, 5, 4, 7, 10

65. Calculate the variance for the following measurements; 20, 17, 14, 12, 18, 15. (**Ans.** 8.4).

66. The following data give heights of boys and girls studying in a college.

	Average Heights	*Variance*
Boys :	68 inchs	9 inches
Girls :	64 inches	4 inches

Is it correct to say that the boy's heights are more variable?

67. Compute standard deviation of the erythrocyte sedimentation rate (ESR) found to be 2, 4, 5, 4, 2, 4, 5 and 3 in eight normal persons.
68. Fifteen days before the annual examination the number of students who consulted central science library each day was; 45, 50, 36, 59, 28, 42, 55, 67, 33, 35, 40, 45, 50, 50, 50 and 36.

    Compute mean mode, median, variance and standard deviation.
69. The variance for the weights of fishes in a sample is calculated to be 20. If all the measurements are multiplied by a constant 5 what will be the variance.
70. Calculate the range for the data given below on the number of patients died during a week in emergency ward of a hospital.

    Number of deaths in a week: 1, 5, 6, 9, 7, 2.

6

The Chi-Square (χ^2) Test

INTRODUCTION

The z, t and the F-test discussed in the previous chapters are based upon an assumption about the distribution in the population studies. Such tests are called *parametric tests*. The other type of tests which are not concerned with any population distribution and its obsrevations are called *non-parametric tests.*

Chi-square (χ^2) is one such statistic and its value is not derived from the observations in a population. The chi-square (χ^2) is computed on the basis of frequencies in a simple. It is used as a test of significance, when we have data that are expressed in frequencies or data that are in terms of percentages or proportions, and that can be reduced to frequencies. It is used as a test statistic in testing a hypothesis that provides a set of theoretical frequencies with which observed frequencies are compared. In general this test is applied to those problems in which we study whether the frequency with which a given event has occurred, is significantly different from the one expected theoretically. *The measure of chi square enables us to find out the degree of discrepancy between observed frequencies and theoretical frequencies and thus to determine whether the discrepancy so obtained between observed frequencies and theoretical frequencies is due to error of sampling or due to chance.*

The χ^2-test was first used in testing statistical hypothesis by Karl Pearson in 1900. Chi-squre is derived from the Greek letter (Chi χ) and pronounced as 'Kye'.

Definition : "Chi-square test is the test of significance of overall deviation square in the observed and expected frequencies divided by expected frequencies".

The following three reasons account for the increasing use of non-parametric test in business research :

(1) These statistical tests are distribution-free (can be used with any shape of population distributions) :

(2) They are usually computationally easier to handle and understand than parametric tests; and

(3) They can be used with types of measurements that prohibit the use of parametric tests.

The increasing popularity of non-parametric tests should not lead the reader to form an impression that they are usually superior to the parametric tests. In fact, in a situation where parametric and non-parametric tests both apply, the former are more desirable than the latter.

χ^2 DEFINED

The χ^2 tests (pronounced as chi-square test) is one of the simplest and most widely used non-parametric tests in statistical work. The symbol χ^2 is the Greek letter Chi. The χ^2 test was first used by Karl Pearson in the year 1990. The quantity χ^2 describes the magnitude of the discrepancy between theory and observation. It is defined as :

$$\chi^2 = \sum \frac{(O-E)^2}{E}$$

Where O is refers to the observed frequencies and E refers the expected frequencies.

Steps. To determine the value of χ^2, the steps required are :

(i) Calculate the expected frequencies. In general the expected frequencies for any call can be calculated from the following equation:

$$\frac{(O-E)^2}{E}$$

E = Expected frequency

RT = The row total for the row containing the cell

CT = The column total for the column containing the cell

N = The total number of observations.

(ii) Take the difference between observed and expected frequencies and obtain the squares of these differences, i.e., obtain the values of $(O-E)^2$.

(iii) Divide the values of $(O-E)^2$ obtained in step (ii) by the respective expected frequency and obtain the total $\Sigma\ [(O-E)^2 / E]$. This give s the value of χ^2 which can range from zero to infinity. If χ^2 is zero to means that the observed and expected frequencies completely coincide. The greater the discrepancy between the observed and expected frequencies, the greater shall be the value of χ^2.

The calculated value of χ^2 is compared with the table value of χ^2 for given degrees of freedom at a certain specified level of significance.

If at the stated level (generally 5% level is selected), the calculated value of χ^2 is more than the table value of χ^2, the differences between theory and observation is considered to be significant, i.e., it could not have arisen due to fluctuations of simple sampling . If, on the other hand , the calculate value of χ^2 is less than the table value, the difference between theory and observation is not considered as significant, i.e., it is regarded as due to fluctuations of simple sampling and hence ignored.

The compute value of χ^2 is a random which takes on different values from sample to sample. That is χ^2 has a sampling distribution just as do the other test statistics discussed in earlier chapter.

Degrees of Freedom

While comparing the calculated value of χ^2 with the table value we have to determine the degrees of freedom. *By degrees of freedom mean the number of classes to which the values can be assigned arbitrarily or at will without the restrictions or limitations placed.* For example, if we are to choose any five numbers whose total is 100, we can exercise our independent choice for any four numbers only, the fifth number is fixed by virtue of the total being 100 as it must be equal to 100 minus the total of four numbers selected . For example , if the four number are 20, 35, 15, 10 the fifth number must be [100 - (20 + 35 + 15 + 10] 20. Thus, though we were to choose any five numbers we could choose any four only. Our choice was reduced by one because of one condition placed in the data, i.e., that of total being 100. Thus, there was only one restraint on our freedom-the degrees of freedom where only four. If more restrictions are placed our freedom to choose will be still curtailed. For example, if there are 10 classes and we want our frequencies to be distributed in such a manner that the number of cases, the mean and the standard deviation agree with the original distribution , we have three constraints (restrictions) and so three degrees of freedom are lost. Hence in this case the degrees of freedom will be $10 - 3 = 7$. Thus the number of degrees of freedom is obtained by subtracting from the number of classes the number of degrees of freedom lost in fitting. Symbolically, the degrees of freedom are denoted by the symbol v (pronounced *nu*) of by *d.f.* and are obtained as follows :

$$\nu = n - k$$

where k refers to the number of independent constraints. We have a constraint or restriction whenever observed or theoretical frequencies are made to agree with one another in some respect in the operations that lead to the calculation of χ^2. Thus a constraint is imposed by the condition $\Sigma f_o = \Sigma f_e$. In general when we fit a binomial distribution, the number of degrees of freedom is one less than the number of classes ; when we fit a Poisson distribution , the degrees of freedom are 2 less than the number of classes ($\therefore$ we use total frequency and arithmetic mean), and when we fit a normal curve, the number of degrees of freedom is small by 3 than the number of classes (because in the fitting we use total frequency, mean and standard deviation).

The following points about the χ^2 test are worth nothing :

(1) The sum of the observed and expected frequencies is always zero. Symbolically,

$\Sigma\ (O - E) = \Sigma\ O - \Sigma\ E = N - N = O$

Needless to say that this provides an important check on the calculation in the computation of χ^2.

(2) The χ^2 tests depend only on the set of observed and expected frequencies and no degrees of freedom ν. It is a non-parametric test, also known as distribution-free test, since no assumption are made about the parameters of the population.

(3) χ^2 Distribution is a limiting approximation of the multinomial distribution.

(4) Though χ^2 distribution is essentially a continuous distribution the χ^2 can be applied to discrete random variables whose frequencies can be counted and tabulated with or without grouping.

THE CHI-SQUARE DISTRIBUTION

For large sample sizes, the sampling (probability) distribution of χ^2 can be closely approximated by a continuous curve known as the Chi-square distribution. The probability function of χ^2 distribution is given by :

$$f(\chi^2) = C(\chi^2)^{(\nu/2-1)} e^{-x^2/2}$$

where $e = 2.71828$

ν = numbers of degrees of freedom

C = a constant depending only on ν.

The χ^2 distribution has only one parameter, ν, the number of degrees of freedom. As in case of t-distribution there is a distribution for each different number of degrees of freedom. For very small number of degrees of freedom , the Chi-square distribution is severely skewed to the right. As the number of degrees of freedom increase , the curve rapidly becomes more symmetrical. For large values of ν the Chi-square distribution is closely approximated by the normal curve.

The Chi-square distribution is a probability distribution and the total area under the curve in each chi-square distribution is 1.0, like the t-distribution so many different chi-square distributions are possible that it is not practical to construct a table that are unduly under the curve for all possible values of the area. In the appendix on tables (for χ^2 only) the areas in the tall most commonly used in significance tests using the χ^2 distribution are given.

Constants of χ^2 Distribution

(1) The mean of the χ^2 distribution is equal to the number of degrees of freedom , *i.e.*, $\bar{\chi} = \nu$.

(2) The variance of the χ^2 distribution is twice the degrees of freedom, variance = 2ν.

(3) $\mu_1 = 0,\ \mu_2 = 2\nu,\ \mu_3 = 8\nu,\ \mu_4 = 48\nu + 12\nu^2$.

(4) $\beta_1 = \frac{\mu_3^2}{\mu_2^3} \frac{64\nu^2}{8\nu^3} = \frac{8}{\nu}$.

(5) $\beta_2 = \frac{\mu_4}{\mu_2^2} \frac{48\nu 12\nu^2}{4\nu^2} = 3 + \frac{12}{\nu}$

The χ^2 Test when the Degrees of Freedom Exceed 30

The table values of χ^2 are available only up to 30 degrees of freedom. For degrees of freedom greater then 30, the distribution of $\sqrt{2\chi^2}$ approximates the normal distribution*. For degrees of freedom greater than 30, the approximation is acceptably close. The mean of the distribution $\sqrt{2\chi^2}$ is $\sqrt{2\chi^2}$, and the standard deviation is equal to 1. Thus the application of the test is simple, for deviation $\sqrt{2\chi^2}$ from $\sqrt{2\chi^2}$ may be interpreted as a normal deviate with unit standard deviation. That is,

$$Z \sqrt{2\chi^2} - \sqrt{2\nu - 1}.$$

Alternative Method of Obtaining the Value of χ^2

In a 2 × 2 table where the cell frequencies and marginal totals are as below :

a	b	(a + b)
c	d	(c + d)
(a + c)	(b + d)	N

N is total frequency and ad the larger cross-product, the value of χ^2 can easily be obtained by the following the formula :

$$\chi^2 = \frac{(ad - bc)^2}{(a + c)(b + d)(c + d)(a + d)} \quad \text{or}$$

with Yate's corrections

$$\chi^2 = \frac{(ad - bc - 1/2\ N)^2\ N}{(a + c)(b + d)\ (c + d)\ (a + d)}$$

CONDITIONS FOR APPLYING χ^2 TEST

The following conditions should be satisfied before applying the χ^2 test:

(1) In the first place *N* must be reasonably large to ensure the similarity between theoretically correct distribution and our sampling distribution of χ^2, the chi-square statistic. It is difficult to say what constitutes largeness, but as a general rule χ^2 test should not be used when *N* is less than 50, however few the cells.

(2) No theoretical cell frequency should be small when the expected frequencies are too small, the value of χ^2 will be overestimated and will result in too many rejections of the null hypothesis. To avoid making incorrect inference, a general rule is followed that expected frequency of less than 5 in one cell of a contingency table is too small to use. When the table contains more than one cell with an expected frequency of less than 5 we "*pool*" the frequencies which are less than 5 with the preceding or succeeding frequency so that the resulting sum is 5 or more. However, in doing so, we reduce the number of categories of data and will gain less information from contingency table.

(3) The constraints on the cell frequencies if any should be linear,* i.e., they should not involve square and higher powers of the frequencies such as $\sum O = \sum E = N$.

YATES' CORRECTIONS

One of the conditions for the application of χ^2 test is that no cell frequency should be less that 5 in any case, though 10 is better. This requirement is to avoid inflated chi-square values due to the division of the squared differences by a small size of the expected frequency. When the theoretical frequency are less than 10 and especially less than 5, the ordinary table values of χ^2 are less reliable. This is especially true for one degree of freedom, it is true to a lesser extent for two or three degrees of freedom. However, the error is negligible for more than three degrees of freedom.

The yates' corrections, also called Yates' corrections for continuity, are introduce because theoretical chi-square distribution is continuous whereas the tabulated values are based on the distribution of discrete χ^2 statistic. The corrections has the effect of reducing the calculated value of χ^2 as continuous compared to the corresponding value without correction.

In a special case of 2×2 contingency table the approximation may be be reduced, by means of a correction proposed by F. Yates in 1934. The correction involves the reduction of the deviation of observed from theoretical frequencies which of course reduces the value of χ^2. The working rule for the application of the correction is : adjust the observed frequency in each cell of the 2 × 2 table in such a way as to reduce the absolute deviation of the observed from the theoretical frequency for that cell by 1/2: adjustments for all the cells are to be made without changing the marginal totals. This operation will increase, *fo,* that is observed frequency, by 1/2 in each of two cells, and will reduce *fo* by 1/2 in each of two cells.

Another method of adjustment which gives the same result as the above procedure is :

χ^2 (corrected)

$$\sum \frac{[O_1 - E_1 - 0.5]^2}{E_1} + \frac{[(O_2 - E_2) - 0.5]^2}{E_2} + \frac{[(O_k - E_k) - 0.5]^2}{E_k}$$

$$\text{or } \sum \frac{(|O - E| - 0.5)^2}{E}$$

In general, correction is made only when the number of degrees of freedom or $\nu = 1$ and N is small. For large samples this yields practically the same results as the uncorrected χ^2. For small samples where each expected frequency is between 5 and 10, it is perhaps best to compare both the corrected and uncorrected values of χ^2 . If both values lead to the same conclusion regarding a hypothesis, such as acceptance or rejection at 0.05 level, difficulties are rarely encountered. If they lead to different conclusions, one can either resort to increasing sample sizes or if this proves impractical, one can employ exact methods of probability the multinomial distribution.

Grouping When individual Frequencies Are Small

If small theoretical frequencies occur (less than 10 and certainly not less than 50, It is generally possible to overcome the difficulty by grouping two or more classes together. In other words, one or ore classes with theoretical frequencies less than 5 may be combined into a single category before calculating the difference between observed and expected frequencies . With this practice it is important to remember that the number of degrees of freedom as determined with the number of classes after the regrouping. For example, if we start with 12 classes and 3 of them have small theoretical frequencies, we may pool these classes into one and shall be left with 10 classes to compare.

USES OF χ^2 TEST

The χ^2 test is one of the most popular statistical inference procedures today. It is applicable to a very large number of problems in practice which can be summed up under the following heads.

(i) χ^2 *tests as a test fo goodness of fit.* χ^2 test is very popularity known as test of goodness of the fit for the reason that is enables us to ascertain how appropriately the theoretical distributions such as Binomial, Poisson, Normal, etc., fit empirical distributions, i.e., those obtained from sample data. When an ideal frequency curve whether normal or some other type is fitted to the data. We are interested in finding out how well this curve fits with the observed facts. A test of the concordance (goodness of fit) of the two can be made just by inspection , but such a test is obviously inadequate. Precision can be secured by applying the χ^2 test.

The following are the steps in testing the goodness of fit :

(1) A null and alternative hypothesis are established , and a significance level is selected for rejection of the null hypothesis.

(2) A random sample of observations is drawn from a relevant statistical population.

(3) A set of expected or theoretical frequencies is derived under the assumption that the null hypothesis is true .This generally takes the form of assuming that a particular probability distribution is applicable to the statistical population under consideration.

(4) The observed frequencies are compared with the expected, or theoretical frequencies.

(5) If the calculated value of χ^2 is less than the table value at a certain level of significance (generally 5% level) and for certain degrees of freedom the fit is considered to be good, i.e., the divergence between the actual and expected frequencies is attributed to fluctuations of simple sampling . On the other hand, if the calculated value of χ^2 is greater than the table value, the fit is considered to be poor, i.e., it cannot be attributed to fluctuations of simple sampling rather it is due to the inadequacy of the theory to fit the observed facts.

It should be borne in mind that in repeated sampling too good a fit is just as likely as too bat a fit. When the computed chi-square value is too close, to zero, we should suspect the possibility that two sets of frequencies have been manipulated in order to force them to agree and therefore, the design of our experiment should be thoroughly checked .

For determining degrees of freedom for a chi-square goodness-of-fit test, we should first count the number of classes (symbolised by k) for which comparison between observed and expected frequencies is going to be made . As a general rule the degrees of freedom would be $(k-1)$. However, if in addition to the total, we have to impose further restrictions the degrees of freedom would be reduced by the number of restrictions imposed. For example, for normal distribution, we have seven classes of observed frequencies $k = 7$, and $\nu = 7 - 1 = 6$. If we have to use the sample mean as an estimate of the population mean, we will have to subtract an addition degree of freedom which leaves us with only 5, and if we have to use the sample standard deviation to estimate the population standard deviation, we will have to subtract one more degree of freedom leaving us with 4. Hence as general rule we should first employ the $(k-1)$ rule and then subtract an additional degree of freedom for each population parameter that has to be estimated from the sample data.

(ii) χ^2 *test as a test of independence.* With the help of χ^2 test we can find out whether two or more attributes are associated or not. Suppose we have N observations classified according to some attributes we may ask whether the attributes are related or independent. Thus, we can find out whether quinine is effective in controlling fever or not, whether there is any association between marriage and failure, or eye colour of husband and wife. In order to test whether or not the attributes are associated we take the null hypothesis that there is no association in the attributes under study or, in other words, the two attributes are independent. If the calculated value of χ^2 is less than the table value at a certain level of significance (generally 5% level), we say that the result of the experiment provide no evidence for doubting the hypothesis or, in other words the hypothesis that the attributes are not associated holds good. On the other hand, if the calculated value of χ^2 is greater than the table value at a certain level of significance. We say that the result of the experiment do not support the hypothesis or, in other words, the attributes are associated. It should be noted that χ^2 is not a measure of the degree or form of relationship, it only tells us whether two principles of classification are or are not significantly related, without reference to any assumptions concerning the form of relationship.

(iii) c^2 *test as a test of homogeneity***. The c^2 test of homogeneity is an extension of the chi-square test of independence. Test of homogeneity are designed to determine whether two or more independent random samples are drawn from the same population or from different populations. Instead of one sample as we use with independence problem we shall now have two or more samples. For example, we may be interested in finding out whether or not university student of various levels, i.e., undergraduates, postgraduate, Ph.D., feel the same in regard to the amount of work required by there professors i.e., too much work, right amount of work or too little work. We shall take the hypothesis that the three samples come from the same population : that is, the three classifications are homogeneous in so far as the opinion of three different groups of students about the amount of work required by their professors is concerned. This also means there exists no difference in opinion among the three classes of people on the issue.

It should be noted that in both the types of tests i.e., test of independence and homogeneity, we are concerned with cross-classified data. The same testing statistic used for tests of independence is used for tests of homogeneity. These two types of tests are, however, different in a number of ways. *First*, they are associated with the different kinds of problems. Tests of independence are concerned with the problem of whether one attribute is independent of another, while tests of homogeneity are concerned with whether different samples come from the same population. *Secondly*, the former involves a single sample taken from one population, but the later involves two or more independent samples one from each of possible population in question.

The following examples will illustrate the various applications of χ^2 test.

Example 1:

Based on information on 1,000 randomly selected fields about the tenancy status of the cultivation of these fields and use of fertilizers, collected in an agro-economic survey, the following classification was noted.

	Owned	***Rented***	***Total***
Using fertilizers	*416*	*184*	*600*
Not using fertilizers	*64*	*336*	*400*
Total	*480*	*520*	*1000*

Would you conclude that owner cultivators are more inclined towards the use of fertilizers at 5% level ?

Carry out chi-square test as per testing procedure.

Solution:

Let us take the hypothesis that ownership of fields and the use of fertilizers are independent attributes.

$$\text{Expectation of (AB)} = \frac{(A) \times (B)}{N}$$

$$= \frac{480 \times 600}{1000}$$

The table of expected frequencies shall be :

288	312	600
192	208	400
480	520	1000

Applying χ^2 test :

O	E	$(O - E)^2$	$(O - E)^2/E$
416	288	16,384	56.889
64	192	16,384	85.333
184	312	16,384	52.513
336	208	16,384	78.769
			$\Sigma [(O - E)^2/E] = 273.504$

$x^2 = \quad \Sigma [O - E)^2 /E] = 273.504$

$\nu = (r - 1)(C - 1) = (2 - 1)(2 - 1) = 1$, For $\nu = 1$, $\chi^2_{0.05} = 3.84$

The calculated value of χ^2 is much more than the table value. The hypothesis is rejected. Hence it can be concluded that owners cultivators are more inclined towards the use of fertilizers.

$$^*E_{11} = \frac{(A) \times (B)}{N} = \frac{60 \times 50}{100} = 30$$

Example 2:

The figures given below are (a) the theoretical frequencies of a distribution, and (b) the frequencies of the normal distribution having the same mean, standard deviation and the total frequency a in (a) :

(a)	*1*	*5*	*20*	*28*	*42*	*22*	*15*	*5*	*2*
(b)	*1*	*6*	*18*	*25*	*40*	*25*	*18*	*6*	*1*

Apply the χ^2 test of goodness of fit :

Solution:

Since the observed and expected frequencies are less than 10 in the beginning and end of the series, we shall group these classes together and then apply the χ^2 test :

O	E	$(O - E)^2$	$(O - E)^2/E$
1, 5 (grouped)	1, 6 (grouped)	1	0.143
20	18	4	0.222
28	25	9	0.360
42	40	4	0.100
22	25	9	0.360
15	18	9	0.500
5, 2 (grouped)	6, 1 (grouped)	0	0.000
			$\Sigma [(O - E)^2/E] = 1.685$

$\chi^2 = \Sigma [(O - E)^2 /] = 1.685$

Hence $v = 4$ because 9 original classes have been reduced to 7 by grouping, thus reducing the degrees of freedom by 2. In addition, the mean, standard deviation and total frequencies of the original distribution have been used in calculating the theoretical frequencies, thus introducing three restrains. The number of degrees of freedom is accordingly 4.

For $v = 4$, $\chi^2_{0.05} = 9.49$. The calculated value of χ^2 is much less than the table and hence the fit is good.

Example 3:

In an anti malarial campaign in a certain area, quinine was administered to 812 persons out of a total population of 3,248. The number of fever cases is shown below :

Treatment	***Fever***	***No Fever***	***Total***
Quinine	*20*	*792*	*812*
No quinine	*220*	*2,216*	*2,436*
Total	*240*	*3,008*	*3,248*

Discuss the usefulness of quinine in checking malaria.

Solution:

Let us take the hypothesis that quinine is not effective in checking malaria. Applying χ^2 test :

$$\text{Expectation of (AB)} = \frac{(A) \times (B)}{N} = \frac{240 \times 812}{3248} = 60$$

or E_1 i.e. expected frequency corresponding to first row and first column is 60.

The bale of expected frequencies shall be :

60	752	812
180	2,256	2,436
240	3,008	3,248

O	E	(O – E)²	(O – E)²/E
20	60	1,600	26.667
220	180	1,600	8.889
792	752	1,600	2.128
2,216	2,256	1,600	0.709
			[Σ [(O – E)²/E] = 38.393

$\chi^2 = [\Sigma\ (O - E)^2/E] = 38.393$

$\nu = (r - 1)\ (c - 1) = (2 - 1)\ (2 - 1) = 1$

For $\nu = 1$, $\chi^2_{0.05} + 3.84$

The calculated value of χ^2 is greater than the table value. The hypothesis is rejected. Hence quinine is useful in checking malaria.

Example 4:

From the data given below about the treatment of 250 patients suffering from a disease, state whether the new treatment is superior to the conventional treatment :

Treatment		*No. of patients*	
	Favourable	*Not favourable*	*Total*
New	*140*	*30*	*170*
Conventional	*60*	*20*	*80*
Total	*200*	*50*	*250*

(Given for degree of freedom = 1, chi-square 5 percent = 3.84)

Solution:

Let us take the hypothesis that there is no significant difference between the new and conventional treatment. Applying χ^2 test :

$$\text{Expectation of (AB)} = \frac{200 \times 170}{250} = 136$$

The table of expected frequencies shall be as follows :

136	34	170
64	16	80
200	50	250

O	E	$(O - E)^2$	$(O - E)^2/E$
140	136	16	0.118
60	64	16	0.250
30	34	16	0.471
20	16	16	1.000
			$\Sigma [(O - E)^2/E] = 1.839$

$\chi^2 = \Sigma (O - E)^2/E] = 1.839$

$\nu = (r - 1)(c - 1) = (2 - 1)(2 - 1) = 1$

For $\nu = 1$, $\chi^2_{0.05} = 3.84$

The calculate value of χ^2 is less than the table value. The hypothesis is accepted. Hence there is no significant difference between the new and conventional treatment.

Example 5:

200 digits are chosen at random form a set of tables. The frequencies of the digits are as follows :

Digit	*0*	*1*	*2*	*3*	*4*	*5*	*6*	*7*	*8*	*9*
Frequency	*18*	*19*	*23*	*21*	*16*	*25*	*22*	*20*	*21*	*15*

Use c^2 test to access the correctness of the hypothesis that the digits were distributed in equal numbers in the tables from which they were chosen .

Solution:

Let us take the hypothesis that the digits were distributed in equal numbers. On the basis of the hypothesis, we should expect 200/10 = 10 as the frequency of 0, 1, 2, ... digits, etc.

Applying the χ^2 test.

	O	E	$(O - E)^2$	$(O - E)^2/E$
	18	20	4	0.20
	19	20	1	0.05
	23	20	9	0.45
	21	20	1	0.05
	16	20	16	0.80
	25	20	25	1.25
	22	20	4	0.20
	20	20	0	0.00
	21	20	1	0.05
	15	20	25	1.25
Total	200	200		$\Sigma (O - E)^2 /E] = 4.3$

$\chi^2 = \Sigma$ (O – E)^2/E] = 4.3

ν = 10 – 1 = 9, For ν = 9, $\chi^2_{0.05}$ =16.22

The calculated value of χ^2 is less than the value and hence the hypothesis that the digits were distributed in equal numbers is accepted.

Example 6:

In an experiment on pea-breeding Mendel obtained the following frequencies of seed : 315 round and yellow, 101 wrinkled and yellow, 108 round and green, 32 wrinkled and green. According to his theory of heredity the numbers should be in proportion 9 : 3 : 3 : 1. Is there any evidence to doubt the theory at 5% level of significance ?

Solution:

Let us take the hypothesis that there is no significant difference in the observed and expected frequencies. According to the theory the expected frequencies should be :

$$\frac{556 \times 9}{16} = 312.75, \frac{556 \times 3}{16} = 104.25 \text{ and } 34.75 \text{ respectively.}$$

Applying χ^2 test :

Category	O	E	(O – E)^2	(O – E)^2 /E
Round & yellow	315	312.75	5.0625	0.016
Wrinkled & yellow	101	104.25	10.5625	0.101
Round & green	108	104.25	14.0625	0.135
Wrinkled & green	32	34.75	7.5625	0.218
			Σ (O–E)^2/E] = 0.47	

ν = n – 1 = 4 –1 = 3

For ν = 3, $\chi^2_{0.05}$ = 7.82

The calculate value is much less than the table value. The hypothesis is accepted. Hence there is no evidence to doubt the theory at 5% level of significance.

ADDITIVE PROPERTY OF χ^2

One of the merits of χ^2 test as an instrument of research is that it is possible to combine the independently derived values of χ^2 relating to samples of similar data by the simple process of addition. It enables a better (because more comprehensive) test than could be made using the data of any one sample by itself. The sum if the χ^2 values thus combined will itself have a χ^2 distribution with degrees of freedom equal to the sum of the degrees of freedom of the separate χ^2 values. However, while adding χ^2 values two points must be remembered :

(1) The combined result in a single inclusive test is appropriate when the samples are independent; and

(2) When χ^2 values are to be added. Yates' corrections should not be applied because the addition theorem holds only for uncorrected constituent items.

The following example will illustrate the additive property:

Example 7:

1,000 students at college are graded according to their I.Q. and their economic conditions. Use chi-square test to find out whether there is any association between economic conditions and the level of I.Q.

Economic Conditions	***I.Q. High***	***Medium***	***Low***	***Total***
Rich	*160*	*300*	*140*	*600*
Poor	*140*	*100*	*160*	*400*
Total	*300*	*400*	*300*	*1,000*

Solution:

Let us take the hypothesis that there is no association between economic conditions and the level of I.Q. On the basis of this hypothesis the expected frequencies corresponding to (a) and (b) are :

$$\beta_{11} \frac{600}{1000} = 300 = 180\,;$$

$$\beta_{12} \frac{600}{100} = 400 = 240$$

The table of expected frequencies is given below :

180	240	180	600
120	160	120	400
300	400	300	1,000

Applying χ^2 test

O	E	$(O - E)^2$	$(O - E)^2/E$
160	180	400	2.222
140	120	400	3.333
300	240	3,600	15.000
100	160	3,600	22.500
140	180	1,000	8.889
160	120	1,000	13.333

$$\Sigma\,(O - E)^2 / E = 65.227$$

$$\chi^2 = \sum \frac{(O - E)^2}{E} = 65.277$$

$\nu = (r - 1)(c - 1) = (2 - 1)(3 - 1) = 2$

For $\nu = 2$, $c^2_{0.05} = 5.99$. The calculated value of χ^2 is much greated than the table value. The hypothesis is rejected. Hence there is association between condition and the level of I.Q.

Example 8:

The number of defects per unit in a sample of 330 units of a manufactured product was found as follows :

Number of defects :	*0*	*1*	*2*	*3*	*4*
Number of units	*214*	*92*	*20*	*3*	*1*

Fit a Poisson distributed to the data and test for goodness of fit.

Solution:

The expected frequencies have been calculated on chapter on. 'Theoretical Distribution'. We now apply the χ^2 test of goodness of fit. Let us take the hypothesis that there is no significant difference between observed frequencies and the frequencies obtained by fitting Poisson distribution.

O	**E**	**$(O - E)^2$**	**$(O - E)^2/E$**
214	212.75	1.5625	0.0073
92	93.4	1.96	0.0210
20	20.5		
3	3	0.0225	0.0009
1	0.35		
			$\Sigma\ [(O - E)^2 / E = 0.0292$

$\nu = 3 - 2 = 1.$

[Since after grouping only 3 classes are left and when Poisson distribution is fitted $\nu = n - 2$].

$\nu = 1$, $\chi^2\ 0.05 = 3.84$

Calculated value of χ^2 is much less than the table value. The hypothesis holds true.

Hence, the fit is good.

Example 9:

A certain drug is claimed to be effective in curing cold. In an experiment on 500 persons with cold, half of them were given the drug and half of them were given the sugar pills. The patients' reactions to the treatment are recorded in the following table :

	Helped	*Harmed*	*No effect*	*Total*
Drug	*150*	*30*	*70*	*250*
Sugar pills	*130*	*40*	*80*	*250*
Total	*280*	*70*	*150*	*500*

On the basis of the data can it be concluded that there is a significant difference in the effect of the drug and sugar pills ?

Solution:

Let us take the null hypothesis that there is a significant difference in the effect of the drug and sugar pills. Applying c^2 test and finding out expected frequencies :

$$\beta_{11} = \frac{250}{500} \times 280 = 140;$$

$$\beta_{12} = \frac{250}{500} \times 70 = 35$$

The table of expected frequencies shall be as follows :

140	35	75	250
140	35	75	250
280	70	150	500

O	E	$(O-E)^2$	$(O-E)^2/E$
150	140	100	0.714
130	140	100	0.714
30	35	25	0.714
40	35	25	0.714
70	75	25	0.333
80	75	25	0.333
			$\Sigma (O-E)^2 / E = 3.522$

$$\chi^2 = \sum \frac{(O-E)^2}{E} = 3.522$$

$\nu = (r - 1)(c - 1) = (2 - 1)(3 - 1) = 2$

For $\nu = 2, \chi^2_{0.05} = 5.99.$

The calculated value of c^2 is less than the table value. Hence the hypothesis is accepted. There is no significant difference in the effect of the drug and sugar pills.

Example 10:

In an experiment on immunization of cattle from tuberculosis, the following results were obtained :

	Affected	*Not affected*
Inoculated	*12*	*26*
Not inoculated	*16*	*6*

Calculate χ^2 and discuss the effect of vaccine in controlling susceptibility to tuberculosis (5% value fo χ^2 for one degree of freedom = 3.84).

Solution:

Let us take the hypothesis that the vaccine is not effective in controlling susceptibility to tuberculosis. Applying χ^2 test :

$$\text{Expectation of (AB)} = \frac{(A)\times(B)}{N} = \frac{38}{60}\times 28 = 177.$$

The table of expected frequencies will be as follows :

17.7	20.3	38
10.3	11.7	22
28	32	60

Since one of the observed frequencies is less than we will use Yates' correction and then apply χ^2 test.

E_{11} denotes expected frequency corresponding to first row and first column. E_{21} will denote expected frequency corresponding to 2nd, row and first column.

In other words, the first denotes row and the second column.

O	**E**	**$(O - E)^2$**	**$(O - E)^2 / E$**
12.5	17.7	27.04	1.528
15.5	10.3	27.04	2.625
25.5	20.4	27.04	1.332
6.5	11.7	27.04	2.311
			$\Sigma (O - E)^2 / E = 7.796$

$\chi^2 = \Sigma (O - E)^2 / E] = 7.796$

$\nu = (r - 1)(c - 1) = (2 - 1)(2 - 1) = 1$

For $\nu = 1; \chi^2_{0.05} = 3.84$

Since the calculated value of χ^2 is greater than the table value, the hypothesis is not true. We therefore, conclude that vaccine is effective in controlling susceptibility to tuberculosis.

Example 11:

The following values of c^2 are obtained in a survey of four districts about inoculation and attack from cholera. What conclusion would you draw for these districts taken together ?

Districts	v	χ^2
A	1	3.62
B	1	3.48
C	1	2.05
D	1	5.43
Total	4	14.58

Solution:

In case of first three districts since the value of χ^2 is less than the 3.84, the results of the tests are not significant. However, in case of district D the result is significant at 5% level. But when the combined test is carried out the sum 14.58 is tested with n = 4. For 4 degrees of freedom $\chi^2_{0.05} = 9.488$. The combined value of χ^2 (14.58) is higher than the table value (9.488) and hence the combined result does not support the hypothesis, i.e., inoculation prevents attack from cholera when these districts are taken together.

CHI-SQUARE TEST FOR SPECIFIED VALUE OF POPULATION VARIANCE

When we want to test if a random sample of $\chi_1, \chi_2......\chi_n$ has been drawn from a normal population with mean μ and a specified variance σ_2^2, the statistic.

$$\chi^2 = \frac{\sum(\chi_1 - \overline{\chi})^2}{\sigma_0^2} = \frac{nS^2}{\sigma_0^2}$$

Where S the standard deviation of the sample follows chi-square distribution with (n – 1) degreed of freedom.

By comparing the calculated value of χ^2 with tabulated value (n – 1) df at certain level of significance we may accept or reject the null hypothesis.

It should be noted that this test can be applied only if the population is normal.

Example 12:

Weights in kg. of 10 students are given below :

38, 40, 45, 53, 47, 43, 55, 48, 52, 49.

Can we say that variance of the distribution of weights of all students from which the above sample of 20 students was drawn is equal to 20 square kg. ? You are given the following table values :

Degrees of freedom	χ^2 *0.05*	χ^2 *0.01*
9	*16.92*	*21.67*
10	*18.31*	*23.21*

Solution:

Applying χ^2 test :

χ	$(\chi - \bar{\chi})$	$(\chi - \bar{\chi})^2$
38	–9	81
40	–7	49
45	–2	4
53	+6	36
47	0	0
43	–4	16
55	+8	64
48	+1	1
52	+5	25
49	+2	4
$\Sigma \chi = 470$	$\Sigma (\chi - \chi) = 0$	$\Sigma (\chi - \chi)^2 = 280$

$$\bar{\chi} \frac{\sum \chi}{N} = \frac{470}{10} = 47$$

$$\chi^2 \frac{\sum(\chi - \bar{\chi})^2}{\sigma^2} = \frac{280}{20} = 14$$

For $\nu = 10 - 1 = 9$

$\nu = 9, \chi^2_{0.05} = 16.92$

The calculated value of c^2 is less than the table value. The null hypothesis holds true and we conclude that the population variance may be 20 sq.kg.

Example 13:

A sample analysis of examination results of 500 students was mad. It was found that 220 students had failed, 170 had secured a third class,90 were placed in second class and 20 got a first class. Are these figues commensurate with the general examination result which is the ratio of 4 : 3 : 2 : 1 for the various categories respectively (the table value of ^{2} for 3 d.f. at 5% level of significance is 7.81) ?

Solution:

Let us take the hypothesis that the observed results are commensura with the general examination result which is the ratio of 4 : 3 : 2 : 1. Th expected number of students, who have failed, obtained a third class, secon class, and first class respectively will be

$$\frac{500 \times 4}{10} = 200, \ \frac{500 \times 3}{10} = 150, \ \frac{500 \times 1}{10} = 100, \ \frac{500 \times 1}{10} = 50$$

Applying the χ^2 test :

Category	O	E	$(O - E)^2$	$(O - E)^2E$
Failed	220	200	400	2.000
3rd class	10	150	400	2.667
2nd class	90	100	100	1.000
1st class	20	50	90	18.000
		$\Sigma\ [(O - E)^2 / E$	$=23.667$	

$\chi^2 = \Sigma\ [(O - E)^2 / E] = 23.667$

For $\nu = 4 - 1 = 3$; $\chi^2_{0.05} = 7.81$

Since the calculated value of χ^2 is greater than the table value, our hypothesis does not hold good.

Example 14:

A random sample of size 25 from a population gives the sample standard deviation to be 8.5. Test the hypothesis that the population standard deviation is 10

Solution:

We set the null hypothesis H_o that the population standard deviation is 10, i.e., $\sigma = 10$.

We are given n = 25; s = 8.5

We know that $\frac{n\ S^2}{\sigma^2}$

Substituting the values

$$\chi^2 \frac{25(8.5)^2}{10^2} = \frac{1806.25}{100} = 18.0625 \quad \nu = n - 1 = (25 - 1) = 24$$

For $\nu = 24$, $\chi^2_{0.05} = 36.415$

Since the calculated value of χ^2 is less than the table value, the hypothesis olds true, i.e., the population standard deviation may be 10.

Example 15:

4 coins were tossed 160 times and the following results were obtained.

No. of heads :	*0*	*1*	*2*	*3*	*4*
Observed frequencies :	*17*	*52*	*54*	*31*	*6*

Under the assumption that coins are balanced, find the expected frequencies of getting 0, 1, 2, 3 or 4 heads and test the goodness of fit.

Solution:

On the assumption that the coins re balanced, the expected frequencies of getting 0, 1, 2, 3 and 4 heads will be given by the binomial :

$$= 160\left[\left(\frac{1}{2}+\frac{1}{2}\right)^4\right] = 160\left[\frac{1}{16}+\frac{1}{4}+\frac{3}{8}+\frac{1}{4}+\frac{1}{16}\right]$$

= 10, 40, 60, 40, 10

Applying χ^2 test :

No. of heads	Observed frequencies	Expected frequencies	$(O - E)^2$	$(O - E)/ E$
0	17	10	49	4.900
1	52	40	144	3.600
2	54	60	36	0.600
3	31	40	81	2.025
4	6	10	16	1.600
			$\Sigma [(O - E)^2 / E$ =12.725	

$\nu = 5 - 1 = 4$

For $\nu = 4$, $\chi^2_{0.05} = 9.49$

The calculated value of χ^2 is greater than the table value and hence the coins are not balanced and the fit is not good.

Example 16:

From the adult male population of four large cities, random samples of size given below were taken and the number of married and single men recorded. Do the data indicate any significant variance among the cities in the tendency of men to marry ?

City	*A*	*B*	*C*	*D*	*Total*
Married	*137*	*164*	*152*	*147*	*600*
Single	*32*	*57*	*56*	*35*	*180*
Total	*169*	*221*	*208*	*182*	*780*

Solution:

Let us take the hypothesis that there is no variance among the cities in the tendency of men to marry. Applying χ^2 test :

The excepted frequencies are :

$$E_{11} = \frac{600}{780} \times 169 = 100; \; E_{12} = \frac{600}{780} \times 221 = 170$$

$$E_{13} = \frac{600}{780} \times 208 = 160.$$

The table of expected frequencies is :

130	170	160	140	600
39	51	48	42	180
169	221	208	182	780

O	**E**	**$(O - E)^2$**	**$(O - E)^2/E$**
137	130	49	0.377
32	39	49	1.256
164	170	36	0.212
57	51	36	0.706
152	160	64	0.400
56	48	64	1.333
147	140	49	0.350
35	42	49	1.167
			$\Sigma (O - E)^2 / E = 5.801$

$\chi^2 = \Sigma [(O - E)^2 / E] = 5.801$

$\nu = (r - 1)(c - 1) = (2 - 1)(4 - 1) = 3$

For $\nu = 3, \; \chi^2_{0.05} = 7.82$

The calculated value of χ^2 is less than the table value. The hypothesis holds true. We, therefore, conduct that the data do not indicates significant variation among the cities in the tendency of men to marry.

Example 17:

1,000 of students at college level were graded according to their I.Q. and the economic conditions of their homes. Use χ^2 test to find out whether there is any association between economic condition at home and I.Q.

	I.Q		
Economic Condition	*High*	*Low*	*Total*
Rich	*460*	*140*	*600*
Poor	*240*	*160*	*400*
Total	*700*	*300*	*1000*

Given for $v = 1,\ \chi^2_{0.05} = 3.84$.

Solution:

Let us taken the hypothesis that there is no association between economic condition at home and I.Q. Applying χ^2 test :

$$E_{11} = \frac{(A) \times (B)}{N} = \frac{600 \times 700}{1000} = 420$$

The table of expected frequencies shall be as follows :

420	180	600
280	120	400
700	300	1,000

O	E	$(O - E)^2$	$(O - E)^2/E$
460	420	1,600	3.810
240	280	1,600	5.714
140	180	1,600	8.889
160	120	1,600	13.333
			$\Sigma (O - E)^2 / E] = 31.746$

$\chi^2 = \Sigma [(O - E)^2 / E] = 31.746$

$v = (v - 1)(c - 1) = (2 - 1)(2 - 1) = 1$

For $v = 1,\ \chi^2_{0.05} = 3.84$

The calculated value of χ^2 is greater than the table value. The hypothesis is rejected. Hence there is association between economic condition at home and I.Q.

Example 18:

The following results were obtained when we sets of items were subjected to two different treatments X and Y, to enhance their tensile strength.

Treatment X was applied on 400 items and 80 wee found to have gained in strength.

Treatment X was applied in 400 items and 20 wee found to have gained in strength.

Is treatment Y superior to treatment X ?

Solution:

Let us take the null hypothesis that there is no difference in the two types of treatments X and Y. The given results and be tabulated as follows:

Treatment	Gained	Not gained	Total
X	80	320	400
Y	20	380	400
Total	100	700	800

$$\text{Expectation of (AB)} = \frac{(A) \times (B)}{N} = \frac{400 \times 100}{800} = 50$$

The table of expected frequencies is given below ;

50	350	400
50	350	400
100	700	800

Applying χ^2 test :

O	E	$(O-E)^2$	$(O-E)^2/E$
80	50	900	18.000
20	50	900	18.000
320	350	900	2.571
380	350	900	2.571
			$\Sigma[(O-E)^2/E] = 41.142$

$\chi^2 = \Sigma[(O-E)^2/E] = 41.142$

$\nu = (r-1)(c-1) = (2-1)(2-1) = 1$

For $\nu = 1, \chi^2_{0.05} = 3.84$

The calculated value of χ^2 is much higher than the table value. Hence the hypothesis is rejected we conclude that there is difference in treatment X and Y.

Example 19:

Verify whether Poisson distribution can be assumed from the data given below :

No. of defects	*0*	*1*	*2*	*3*	*4*	*5*
f_o	*6*	*13*	*13*	*8*	*4*	*3*
f_o	*6.24*	*13.52*	*13.52*	*9.01*	*4.50*	*1.80*

Solution:

Applying χ^2 test :

No. of defects	O	E	$(O - E)^2$	$(O - E)^2 / E$
0	6	6.24	0.0576	0.009
1	13	13.52	0.2704	0.020
2	13	13.52	0.2704	0.020
3	8	9.01	1.0201	0.113
4	4	4.50	0.2500	0.056
5	3	1.80	1.4400	0.800
		$\chi^2 = \Sigma\,[(O - E)^2 / E] = 1.018$		

$\nu = n - 1 = 6 - 2 = 4$

For $\nu = 4$, $\chi^2_{0.05} = 9.49$.

The calculated value of c^2 is much less than the table value. Hence the Poisson contribution provides a good fit to the data.

Example 20:

A movie producer is bringing out a new movie. In order to map out his advertising campaign, he wants to determine whether the movie will appeal most to particular age group or whether it will appeal equally to all age groups, the producer takes a random sample from persons attending preview of the new movie, and obtains the following results :

	Age Groups				
	Under 20	*20-39*	*40-59*	*60 & over*	*Total*
Liked the movie	*146*	*78*	*48*	*28*	*300*
Disliked the movie	*54*	*22*	*42*	*22*	*140*
Indifferent	*20*	*10*	*10*	*20*	*60*
Total	*220*	*110*	*100*	*70*	*500*

What inference will you draw from this data ?

Solution:

Let us take the hypothesis that the movie appeals equally to all age groups.

$$E_{11} = \frac{300}{500} \times 200 = 132; \quad E_{21} = \frac{140}{500} \times 220 = 61.6$$

$$E_{12} = \frac{300}{500} \times 110 = 66; \quad E_{22} = \frac{140}{500} \times 1.0 = 30.8$$

$$E_{13} = \frac{300}{500} \times 100 = 60; \quad E_{23} = \frac{140}{500} \times 100 = 28$$

The expected frequencies can be obtained below :

				Total	
132.0	66.0	60	42	300	
61.6	30.5	28	19.6	140	
26.4	13.2	12	8.4	60	
220	110	100	70	500	← Total

Applying c² test :

O	E	$(O - E)^2$	$(O - E)^2/E$
146	132.0	196.0	1.485
54	61.6	57.76	0.938
20	26.4	40.96	1.552
78	66.0	144.00	2.182
22	30.8	77.44	2.514
10	13.2	10.24	0.776
48	60.0	144.00	2.400
42	28.0	196.00	7.000
10	12.0	4.00	0.333
28	42.0	196.00	4.667
22	19.6	5.76	0.294
20	8.4	134.56	16.019
			$\Sigma [(O - E)^2 / E] = 40.16$

$\chi^2 + \Sigma [(O - E)^2 / E] = 40.16$

$v = (r - 1)(c - 1) = (3 - 1)(4 - 1) = 6$

For $v = 6, \chi^2_{0.05} = 12.59$

The calculated value of χ^2 is much greater than the table value. The hypothesis does not hold good and we, therefore, include that the movie does not appeal equality to all groups.

Example 21:

A set of 5 coins is tossed 3,200 times, and the number of heads appearing each time is noted. The results are given below :

No. of heads	*0*	*1*	*2*	*3*	*4*	*5*
Frequency	*80*	*570*	*1,100*	*900*	*50*	*50*

Test the hypothesis that the coins are unbiased.

Solution:

Let us take the hypothesis that the coins are unbiased. If this is true, then the chances of getting heads, etc., in a toss of 5 coins are the successive terms of the binomial expansion $\left(\frac{1}{2}+\frac{1}{2}\right)^5$. So the theoretical frequencies in 3,200 tosses are the terms in the expansion $3200\left(\frac{1}{2}+\frac{1}{2}\right)^2$ and are as follows :

No. of heads	0	1	2	3	4	5
Frequencies	100	500	1,000	1,000	500	100

Applying χ^2 test :

O	E	$(O-E)^2$	$(O-E)^2/E$
80	100	400	4.00
	570	500	4,900 9.80
1,100	1,000	10,000	10.00
900	1,000	10,000	10.00
500	500	0	0.00
50	100	2,500	25.00
			$\Sigma\,[(O-E)^2/E] = 58.80$

$\chi^2 = \Sigma\,[(O-E)^2/E] = 58.80$

$\nu = n - 1 = 6 - 1 = 5$

For $\nu = 5$, $\chi^2_{0.05} = 11.07$

The calculated value of χ^2 is much greater than the table value. Hence the hypothesis. We, therefore, conclude that the coins are biased.

Example 22:

The demand for a particular spare part in a factory as found to vary from day to day. In a sample study the following information was obtained:

Days *:*	*Mon.*	*Tues.*	*Wed.*	*Thurs.*	*Fri.*	*Sat.*
No. of parts : demanded :	*1,124*	*1,125*	*1,110*	*1,120*	*1,126*	*1,115*

Test the hypothesis that the number of parts demanded and on the day of the week. (The table value of χ^2 for 5 d.f. at 5% level of significance is 11.07).

Solution:

The number of parts demanded during six days = 6,720, we should expect $\frac{6{,}720}{6} = 1{,}120$ parts to be demanded each day of the week.

Applying χ^2 test :

Days	O	E	$(O - E)^2$	$(O - E)^2 / E$
Monday	1,124	1,120	16	0.014
Tuesday	1,125	1,120	25	0.022
Wednesday	1,110	1,120	100	0.089
Thursday	1,120	1,120	0	0.0
Friday	1,126	1,120	36	0.032
Saturday	1,115	1,120	25	0.022
				$\Sigma [(O-E)^2 / E]=0.179$

$\chi^2 = [\Sigma (O - E)^2 / E] = 0.179$

$\nu = n - 1 = 6 - 1 = 5$

For $\nu = 5$, $\chi^2_{0.05} = 11.07$

The calculated value of χ^2 is less than the table value and hence the hypothesis is true. The number of parts demanded does not depend on the day of the week.

Example 23:

From the adult male population of seven cities random samples of married and unmarried men as given below were taken. Can it be said that there is a significant variation among the people of different cities in the tendency to marry ?

City	*A*	*B*	*C*	*D*	*E*	*F*	*G*	*Total*
Married	*170*	*285*	*165*	*106*	*153*	*125*	*146*	*1,150*
Unmarried	*40*	*125*	*35*	*37*	*55*	*35*	*33*	*360*
	210	*410*	*200*	*143*	*208*	*160*	*179*	*1,510*

(Given for $\nu = 6$, χ^2 0.05 = 12.6)

Solution:

Let us take the hypothesis that their is no significant difference in the tendency of men to marry.

Expected number of married people in city A :

i.e. $E_{11} = \frac{1150}{1510} \times 210 = 159.93$; $E_{12} = \frac{1150}{1510} \times 410 = 312.25$

$E_{13} = \frac{1150}{1510} \times 200 = 152.32$; $E_{14} = \frac{1150}{1510} \times 143 = 108.91$

$$E_{15} = \frac{1150}{1510} \times 208 = 158.41; \quad E_{16} = \frac{1150}{1510} \times 160 = 121.85;$$

$$E_{17} = \frac{1150}{1510} \times 179 = 136.33$$

The table of expected frequencies is given below :

159.93	312.25	152.32	108.91	158.41	121.85	136.33
50.07	97.75	47.68	34.09	49.59	38.15	42.67
210	410	200	143	208	160	179

Applying c² test :

O	E	$(O-E)^2$	$(O-E)^2/E$
170	159.93	101.405	0.624
285	312.25	742.562	2.378
165	152.32	160.782	1.056
106	108.91	6.812	0.062
153	158.41	29.268	0.185
125	121.85	9.922	0.081
146	136.33	93.509	0.686
40	50.07	101.405	2.025
125	97.75	742.562	7.597
35	47.68	160.782	3.372
37	34.09	8.468	0.248
55	49.59	29.268	0.590
35	38.15	9.922	0.260
33	42.67	93.509	2.191
			$\Sigma [(O-E)^2 / E] = 21.365$

$$\chi^2 = \Sigma [(O-E)^2 / E] = 21.365$$

$\nu = (r-1)(c-1) = (2-1)(7-1) = 6$

For $\nu = 6$, $\chi^2_{0.05} = 12.6$

The calculated value of χ^2 is much greater than the table value. The hypothesis is rejected and we conclude that there is difference in the tendency of people to marry in different cities.

Example 24:

The following table shows the number of people interviewed by age group and the number in each age group estimated to have peptic ulcer.

Age group	*20—25*	*25—30*	*30—35*	*35—40*	*40—45*
Nos. Interviewed	*50*	*100*	*200*	*350*	*400*

P.U. Cases	*5*	*12*	*25*	*28*	*40*
Age Group	*45—50*	*50—55*	*55—60*	*Total*	
Nos. Interviewed	*250*	*150*	*100*	*1,600*	
P.U. Cases	*27*	*12*	*11*	*160*	

Do the data reveal an association between age groups, and peptic ulcer?

Solution:

Let us take the hypothesis that peptic ulser is not associated with age groups, i.e., the two attributes are independent.

If this is true then $\frac{100}{1,000} \times 100 = 100\%$ of people should suffer from peptic ulser in each age group. The observed and expected number of case suffering from peptic ulser shall be as follows :

Age group	Observed cases	Expected cases
20—25	5	5
25—30	12	10
30—35	25	20
35—40	28	35
40—45	40	40
45—50	27	25
50—55	12	15
55—60	11	11

Applying c^2 test :

O	E	$(O - E)^2$	$(O - E)^2/E$
5 ⎤ 12 ⎦	5 ⎤ 10 ⎦	4	0.267
25	20	25	1.250
28	35	49	1.400
40	40	0	0.000
27	25	4	0.160
12	15	9	0.600
11	10	1	0.100
			$\Sigma [(O - E)^2 / E] = 3.777$

$\chi^2 = \Sigma (O - E)^2 / E] = 3.777$

$\nu = 8 - 1 - 1 = 6$

For $\nu = 6$, $\chi^2_{0.05} = 12.59$

The calculated value of χ^2 is much greater than the table value. Hence the hypothesis is true. We therefore, conclude that peptic ulcer is not associated with age groups.

Example 25:

The following contingency table shows the classification of 1,000 workers in a factory, according to the disciplinary action taken by the management and their promotional experience :

Disciplinary action	*Promotional Experience*		*Total*
	Promoted	*Not promoted*	
Offenders	*30*	*670*	*700*
Non-offenders	*70*	*230*	*300*
Total	*100*	*900*	*1,000*

Use χ^2 -test to ascertain whether the disciplinary action and promotional experience are associated.

(Given for ν = =, $\chi^2_{0.05} = 3.84$).

Solution:

Let us take the hypothesis that the disciplinary action taken and promotional experience are not associated.

Expected frequency corresponding to

$$(AB) = \frac{300}{1000} \times 100 = 30$$

The table of expected frequencies shal! be as follows :

70	630	700
30	270	300
100	900	1000

O	**E**	**$(O - E)^2$**	**$(O - E)^2/E$**
30	70	1600	22.857
70	30	1600	53.333
670	630	1600	2.539
230	270	1600	5.926
			Σ [(O–E)² / E]=84.655

$\nu = (r - 1)(c - 1) = (2 - 1)(2 - 1) = 1$

For $\nu = 1, \chi^2_{0.05} = 3.84$

The calculated value of χ^2 is much more than the table value. Hence the hypothesis is rejected. We, therefore, conclude that the disciplinary action taken and promotional experience are associated.

Example 26:

The number of scooter accidents per month is a certain town wos as follows :

12, 8, 20, 2, 14, 10, 15, 6, 9, 4

Use chi-square test to determine if these frequencies are in agreement with the belief that accident conditions were the same during 10-month period.

Solution:

On the hypothesis that accident conditions were the same during the period we should expect

$$\frac{12 + 8 + 20 + 2 + 14 + 10 + 15 + 6 + 9 + 4}{10} = 10 \text{ accident per month}$$

Applying χ^2 test :

O	E	$(O - E)^2$	$(O - E)^2/E$
12	10	4	0.40
8	10	4	0.40
20	10	100	10.00
2	10	64	6.40
14	10	16	1.60
10	10	0	0.00
15	10	25	2.50
6	10	16	1.60
9	10	1	0.10
4	10	36	3.60
			$\Sigma [(O - E)^2 / E] = 26.60$

For $\nu = (n - 1) = (10 - 1) = 9$

$\nu = 9, \chi^2_{0.06} = 16.92$

The calculated value of χ^2 is much greater than the table value. The hypothesis is rejected. The frequency are not in agreement with the belief that the accident conditions were the same during 10-month period.

Example 27:

The contingency table below summarise the results obtained in a study conducted by a research organisation with respect to the performance of four competing brands of toothpaste among the users :

Brand	*Brand A*	*Brand B*	*Brand C*	*Brand D*	*Total*
No Cavities	*9*	*13*	*17*	*11*	*50*
One to five Cavities	*63*	*70*	*85*	*82*	*300*
More than five Cavities	*28*	*37*	*48*	*37*	*150*
Total	*100*	*120*	*150*	*130*	*500*

Test the hypothesis that incidence of cavities is independent of the barbed of the toothpaste used. (The table value of χ^2 for 6 df. are 12.59 and 16.81 at 5 percent levels of significant respectively).

Solution:

Our hypothesis is that incidence of cavities is independent of the brand of the toothpaste used. Applying χ^2 test.

$$E_{11} = \frac{50}{500} \times 100 = 10; \quad E_{21} = \frac{300}{500} \times 100 = 60$$

$$E_{12} = \frac{50}{500} \times 120 = 12; \quad E_{22} = \frac{300}{500} \times 120 = 72$$

$$E_{13} = \frac{50}{500} \times 150 = 15; \quad E_{23} = \frac{300}{500} \times 150 = 90$$

The table of expected frequencies shall be :

10	12	15	13	50
60	72	90	78	300
30	36	45	39	150
100	120	150	130	500

O	E	$(O - E)^2$	$(O - E)^2/E$
9	10	1	0.100
63	60	9	0.150
28	30	4	0.133
13	12	1	0.083
70	72	4	0.056
37	36	1	0.028
17	15	4	0.267
85	90	25	0.278
48	45	9	0.200
11	13	4	0.308
82	78	16	0.205
37	39	4	0.103
			$\Sigma [(O-E)^2 / E]=1.911$

$\nu = (r - 1)(c - 1) = (3 - 1)(4 - 1) = 6$

For $\nu = 6$, $\chi^2_{0.05} = 12.59$

The calculated value of χ^2 is much less than the table value. Our hypothesis holds true. Hence, incidence of cavities is independent of the brand to the toothpaste used.

Example 28:

An automobile company gives you the following information about age groups and the liking for particular model of car which it plans to introduce.

	Age Group				
	Below 20	*20—39*	*40—59*	*60 & above*	*Total*
Persons who :					
Liked the car	*140*	*80*	*40*	*20*	*280*
Disliked the car	*60*	*50*	*30*	*80*	*220*
Total	*200*	*130*	*70*	*100*	*500*

On the basis of this data can it be conducted that the model appeal is independent of the age group(given for $\nu = 3, \chi^2 0.05 = 7.815$.

Solution:

Let us take the hypothesis that the model appeal is independent of the age group. Applying χ^2 test :

$$E_{11} = \frac{280}{500} \times 200 = 112; \qquad E_{12} = \frac{280}{500} \times 130 = 72.8$$

$$E_{13} = \frac{280}{500} \times 70 = 39.2$$

The table of expected frequencies will be as follows :

112	72.8	39.2	56	280
88	57.2	30.8	44	220
200	130	70	100	500

O	E	$(O - E)^2$	$(O - E)^2/E$
140	112.0	784.00	7.000
60	8	784.00	8.910
80	7208	51.84	0.712
50	57.2	51.84	0.906
40	39.2	0.64	0.016
30	30.8	0.64	0.021
20	56.0	1296.00	23.143
80	44.0	1296.00	29.454
			$\Sigma [(O - E)^2 / E] = 70.162$

$$\chi^2 = \Sigma\,[(O - E)^2 / E] = 70.162$$

For $\nu = (r - 1)(c - 1) = (2 - 1) = 3$

$\nu = 3,\ \chi^2 0.05 = 7.815$

The calculated value of χ^2 is much greater than the table value. Hence the hypothesis is rejected. We, therefore, conclude that the model appeal is not independent of the age groups.

Example 29:

A survey of 800 families with 4 children each revealed following distribution.

No. of boys	*0*	*1*	*2*	*3*	*4*
No. of girls	*4*	*3*	*2*	*1*	*0*
No. of families	*32*	*178*	*290*	*236*	*64*

Is this result consistent with the hypothesis that the male and female births are equally probable ?

Solution:

On the hypothesis that the male and female births are equally probable, the expected number of families would be obtained by the expansion of :

$$800\left(\frac{1}{2} + \frac{1}{2}\right)^4$$

$$= 800\left(\frac{1}{16}, \frac{4}{16}, \frac{6}{16}, \frac{4}{16}, \frac{1}{16}\right)$$

$$= 50,\ 200,\ 300,\ 200,\ 50$$

Applying χ^2 test :

O	E	$(O - E)^2$	$(O - E)^2/E$
32	50	324	6.480
178	200	484	2.420
290	300	100	0.333
236	200	1,296	6.480
64	50	196	3.920
			$\Sigma\,[(O - E)^2 / E] = 19.633$

$$\chi^2 = \Sigma\,[(O - E)^2 / E] = 19.633$$

$$\nu = 5 - 1 = 4$$

For $\nu = 4,\ \chi^2_{0.05} = 9.488$

The calculated value or χ^2 is greater than the table value. The hypothesis is rejected. Hence the male and female births don't seem to have been equally probable.

Example 30:

A book has 700 pages. The number of pages with various numbers of misprints is recorded below. At 5% significant level are the misprints distributed according to Poisson law ?

No. of misprints (X) :	*0*	*1*	*2*	*3*	*4*	*5*
No. of pages with X misprints :	*616*	*70*	*10*	*2*	*1*	*1*

Solution:

By applying Poisson law find expected values :

x	f	f_x
0	616	0
1	70	70
2	10	20
3	2	6
4	1	4
5	1	5
	N = 700	Σ fx = 105

$$\overline{X} \text{ or } m = \frac{\sum fX}{N} = \frac{105}{700} = 0.15$$

$P_{(0)} = e^{-m} = e^{-.15} = 0.8607$ (as per table)
$NP_{(0)} = .8607 \times 700 = 602.5$
$NP_{(1)} = NP_{(0)} \times m = 602.5 \times .15 = 90.4$
$NP_{(2)} = NP_{(1)} \times m/2 = 90.4 \times .15/2 = 6.8$
$NP_{(3)} = NP_{(2)} \times m/3 = 6.8 \times .15/3 = 0.4$
$NP_{(4)} = NP_{(3)} \times m/4 = 0.4 \times .15/4 = 0$
$NP_{(5)} = 0$

Applying χ^2 test

O	**E**	**$(O - E)^2$**	**$(O - E)^2/E$**
616	602.5	182.25	0.302
70	90.4	416.16	4.604
10	6.8	10.24	1.506
2	0.4	12.096	32.400
1	0	–	–
1	0	–	–
			$\Sigma [(O - E)^2 / E] = 38.812$

$\nu = 4 - 2 = 2$ For $= 2$, $\chi^2_{0.05} = 5.99$

The calculated value of χ^2 is greater than the table value. Hence at 5% level the misprints are not distributed according to Poisson law.

Example 31:

The following data relate to the sales, in a time of trade depression of a certain article in wide demand. Do the data suggest that the sales are significance affected by depression ?

District where sates are :

Not hit by Depression	***Districts Hit***		***Total***
Satisfactory	*140*	*60*	*200*
Not satisfactory	*40*	*60*	*100*
Total	*180*	*120*	*300*

Solution:

Let us take the hypothesis that the sales are not significantly affected by depression.

Applying χ^2 test

$$\text{Expectation of (AB)} = \frac{200}{300} \times 180 = 120$$

The table of expected frequencies shall be as follows :

120	80	200
60	40	100
180	120	300

O	E	$(O-E)^2$	$(O-E)^2/E$
140	120	400	3.333
40	60	400	6.667
60	80	400	5.000
60	40	400	10.000
			$\Sigma\,[(O-E)^2/E] = 25$

$\nu = (r-1)(c-1) = (2-1)(2-1) = 1$

For $= \nu = 1;\ \chi^2_{0.05} = 3.84.$

The calculated value of χ^2 is less than the table value. The hypothesis is rejected. Hence the sales are significantly affected by depression.

200	100	300
200	100	300
400	200	600

O	E	$(O-E)^2$	$(O-E)^2/E$
190	200	100	0.50
210	200	100	0.50
110	100	100	1.00
90	10	100	1.00
			$\Sigma[(O-E)^2/E] = 3.0$

$$\nu = (r-1)(c-1) = (2-1)(2-1) = 1$$

The calculated value of χ^2 is less than the table value. The hypothesis accepted. Hence the opinion of the Under-graduate and Post-graduate students on autonomous status of colleges are independent.

Example 32:

1,000 families were selected at random in a city to test the belief that high incomes families usually send their children to public schools and the low income families often sent their children to government schools. The following results were obtained.

	School		
Income	***Public***	***Govt.***	***Total***
Low	*370*	*430*	*800*
High	*130*	*70*	*200*
Total	*500*	*500*	*1,000*

Test whether income and type of schooling are independent.

Solution:

Let us take the hypothesis that the income and type of schooling are independent.

$$\text{Expectation of } (AB) = \frac{(A)\times(B)}{N} = \frac{50\times 800}{1{,}000} = 400$$

The table of expected frequencies shall be :

400	400	800
100	100	200
500	500	1000

Applying χ^2 test :

O	E	$(O-E)^2$	$(O-E)^2/E$
370	400	900	2.25
130	100	900	9.00
430	400	900	2.25
70	100	900	9.00
			$\Sigma[(O-E)^2/E] = 22.5$

$$\chi^2 = \Sigma [(O - E)^2 / E] = 22.5$$
$$\nu = (r - 1)(c - 1) = (2 - 1)(2 - 1) = 1$$
$$\text{For} = \nu = 1; \chi^2_{0.05} = 3.84.$$

The calculated value of χ^2 is more than the table value. The hypothesis is rejected. Hence income and type of schooling are not independent.

Example 33:

Two researches adopted different sampling techniques while investing the same group of students to find the number of students failing in different intelligence level. The results are as follows :

No. of students in each level

Researcher	*Below average*	*Average*	*Above average*	*Genius*	*Total*
X	*86*	*60*	*44*	*10*	*200*
Y	*40*	*33*	*25*	*2*	*100*
Total	*126*	*93*	*69*	*12*	*300*

Would you say that the sampling techniques adopted by two researches are significantly different ?[For $\nu = 3$, $\chi^2 0.05 = 7.82$]

Solution:

Let us take the hypothesis that the sampling techniques adopted by the two researches are not significantly different.

Applying χ^2 test :

$$E_{11} = \frac{200}{300} \times 126 = 84; \qquad E_{12} = \frac{200}{300} \times 93 = 62$$

$$E_{13} = \frac{200}{300} \times 69 = 46;$$

Table of expected frequencies

84	62	46
42	31	23
126	93	69

O	**E**
86	84
40	42
60	62
33	31
44	46
25	23

10 8

2 4

For = ν = 3; $\chi^2_{0.05}$ = 7.82

The calculated value of χ^2 is more than the table adopted by the two researches are not significantly different

Example 34:

In the accounting department of a bank 100 accounts are selected at random and examined for errors. Suppose the following results have been obtained :

No. of errors	*0*	*1*	*2*	*3*	*4*	*5*	*6*
No. of accounts	*35*	*40*	*19*	*2*	*0*	*2*	*2*

On the basis of this information can it be concluded that the errors are distributed according to Poisson probability law ?

Solution:

To solve this question we have to obtain the expected frequencies by supplying Poisson distribution and test of goodness of fit by χ^2 test.

Fitting Poisson Distribution

x	f	fx
0	35	0
1	40	40
2	19	38
3	2	6
4	0	0
5	2	10
6	2	12
	N = 100	Σ.f x = 106

$$X = \frac{\sum fX}{N} = \frac{106}{100} = 1.06$$

$$P(r) = \frac{e^{-m}\, m^r}{r^i} = e^{-1.06}$$

$e^{-1.06} = (e^{-1})\,(e^{0.06}) = .36788 \times .9418 = .3465$ (from the table)

$N\,(P_0) = .3465 \times 110 = 34.65$

$N\,(P_1) = N\,(P_0) \times m = 34.65 \times 1.06 = 36.73$

$$N\,(P_2) = N\,(P_1)\,\frac{m}{2} = 36.73 \times \frac{1.06}{3} = 19.47$$

$$N(P_3) = N(P_2)\frac{m}{3} = 19.47 \times \frac{1.06}{4} = 6.88$$

$$N(P_4) = N(P_3)\frac{m}{4} = 6.88 \times \frac{1.06}{4} = 1.82$$

$$N(P_5) = N(P_4)\frac{m}{5} = 1.82 \times \frac{1.06}{5} = 0.39$$

$$N(P_6) = N(P_5)\frac{m}{6} = 0.39 \times \frac{1.06}{6} = 0.06$$

Applying χ^2 test :S

O	E		$(O-E)^2$	$(O-E)^2/E$
35	34.65		0.1225	0.004
40	36.73		10.6929	0.291
19	19.47		0.2209	0.011
2	6.88			
6	1.82			
2	0.39	9.15	9.925	1.084
2	0.06			
				$\Sigma[(O-E)^2/E]=1.39$

After grouping, only 4 classes are left.

$\nu = 4 - 2 = 2$

For $\nu = 2$, c^2 0.05 = 5.59

The calculate value of χ^2 is less than the table value. Hence the given information verify that the errors are distributed according to normal distribution

Example 35:

Three samples are taken comprising university teachers. Each person chosen is asked to best represent his feeling toward a certain national favour of policy (F), against the policy (A), and indifferent of the interviews are given below :

Reaction Occupation	*F*	*A*
Doctors	*80*	*30*
Advocates	*70*	*40*
University Teachers	*50*	*50*
Total	*200*	*120*

On the basis of this data can it be concluded that University teachers are homogeneous in so far as National

Solution:

Let us take the hypothesis that the three far as the opinion of three different groups of people is concerned.

Applying χ^2 test :

$$E_{11} = \frac{120}{400} \times 200 = 60; \qquad E_{12} = \frac{120}{400};$$

$$E_{21} = \frac{150}{400} \times 200 = 75; \qquad E_{22} = \frac{150}{400};$$

The table of expected frequencies as shall be as follows :

60	36	24
75	45	30
65	39	26
200	120	80

O	E	$(O - E)^2$	$(O - E)^2/E$
80	60	400	6.667
70	75	25	0.333
50	65	225	3.462
30	36	36	1.000
40	45	25	0.556
50	39	121	3.103
10	24	196	8.167
40	30	100	3.333
30	26	16	0.616
			$\Sigma [(O - E)^2 / E] = 27.237$

$$\nu = (r - 1)(c - 1) = (3 - 1)(3 - 1) = 4$$

For = $\nu = 4$; $\chi^2_{0.05} = 9.488$

The calculated value of χ^2 is greater than the table value. The hypothesis is rejected. The three classifications are not homogenous in so far as the opinion of the three different groups of people about national policy under consideration is concerned.

Example 36:

From the following data find out whether there is any relationship between sex and preference of colour :

Colour	*Males*	*Females*	*Total*
Red	*10*	*40*	*50*
White	*70*	*30*	*100*
Green	*30*	*20*	*50*
Total	*110*	*90*	*200*

Solution:

Let us take the hypothesis that there us no relationship between sex and colour preference of colour. Applying χ^2 test :

$$E_{11} = \frac{50}{200} \times 110 = 27.5; \qquad E_{21} = \frac{100}{200} \times 110 = 55$$

27.5	22.5	50
55.0	45.0	100
27.5	22.5	50
110	90	200

O	E	$(O - E)^2$	$(O - E)^2/E$
10	27.5	306.25	11.14
70	55.0	225.00	4.09
30	27.5	6.25	0.23
40	22.5	306.25	13.61
30	45.0	225.00	5.00
20	22.5	6.25	.028
			$\Sigma [(O - E)^2 / E] = 34.35$

$$\chi^2 = \Sigma [(O - E)^2 / E] = 34.35$$

$$\nu = (r - 1)(c - 1) = (3 - 1)(2 - 1) = 2$$

$$\text{For} = \nu = 2; \ \chi^2_{0.05} = 5.99$$

The calculated value of χ^2 is greater than the table value. The hypothesis is rejected. Hence there is relationship between sex and preference of colour.

Example 37:

In across-breeding experiment with plants at certain species 240 offspring were classified into 4 classes with respect to the structure of their leaves as follows :

Class	*I*	*II*	*III*	*IV*	*Total*
Frequency	*21*	*127*	*40*	*52*	*240*

According to theory of heredity, the probabilities of the four classes should be in the ratio 1 : 9 : 3 : 3. Are these data consistent with theory?

Solution:

Applying χ^2 test :

O	E	$(O - E)^2$	$(O - E)^2/E$
21	15	36	2.400
127	135	64	0.474
40	45	25	0.556
52	45	49	1.089
		$\Sigma [(O - E)^2 / E] = 4.519$	

From 5% value of χ^2 for 3 d.f. = 7.82. The calculate value is less than the table value. Hence hypothesis that these data are consistent with theory is accepted.

Example 38:

"A sample of 300 students of Under-graduate and 300 students of Post-graduate classes of a university were asked to give their opinion towards the autonomous college. 190 of the Under-graduate and 210 of the Post-graduate students favoured the autonomous status".

Present the above fact in the form of a frequency table and test, at 5% level, that opinions of Under-graduate and Post-graduate students on autonomous status of colleges are independent. (The value of chi-square at 5% level of 1 d.f. is 3.84).

Solution:

The given data can be put in the form of frequency table :

	Favoured	Not favoured	Total
Under-graduates	190	110	300
Post-graduates	210	90	300
Total	400	200	600

Let us take the opinions of the Under-graduate and Post-graduate students on autonomous status of college are independent.

Applying χ^2-test :

$$\text{Expectation of (AB)} = \frac{300}{600} \times 400 = 200$$

Example 39:

Given below is the contingency table for production in three shifts and the number of defective goods turnout. Find the value of C. Is it possible that the number of defective goods depend on the shift run by the factory?

Shift	*No. of defective goods in 3 week*			
	Ist week	***2nd week***	***3rd week***	***Total***
I	*15*	*5*	*20*	*40*
II	*20*	*10*	*20*	*50*
III	*25*	*15*	*20*	*60*
Total	*60*	*30*	*60*	*150*

Solution:

$$C \quad \frac{\chi^2}{\sqrt{N + \chi^2}}$$

Let us take the hypothesis that the number of defective goods does not depend upon the shift run by the factory.

$$E_{11} = \frac{40}{150} \times 60 = 16; \qquad E_{21} = \frac{50}{150} \times 60 = 20$$

$$E_{12} = \frac{40}{150} \times 30 = 8; \qquad E_{22} = \frac{50}{150} \times 30 = 10$$

The expected frequencies are tabulated below :

16	8	16	40
20	10	20	50
24	12	24	60
60	30	60	150

O	*E*	*$(O - E)^2$*	*$(O - E)^2/E$*
15	16	1	0.063
20	20	0	0.000
25	4	1	0.042
5	8	9	1.125
20	16	16	1.000
20	20	0	0.000
20	4	16	0.667
			$\Sigma [(O - E)^2 / E] = 3.647$

$$c = \frac{\chi^2}{\sqrt{N + \chi^2}} = \frac{3.647}{\sqrt{150 + 3.647}} = \frac{3.647}{12.395} = 0.294$$

$$\nu = (3 - 1)(3 - 1) = 4,$$

For $\nu = 4$, $\chi^2_{0.05} = 9.488$

Since the calculated value of χ^2 is less than the table value, the hypothesis is accepted. Hence the number of defective goods does not depend on the shift run by the factory.

Example 40:

From the following data find out whether there is any relationship between sex and preference of colours :

Colour	Males	Females	Total
Green	*40*	*60*	*100*
White	*35*	*25*	*60*
Yellow	*25*	*15*	*40*
Total	*100*	*100*	*200*

Solution:

Let us take the null hypothesis that there is no relationship between sex and preference of colours.

Applying χ^2 test :

$$E_{11} = \frac{100}{200} \times 100 = 50; \qquad E_{12} = \frac{60}{200} \times 100 = 30$$

The table pf expected frequencies shall be as follows :

50	50	100
30	30	60
20	20	40
100	100	200

O	E	$(O - E)^2$	$(O - E)^2/E$
40	50	100	2.000
35	30	25	0.833
25	20	25	1.250
60	50	100	2.000
25	30	25	0.833
15	20	25	1.250
			$\Sigma [(O - E)^2 / E] =$ **8.166**

$$\nu = (r - 1)(c - 1) = (3 - 1)(2 - 1) = 2$$

For $\nu = 2, \chi^2_{0.05} = 5.99$

The calculated value of χ^2 is less than the table value, the hypothesis is rejected. Hence there is relationship between sex and preference of colours.

Example 41:

A firm selling four products is interested in finding out whether the sales are distributed similarly among four general classes of customers. A random sample of 400 sales records provides the following information :

	Products				
Customer's group	***1***	***2***	***3***	***4***	***Total***
Partners	*25*	*10*	*30*	*15*	*80*
Factory workers	*32*	*20*	*10*	*28*	*90*
Businessmen	*35*	*48*	*25*	*40*	*148*
Professionals	*28*	*22*	*15*	*17*	*82*
Total	*120*	*100*	*80*	*100*	*400*

Formulate a suitable hypothesis. Applying χ^2-test. What conclusion can you draw from the test results ?

Solution:

Let us take the hypothesis that the four samples representing farmers, factory workers, businessmen and professionals are drawn from the same population, i.e., the four products are homogeneously distributed amongst four general classes of customers.

The expected frequencies corresponding to each group and product can be obtained as follows :

Expected Frequencies

Group	1	2	3	4	Total
Farmers	$\frac{80}{400}\times120=240$	$\frac{80}{400}\times100=20.0$	$\frac{80}{400}\times80=16.0$	$\frac{80}{400}\times200=20.0$	86
Factory Workers	$\frac{90}{400}\times120=27.0$	$\frac{90}{400}\times100=22.5$	$\frac{90}{400}\times80=18.0$	$\frac{90}{400}\times100=22.0$	90
Business -men	$\frac{148}{400}\times120=44.4$	$\frac{148}{400}\times100=37.0$	$\frac{148}{400}\times80=29.6$	$\frac{148}{400}\times100=37.0$	148
Professio- nals	$\frac{82}{400}\times120=24.6$	$\frac{82}{400}\times100=20.5$	$\frac{148}{400}\times80=16.4$	$\frac{82}{400}\times100=205.5$	82
Total	120	100	80	100	400

Applying χ^2 test :

O	E	$(O - E)^2$	$(O - E)^2/E$
25	24.0	1.00	0.042
32	27.0	25.00	0.926
35	44.4	88.36	1.990
28	24.6	11.56	0.470
10	20.0	100.00	5.000
20	22.5	6.25	0.278
48	37.0	121.0	3.270
22	20.5	2.25	0.110
30	16.0	196.00	12.250
10	18.0	64.00	3.556
25	29.6	21.16	0.715
15	16.4	1.96	0.119
15	20.00	25.00	1.250
28	22.5	30.25	1.344
40	37.0	9.00	0.243
17	20.5	12.25	0.598
			$\Sigma [(O - E)^2 / E] = 32.161$

$$\chi^2 = \Sigma [(O - E)^2 / E] = 32.161$$

$$\nu = (r - 1)(c - 1) = (4 - 1)(4 - 1) = 9$$

For $\nu = 9$, $\chi^2_{0.05} = 16.919$

Since the calculated value of greater than the table value, the hypothesis is rejected. Hence the four given products are not homogeneously distributed amongst different classes of customers.

Example 42:

Following information is obtained in a sample survey :

Condition of child	*Condition of Home*		
	Clean	*Dirty*	*Total*
Clean	*70*	*50*	*120*
Fairy Clean	*80*	*20*	*100*
Dirty	*35*	*45*	*80*
Total	*185*	*115*	*300*

State the whether the two attributes i.e., condition of home and condition of child are independent. Use χ^2 test for the purpose.

Solution:

Let us take the hypothesis that the two attributes in i.e., condition of home and condition of child are independent.

Applying χ^2 test :

$$E_{11} = \frac{120}{300} \times 185 = 74; \qquad E_{12} = \frac{100}{300} \times 185 = 61.67 \text{ or } 62$$

The table of expected frequencies shall be as follows :

74	46	120
62	38	100
49	31	80
185	31	80
185	115	300

O	E	$(O-E)^2$	$(O-E)^2/E$
70	74	16	0.216
80	62	324	5.226
35	49	196	4.000
50	46	16	0.348
20	38	324	8.526
45	31	196	6.322
			$\Sigma[(O-E)^2/E] = 24.638$

$$\nu = (3-1)(2-1) = 2$$

For $\nu = 2, \chi^2_{0.05} = 5.99$

The calculated value of χ^2 is much greater than the table value. The hypothesis is rejected. Hence the attributes condition of home and condition of child are associated

Example 43:

A marketing agency gives following information about the age-groups of the sample informants and their liking for a particular model of scooter which a company plans to introduce :

	Age Group of Informants			
	Below 20	*20-39*	*40-59*	*Total*
Liked :	*125*	*420*	*60*	*605*
Disliked :	*75*	*220*	*100*	*395*
Total	*200*	*640*	*160*	*1,000*

On the basis of the above data, can it be concluded that the model appeal is independent of the age-group of the informants ?

Solution:

Let us take the hypothesis that the model appeal is independent of the age group. Applying χ^2 test.

$$E_{11} = \frac{605 \times 200}{1{,}000} = 121.0; \qquad E_{12} = \frac{605 \times 640}{1{,}000} = 387.2$$

$$E_{13} = \frac{605 \times 160}{1{,}000} = 96.8$$

The table of expected frequencies is as follows :

121	387.2	96.8	605.0
79	252.8	63.2	395.0
200	640	160	1,000.0

O	E	$(O - E)^2$	$(O - E)^2/E$
125	121	16.00	0.132
75	79	16.00	0.203
420	387.2	1075.84	2.779
220	252.8	1075.84	4.256
60	96.8	1354.25	13.990
100	63.2	1354.24	21.428
			$\Sigma [(O - E)^2 / E] = 42.788$

$$\nu = (2 - 1)(2 - 1) = 2$$

For $\nu = 2, \chi^2_{0.05} = 5.99$

The calculated value is mush higher than the table value. Hence, the hypothesis that model appeal is independent of the age group is rejected and it is concluded that model appeal is not independent of the age group of the informants.

Example 44:

A controlled experiment was conducted to test the effectiveness of a new drug. Under this experiment 300 patients were treated with new drug and 200 were not treated with the drug. The results of the experiment are given below :

Details	***Cured***	***Condition Worsened***	***No effect***	***Total***
Treated with the drug	*200*	*40*	*60*	*300*
Not treated with the drug	*120*	*30*	*50*	*200*
Total	*320*	*70*	*110*	*500*

Use χ^2 test and comment on the effectiveness of the drug.

Solution:

Let us take the hypothesis that is no significant different in the patients treated with the new drug and those not treated. Applying χ^2 test:

$$E_{11} = \frac{300}{500} \times 320 = 192; \qquad E_{12} = \frac{300}{500} \times 70 = 42$$

The table of expected frequencies shall be as follows :

192	42	66	300
128	28	44	200
32	70	110	500

O	E	$(O - E)^2$	$(O - E)^2/E$
200	192	64	0.333
120	128	64	0.500
40	42	4	0.095
30	28	4	0.143
60	66	35	0.545
50	44	36	0.818
			$\Sigma [(O - E)^2 / E] = 2.434$

$$\chi^2 = \Sigma (O - E)^2 = 2.434$$

$$\nu = (r - 1)(c - 1) = (2 - 1)(3 - 1) = 2$$

For $\nu = 2, \chi^2_{0.05} = 5.49$

The calculated value of χ^2 is less than the table value. The hypothesis is accepted. There is no significant difference in the patients treated with the new drug and those not treated.

Using the Computer for Chi-Square Test

The χ^2-test is a very simple test and does not involve much of calculation work. However, for large sets of data calculations may become time consuming. In such a situation it is possible to make use of the computer. Most commonly used computer statistics packages contain routine for doing these tests.

MISUSE OF CHI-SQUARE TEST

Probably, one of the most frequently used statistics is the chi-square. Unfortunately, it is also one of the most frequently is not so easily learnt.

The most common mistake of the chi-square statistics and yet the most critical for its correct application is the violation of independence between measures or events. This assumption of independence is not to be confused with the chi-square as test of independence. The assumption of independence refers to the individual observations or frequencies and means that the occurrence of one event has no effect upon the occurrence of any other event. Another way of stating this meaning of independence is that the probability of all other events. In statistical terms we say that the joint probability of two random events is equal to the product of the probabilities of these events.

The chi-square as a test of independence refers to the statistical test of the possibility of a relationship between two variables. This is often called a test of association, the question tested is whether the frequencies observed of one category are contingent upon the another category. For example, is the number of "yes" answers to some questions contingent upon the age of the respondent ?

Some other sources of error in the application of χ^2 test as revealed in a survey of research papers published in the *Journal of experimental Psychology* are :

(i) Small theoretical frequencies.

(ii) Neglect of frequencies of non-occurrence.

(iii) Failure to equals the sum of observed frequencies and the sum of the theoretical frequencies.

(iv) Indeterminate theoretical frequencies.

(v) Incorrect or questionable categorizing.

(vi) Use of non-frequency data.

(vii) Incorrect determination of the numbers of degrees of freedom.

(viii) Incorrect computations.

The number of applications of chi-square test does not seem to be increasing but the number of misuses of χ^2 has become surprisingly large. A lesson can be drawn from the findings of the Lewis and Burke* article. You cannot simply use a Statistic because you know how to calculate it; you must understand the rationale behind its development and the limitations on its applications imposed by the assumption underlying it.

LIMITATIONS ON THE USE OF χ^2 TEST

χ^2 test is very widely used in practice. However, in order to avoid the misapplication of the test its following limitations should be kept in mind;

(1) Frequencies of non-occurrence should not be omitted for binomial or multinomial events. For example, if 5 drugs were tried out on 5 separate groups of 200 patients each, the number of cures per drug might be shown in one-way table as follows ;

Data for Five Drugs

	Drug					
	1	**2**	**3**	**4**	**5**	**Total**
Number Cured	80	120	40	60	20	320

However, the χ^2 test should not be applied to these data until the alternative outcome (i.e., "not cured") is represented in the table.

(2) The formula presented for χ^2 statistics is in terms of frequencies. Hence an attempt should not be made to computer on the basis of proportions or other derived measures.

(3) The formula presented in this chapter is not appropriate for cases in which repeated measurements on the same or matched groups are represented in one table. When data from questionnaires and similar devices are analysed. The reader should be careful that he does not set up the tables incorrectly. For example, it nay seem reasonable to set up a table as follows :

	Agree	**Neutral**	**Disagree**	**Total**
Item X :	140	190	170	500
Item Y :	180	150	170	500

However, a χ^2 contingency test should not be performed on the basis of this table, since it is not really a contingency table because each student is classified twice in the table.

EXERCISES

1. What is χ^2 distribution? Mention its important properties.
2. Define χ^2 (Chi-square) and discuss its uses in testing of hypothesis.
3. An automobile manufacturing company is bringing out a new car model. The company is interested to known whether the model will appeal most to a particular age-group or equally to all age-groups. The company takes a random sample from persons attending a demonstration show of the new model and obtained the following information:

	Age group			
Persons who	Under 20	30-39	40-59	60 and above
Liked the car	146	78	41	28
Disliked the car	54	52	32	62

What conclusion can be drawn from the above data of 5% level of significance?

[χ^2 = 45.38]

4. The following grades were given to a class of 100 students :

Grade :	A	B	C	D	E
Frequency :	14	18	32	20	16

Test the hypothesis at the 0.05 level that the distribution of grades is uniform.

[χ^2 = 10. Yes]

5. The following table gives the classification of 100 workers according to sex and the nature of work. Test whether nature of work is independent of the sex of the worker.

	No. of patients		
	Favourable	*Not Favourable*	*Total*
New	280	60	340
Conventional	120	40	160
Total	400	100	500

(Given for $u = 1 . \chi^2_{20.05} = 3.84$)

[$\chi^2 = 3.676$]

6. What are the basic conditions for the application of χ^2 test?
7. χ^2 test is a test of independence homogeneity and goodness of fit. Discuss breifly.
8. A book has 700 pages. The number of pages with various number of misprints is recorded below. At 5% significance level are the misprints distributed according to Poisson Law?

Number of misprints X :	0	1	2	3	4	5	Total
Number of pages with X mistrpints :	616	70	10	2	1	1	700

9. A production supervisor is interested in knowing if the number of breakdowns of four machines is independent of the shift using the machines. Test this hypothesis based on the following sample information :

	Machine			
Shift	A	B	C	D
Morning	15	10	18	12
Evening	12	8	15	10

[$\chi^2 = 0.01$]

10. The distribution of typing mistakes by a typist is given below: Would you say that typing mistakes are subject to Poisson model at 5% level of significance?

Mistakes per page :	0	1	2	3	4	5
No. of pages :	142	156	69	27	5	1

11. For the following contingency table :
    (a) Construct a table of observed and expected frequencies.
    (b) Calculate the Chi-square statistic.
    (c) State the null and alternate hypothesis.
    (d) Using 0.05 level of significance derive the analysis:

		Income level	
Attendance	*Law*	*Middle*	*High*
Never	27	48	15
Occasional	25	63	14
Regular	22	74	12

12. A certain drug is claimed to be effective in curing cold. In an experiment on 500 persons with cold half of them were given the drug and half of them were given the sugar pills. The patients reactions to the treatment are recorded in the following table :

	Helped	*Harmed*	*No effect*	*Total*
Drug	150	30	70	250
Sugar pills	130	40	80	250
Total	280	70	150	500

On the basis of the data, can it be concluded that there is a significant difference in the effect of the drug and sugar pills?

13. Discuss briefly the uses of χ^2 test as a test of goodness of fit. State the conditions to be satisfied for the applicability of the test.
    (b) Discuss the Chi-square test of goodness of fit of a theoretical distribution to an observed frequency distribution. State the conditions for the validity of χ^2 test.
14. What is χ^2 test? Given various uses of χ^2 test. What are the limiting values of χ^2 test? How will you determine the degrees of freedom for χ^2 test?
15. What are Yate's corrections?
16. Theory gives proportion $p_1 . p_2 ... p_n$ whereas the observe proportions are $a_1, a_2 .... a_n$. How to test the consistency?
17. What is Chi-square test? What are the assumptions for the test? What type of conclusions you can draw using it? Explain.
18. Describe the uses of Chi-square test
19. What is χ^2 test? What is it called non-parametric test?
20. What is the Chi-square test for independence? State two business situations.

7

Partial and Multiple Correlation

INTRODUCTION

The correlation and regression coefficients discussed earlier measure the degree and nature of the effect of one variable on another. While it is useful to know how one phenomenon is influenced by another, it is also important to know how one phenomenon is affected by several other variables. Its nature, relationship tends to be complex rather than simple. One variable is related to a great number of others, many of which may be interrelated among themselves.

The basic distinction between multiple and partial correlation analysis is that whereas in the former we measure the degree of relationship between the variable Y and all the variables. X_1, X_2, X_3 .... X_n, taken together, in the latter we measure the degree of relationship between Y and one of the variables X_1, X_2, X_3 ..... X_n with te effect of all the other variables removed.

PARTIAL CORRELATION

It is often important to measure the correlation between a dependent variable and one particular independent variable when all other variables involved are kept constant i.e.. when the effects of all other variables are removed (often indicated by the phrase "other things being equal"). This can be obtained by calculating coefficient of partial correlation. Thus partial correlation analysis measures the strength of the relationship between Y and one independent variable in such a way that variations in te other independent variables are taken into account. A partial correlation coefficient is analogous to a partial regression coefficient in that all other factors are "held constant". Simple correlation, on the other hand, ignores the effect of all other variables even though these variables might be quite closely related to the independent variable, on to one another.

Partial Correlation Coefficient

Partial correlation coefficient provides a measure of the relationship between the dependent variable and other variables, with the effect of the most of the variables eliminated.

If we denote by r_{123} the coefficient of partial correlation between X_1 and X_2 keeping X_3 constant, we find that

$$r_{12.3} = \frac{r_{12} - r_{13}\ r_{23}}{\sqrt{(1-r_{13}^{2})}\sqrt{(1-r_{23}^{2})}}$$

Similarly,

$$r_{13.2} = \frac{r_{13} - r_{12}\ r_{23}}{\sqrt{(1-r^{2}{}_{12})}\sqrt{(1-r^{2}{}_{23})}}$$

where $r_{13.2}$ is the coefficient of partial correlation between X_1 and X_2 keeping X_2 constant.

$$r_{23.1} = \frac{r_{23} - r_{12}\ r_{13}}{\sqrt{(1-r^{2}{}_{12})}\sqrt{(1-r^{2}13)}}$$

where $r_{23.1}$ is the coefficient of partial correlation between X_2 and X_3 keeping X1 constant.

Thus, for three variables, X_1, X_2 and X_3, there will be three coefficients of partial correlation each studying the relationship between two variables when the third is held constant.

The partial correlation coefficient helps us to answer questions such as this. Is the correlation between, say, X_1 and X_2, merely due to the fact that both are affected by X_2 or is there a no covariation between X_1 and X_2 over and above the association due to the common influence of X_3? Thus, in determining a partial correlation coefficient between X_1 and X_2, we attempt to remove the influence of X_3 from each of the two variables so as to ascertain whether net relationship exists between the "unexplained" residuals that remain.

Zero Order, First Order and Second Order Coefficients

Partial coefficients such as $r_{12.3}$, $r_{13.2}$ are often referred to as first order coefficients, since one variable has been held constant. Simpie coefficients (correlation between two variables only) are called zero order coefficients, since no variables are held constant. $r_{12.34}$, $r_{13.24}$, etc., are called second order coefficients since two variables are kept constant. Stated generally, the order of designation indicates the number of variables that have been held constant statistically.

Example 1:

On the basis of the following information compute:

(i) $r_{23.1}$ *(ii)* $r_{13.2}$ *and (iii)* $r_{12.3}$:

$r_{12} = 0.70;$

$r_{13} = 0.61;\ r_{23} = 0.40$

Solution:

$$r_{23.1} = \frac{r_{23} - r_{12}\ r_{13}}{\sqrt{\left(1-r^2 12\right)}\sqrt{\left(1-r^2 13\right)}}$$

Substituting te given values

$$r_{23.1} = \frac{0.4 - 0.7 \times 0.61}{\sqrt{1-(0.7)^2}\sqrt{1-(0.61)^2}}$$

$$= \frac{0.4 - 0.427}{\sqrt{0.51}\sqrt{1-0.3721}} = \frac{-0.027}{0.714 \times 0.792} = 0.048$$

$$r_{13.2} = \frac{r_{13} - r_{12}\ r_{23}}{\sqrt{\left(1-r^2 12\right)}\sqrt{\left(1-r^2 23\right)}}$$

$$= \frac{0.61-(0.7)\ (0.4)}{\sqrt{1-(0.7)^2}\sqrt{1-(0.4)^2}} = \frac{0.61-0.28}{\sqrt{1-0.49}\sqrt{1-0.16}}$$

$$= \frac{0.33}{\sqrt{0.51}\sqrt{0.84}} = 0.504$$

$$r_{12.3} = \frac{r_{12} - r_{13}\ r_{23}}{\sqrt{1-r^2 13}\sqrt{1-r^2 23}}$$

$$= \frac{0.7-(0.61 \times 0.4)}{\sqrt{1-(0.61)^2}\sqrt{1-(0.4)^2}}$$

$$= \frac{0.7-0.244}{\sqrt{1-0.3721}\sqrt{1-0.16}} = 0.629$$

Example 2:

Is it possible to get the following from a set of experimental data:

(a) $r_{23} = 0.8\ r_{31} = -\ 0.5,\ r_{12} = 0.6$

(b) $r_{23} = 0.7,\ r_{31} = -\ 0.4,\ r_{12} = 0.6$

Solution (a):

$$r_{12.3} = \frac{r_{12} - r_{13}r_{23}}{\sqrt{\left(1-r^2{}_{13}\right)}\sqrt{\left(1-r^2 23\right)}}$$

$$= \frac{0.6-(-.05)\ (.8)}{\sqrt{0.75}\sqrt{0.36}} = \frac{1}{0.52} = 1.923$$

Since the value of $r_{12.3}$ is greater than one, there is some inconsistency in the given data.

(b) $$r_{12.3} = \frac{r_{12} - r_{13}r_{23}}{\sqrt{(1-r^2_{13})}\sqrt{(1-r^2_{23})}}$$

$$= \frac{0.6-(-0.4)(0.7)}{\sqrt{1-(-0.4)^2}\sqrt{1-(0.7)^2}}$$

$$= \frac{0.6+0.28}{\sqrt{0.84}\sqrt{0.51}} = \frac{0.88}{0.655} = 1.344$$

This again is greater than one which is not possible.

Second-order Partial Correlation Coefficients

Second order coefficients may be obtained from first order coefficients. In case of four variables. If $r_{12.34}$ is the coefficient of partial correlation between X_1 and X_2 keeping X_3 and X_4 constant, then

$$r_{12.34} = \frac{r_{12.4} - r_{13.4}\, r_{23.4}}{\sqrt{(1-r^2_{13.4})}\sqrt{(1-r^2_{23.4})}}$$

Similarly $$r_{13.24} = \frac{r_{13.4} - r_{12.4}\, r_{23.4}}{\sqrt{(1-r^2_{12.4})}\sqrt{(1-r^2_{23.4})}}$$

and $$r_{14.23} = \frac{r_{14.3} - r_{12.3}\, r_{24.3}}{\sqrt{(1-r^2_{12.3})}\sqrt{(1-r^2_{24.3})}}$$

Alternative formulae giving the same results are available for all three of the second-order coefficients. They are:

$$r_{12.34} = \frac{r_{12.3} - r_{14.3}\, r_{24.3}}{\sqrt{(1-r^2_{14.3})}\sqrt{(1-r^2_{24.3})}}$$

$$r_{13.24} = \frac{r_{13.2} - r_{14.2}\, r_{34.2}}{\sqrt{(1-r^2_{14.2})}\sqrt{(1-r^2_{34.2})}}\cdot$$

Example 3:

On te basis of observations made on 39 cotton plants, the total correlation of yield of cotton (X_1), number of bolls, i.e., seed vessels (X_2) and height (X_3) and height (X_3) are fod to be:

$$r_{12} = 0.8,\ r_{13} = 0.65 \text{ and } r_{23} = 0.7$$

Commenting the partial correlation between yield of cotton and the number of bolls, eliminating the effect of height.

Solution:

We have to find the partial correlation between yield of cotton and te number of bolls, eliminating the effect of height, i.e., in terms of symbols, we have to calculate $r_{12.3}$.

$$r_{12.3} = \frac{r_{12} - r_{13}\ r_{23}}{\sqrt{(1-r^2 13)}\sqrt{(1-r^2 23)}}$$

$r_{12} = 0.8$, $r_{13} = 0.65$ and $r_{23} = 0.7$

Substituting the values

$$r_{12.3} = \frac{0.8-(0.65\times 0.7)}{\sqrt{1-(0.65)^2}\sqrt{1-(0.7)^2}}$$

$$= \frac{0.8-0.455}{\sqrt{1-.4225}\sqrt{1-0.49}}$$

$$= \frac{0.345}{0.76\times 0.714} = \frac{0.345}{0.543} = 0.635$$

Example 4:

If $r_{12} = 0.86$, $r_{13} = 0.65$ and $r_{23} = 0.72$, find the partial correlation coefficient $r_{12.3}$.

Solution:

$$r_{12.3} = \frac{r_{12} - r_{13}\ r_{23}}{\sqrt{\left(1.r^2 13\right)}\sqrt{\left(1-r^2 23\right)}}$$

$r_{12} = 0.86$, $r_{13} = 0.65$, $r_{23} = 0.72$

Substituting the values

$$r_{12.3} = \frac{.86-.65\times .72}{\sqrt{1-(.65)^2}\sqrt{1-(.72)^2}}$$

$$= \frac{.86-.468}{\sqrt{.5775\times 4816}} = \frac{.392}{.5274} = 0.743$$

Partial Correlation Coefficients in case of Four Variables

When four variables are involved in a correlation problem, there are twelve possible first-order coefficients. Some of these are:

$$r_{14.2} = \frac{r_{14} - r_{12}\ r_{24}}{\sqrt{\left(1-r^2{}_{12}\right)}\sqrt{\left(1-{}^2{}_{24}\right)}}$$

$$r_{14.3} = \frac{r_{14} - r_{13}\, r_{34}}{\sqrt{\left(1-r^2{}_{13}\right)}\sqrt{\left(1-{}^2{}_{34}\right)}}$$

$$r_{13.4} = \frac{r_{13} - r_{14}\, r_{34}}{\sqrt{\left(1-r^2{}_{14}\right)}\sqrt{\left(1-{}^2{}_{34}\right)}}$$

$$r_{12.4} = \frac{r_{12} - r_{14}\, r_{24}}{\sqrt{\left(1-r^2{}_{14}\right)}\sqrt{\left(1-{}^2{}_{24}\right)}}$$

$$r_{24.3} = \frac{r_{24} - r_{23}\, r_{34}}{\sqrt{\left(1-r^2{}_{23}\right)}\sqrt{\left(1-{}^2{}_{34}\right)}}$$

$$r_{34.2} = \frac{r_{34} - r_{23}\, r_{24}}{\sqrt{\left(1-r^2{}_{23}\right)}\sqrt{\left(1-{}^2{}_{24}\right)}}$$

$$r_{23.4} = \frac{r_{23} - r_{24}\, r_{34}}{\sqrt{\left(1-r^2{}_{24}\right)}\sqrt{\left(1-{}^2{}_{34}\right)}}$$

In a similar manner, the formulae for other partial correlation coefficients, i.e. $r_{12.3}$, $r_{13.2}$, $r_{24.1}$, $r_{34.1}$ can·also be written.

$$r_{14.23} = \frac{r_{14.2} - r_{13.2}\, r_{34.2}}{\sqrt{\left(1-r^2{}_{12.3}\right)}\sqrt{\left(1-{}^2{}_{34.1}\right)}}$$

The value of a partial correlation correlation coefficient is usually interpreted via the corresponding coefficient of partial determination, which is merely the square of the former. Thus if $r_{12.3} = 0.4$, $r^2{}_{12.3} = 0.16$.

The t-test employed to test the significance of a sample correlation can be employed to test the significance of a partial correlation when the number of degrees of freedom is reduced by the number of variables eliminate.

Characteristic and Uses of Partial Correlation Analysis

The function of partial correlation analysis is the measurement of relationship between two factors, with the effects of one or more other factors eliminated. If the assumptions of the method are true for a series of data, the power f partial analysis is great. The problem of holding certain variables constant, while the relationship between the others is measured, often presents difficulty itself in statistical analysis. Partial correlation is especially useful in the analysis of interrelated series. It is particularly pertinent to uncontrolled experiments of various kinds, in which such interrelationships usually exist. Most economic data fall in this category.

Partial correlation is of greatest value when used in conjunction with gross and multiple correlation in the analysis of factors affecting variations in many kinds of phenomena.

Partial analysis, like all correlation, has the advantage that the relationships are expressed concisely in a few well-defined coefficients. Also it is adaptable to small amounts of data and the reliability of the results can be rather easily tested.

Limitations of Partial Correlation Analysis

(1) The usefulness of the partial analysis is somewhat limited by the following basic assumptions of the method:

  (i) The gross or zero-order correlation must have linear regressions.

  (ii) The effects of the independent variables must be additively and not jointly related.

  (iii) Because the reliability of partial coefficients decreases as its order increases, the number of observations in gross correlations should be fairly large. After the student carries the analysis beyond the limits of the data. Thus weakness to some extent can be guarded against by test of reliability.

(2) When the above assumptions have been satisfied, partial analysis still possesses the disadvantages of laborious calculations and difficult interpretation even for statisticians.

The interpretation of the partial and multiple correlation results tends to assume that the independent variables have causal effects on dependent variable. This assumption is sometimes true, but more often untrue in varying degrees.

The Significance of a Partial Correlation Coefficient

The significance of a partial r may be determined, most readily, by the *Z* transformation. The S.E. $= \frac{1}{\sqrt{N-3}}$ and the S.E. of the *Z* corresponding to $r_{12.3}$ is $= \frac{1}{\sqrt{N-4}}$. One degree of freedom is subtracted from *N* for each variable eliminated, in addition to the 3 already lost. So for $r_{12.34}$ the S.E. of the corresponding *Z* is $\frac{1}{\sqrt{N-3-2}} = \frac{1}{\sqrt{N-5}}$.

Example 5:

The following zero-order correlation coefficients are given

$r_{12} = 0.98,\ r_{13} = 0.44$ *and* $r_{23} = 0.54.$

Calculate multiple correlation coefficient treating first variable as dependent and second and third variables as independent.

Solution:

We have to calculate the multiple correlation coefficient treating first variable as dependent and second and third variables as independent, i.e., we have to find $R_{1.23}$.

$$R^2 = \frac{\text{Explained variation}}{\text{Total variation}}$$

$$R_{1.23} = \frac{\sqrt{r^2_{12} + r^2_{13} - 2 r_{12} r_{13} r_{23}}}{1 - r^2_{23}}$$

Substituting the given values $R_{1.23}$

$$= \frac{\sqrt{(.98)^2 + .(44)^2 - 2(.98)(.44)(.54)}}{1 - (.54)^2}$$

$$= \frac{\sqrt{.9604 + .1936 - .4657}}{.7084} = 0.986.$$

MULTIPLE CORRELATION

In problems of multiple correlation we are dealing with situations that involve three or more variables. For example, we may consider the association between the yield of wheat per acre and both the amount of rainfall and the average daily temperature. We are trying to make estimates of the value of one of the these variables based on the values of all the others. The variables whose value we are trying to estimate is called the dependent variable and the other variables on which our estimates are based are known as independent variables. The statistician himself chooses which variable is to be dependent and which variables are to be independent. It is merely a question of problem being studies. If we are trying to determine the most problem weight of men, we make weight of men, we make weight the dependent variable and height, age, etc., independent variables. If on the other hand, we are interested in estimating height, we will make height the dependent variable anu weight, age, etc., the independent variables. Thus in problems of multiple correlation we always have three or more variables (one dependent and the other independent). In order that we may distinguish them easily we follow the custom of representing them by the letter X with subscript. The dependent variable is always denoted by X_1 and the other by X_2, X_3, etc. Thus, in the height, age and weight problem, if we are trying to estimate men's weight (that is, if weight be a dependent variable), we might denote.

$X_1 \rightarrow$ weight in lbs.

$X_2 \rightarrow$ height in inches.

$X_3 \rightarrow$ age in years.

Coefficient of Multiple Correlation

the coefficient of multiple linear correlation is represented by R_1 and it is common to add subscripts designating the variables involved. Thus, $R_{1.234}$ would represent the coefficient of multiple linear correlation between X_1, on the one hand and X_2, X_3 and X_4 on the other. The subscript of the dependent variable is always to the left of the point.

The coefficient of multiple correlation can be expressed in terms of r_{12}, r_{13} and r_{23} as follows:

$$R_{1.23} = \frac{\sqrt{r^2_{12} + r^2_{13} - 2r_{12}\, r_{13}\, r_{23}}}{1 - r^2_{23}}$$

$$R_{2.13} = \frac{\sqrt{r^2_{12} + r^2_{23} - 2r_{12}\, r_{13}\, r_{23}}}{1 - r^2_{13}}$$

$$R_{3.12} = \frac{\sqrt{r^2_{13} + r^2_{23} - 2r_{12}\, r_{13}\, r_{23}}}{1 - r^2_{12}}$$

It should be noted that $R_{1.23}$.

A coefficient of multiple correlation such as $R_{1.23}$ lies between 0 and 1. The closer it is to 1, the better is the linear relationship between the variables. The closer it is to 0, the worse is the linear relationship. If the coefficient of multiple correlation is 1, the correlation is called perfect. Although a correlation coefficient of 0 indicates no linear relationship between the variables, it is possible that a non-linear relationship may exist. It should be noted that whereas the correlation coefficients range from + 1.0 to 0 to – 1.0, the coefficients of multiple correlation are always positive in sign and range from + 1.0 to 0. Since some of the individual variables may be positively correlated with the dependent variables and other negatively correlated, no useful purpose would be served in distinguishing between a positive and negative value of R.

By squaring $R_{1.23}$, we obtain te coefficient of multiple determination.

An alternate formula for obtaining the value of $R_{1.23}$ is follows:

$$R_{1.23} = \sqrt{r^2_{12} + r_{13.2}\left(1 - r^2_{12}\right)}$$

or $$R^2_{1.12} = r^2_{12} + r_{13.2}\left(1 - r^2_{12}\right)$$

Similarly,

$$R_{1.24} = \frac{\sqrt{r^2_{12} + r^2_{14} - 2r_{12}\,r_{14}\,r_{24}}}{1 - r^2_{24}}$$

$$\Rightarrow \quad R_{1.24} = \sqrt{r^2_{12} + r^2_{14.2}\left(1 - r^2_{12}\right)}$$

and $$R_{1.34} = \frac{\sqrt{r^2_{13} + r^2_{14} - 2r_{13}r_{14}r_{34}}}{1 - r^2_{34}}$$

$$\Rightarrow \quad R_{1.34} = \sqrt{r^2_{12} + r^2_{14.2}\left(1 - r^2_{13}\right)}$$

To determine a multiple coefficient with three independent variables, the following formula shall be used:

$$R_{1.234} = \sqrt{1 - \left(1 - r^2_{14}\right)\left(1 - r^2_{13.4}\right)\left(1 - r^2_{12.34}\right)}$$

Example 6:

Suppose that $r_{12.34}$ = 0.5 and N = 41. Is the partial r significant? Also determine 95% confidence limits.

Solution:

$$\text{S.E.} = \frac{1}{\sqrt{N-5}} = \frac{1}{\sqrt{41-5}} = 0.167$$

The Z corresponding to r of 0.50 is 0.55. Hence $r_{12.34}$ is significant. The 95% confidence interval for the population Z is

$0.55 \pm 1.96 \times 0.167$ or from 0.223 to 0.877.

Advantages of Multiple Correlation Analysis

The coefficient of multiple correlation serves the following purposes:

(1) It also serves as a measure of goodness of fit of the calculated plane of regression and consequently as a measure of the general degree of accuracy of estimates made by reference to equation for the plane of regression.

(2) It serves as a measure of the degree of association between one variable taken as the dependent variable and a group of other variables taken as the independent variables.

Limitations of Multiple Correlation Analysis

(1) Linear multiple correlation involves a great deal of work relative to the results frequently obtained. When the result are obtained, only

a few students, well-trained in the method are able to interpret them. The misuse of correlation results has probability led to more doubt on the method than is justified. However, tis lack of understanding and resulting misuse are due to the complexity of the method.

(2) Multiple correlation analysis is based on the assumption that the relationship between the variables is linear. In other words, the rate of change in one variable in terms of another is assumed to be constant for all values. In practice, most relationships are not linear but follow some other pattern. This limits somewhat the use of multiple correlation analysis. The linear regression coefficients are not accurately descriptive of curvilinear data.

(3) A second important limitation is te assumption that effects of independent variables on the dependent variables are separate, distance and additive. When the effects of variables are additive, a given change in one has the same effect on the dependent variable regardless of the sizes of the other two independent variables.

MULTIPLE REGRESSION ANALYSIS

Multiple Regression and Correlation Analysis

Multiple regression analysis represents a logical extension of two-variable regression analysis. Instead of a single independent variable, two or more independent variables are used to estimate the values of a dependent variable. However, the fundamental concept in the analysis remains the same. The following are the three main objectives of multiple regression and correlation analysis:

(1) To derive an equation which provides estimates of the dependent variable from values of the two or more independent variables.

(2) To obtain a measure of the error involved in using this regression equation as a basis for estimation.

(3) To obtain a measure of the proportion of variance in the dependent variable accounted for or "explained by" the independent variables.

The first purpose is accomplished by deriving an appropriate regression equation by the method of least squares. The second purpose is achieved through the calculation of a standard error of estimate. The third purpose is accomplished by computing the multiple coefficient of determination.

Multiple Regression Equation

The multiple regression equation describes the average relationship between these variables and this relationship is used to predict or control the dependent variable.

A regression equation is an equation for estimating a dependent variable, say, X_1 from the independent variables X_2, X_3, ... and is called a regression equation of X_1 on X_2, X_3, ... In functional notation, this is sometimes written briefly as X1 = F(X_2, X_3...) read as "X_1 is a function of X_2, X_3 and so on".

In case of three variables, the regression equation of X_1 on X_2 and X_3 has the form

$$X_{1.23} = a_{1.23} + b_{12.3} X_2 + b_{13.2} X_3 \quad ...(i)$$

$X_{1.23}$ is the computed or estimated value of the dependent variable and X_2. X_3 are the independent variables.

The constant $a_{1.23}$ is the intercept made by the regression plane. It gives the value of the dependent variable when all the independent variables assume a value equal to zero. $b_{12.3}$ and $b_{13.2}$ are called partial regression coefficients or the net regression coefficients. $b_{12.3}$ measures the amount by which a unit change in X_2 is expected to affect X_1 when X_3 is held constant and $b_{13.2}$ measures the amount of change in X_1 per unit change in X_3 when X_2 is held constant.

Due to the fact the X_1 varies partially because of variation in X_2 and partially because of variation in X_3, we call $b_{12.2}$ and $b_{13.2}$ the partial regression coefficients of X_1 on X_2 keeping X_3 constant and of X_1 on X_3 keeping X_2 constant.

Normal Equation for the Least Square Regression Plane

Just as there exist least square regression lines approximating a set of N data points (X', Y) in a two - dimensional scatter diagram so also there exist least square regression planes fitting a set of N data points (X_1, X_2, X_3) in a three - dimensional scatter diagram.

The least square regression plane of X_1 on X_2 has the equation (i), where $b_{12.3}$ and $b_{13.2}$ are determined by solving simultaneously the normal equations

$$\Sigma X_1 = N a_{1.23} + b_{12.3} \Sigma X_2 + b_{13.2} \Sigma X_3$$

$$\Sigma X_1 X_2 = a_{1.23} \Sigma X_2 + b_{12.3} \Sigma X_2^2 + b_{13.2} \Sigma X_2 X_3$$

$$\Sigma X_1 X_3 = a_{1.23} \Sigma X_3 + b_{12.3} \Sigma X_2 X_3 + b_{13.2} \Sigma X_2^2$$

These equations can be obtained by multiplying both sides of equation (i) by 1, X_2 and X_3 successively and summing on both sides.

When the number of variables is 4 or more, solving the above system of normal equation becomes a very tedious procedure. Efficient methods of solving simultaneous equations require a knowledge of matrix algebra, which is not assumed for the reader of this text. Computer programmes are widely available for determining the variables of the constants in a multiple regression equation. In our discussion that follows, we shall confine ourselves to the two independent variables case which of course, can be extended to cover case with three or more independent variables.

Assumptions of Linear Multiple Regression Analysis

For point estimation, the principle assumptions of linear multiple regression analysis are:

(1) The dependent variable is a random variable whereas the independent variable need not be random variable.

(2) The relationship between the several independent variables and the one dependent variable is linear, and

(3) The variances of the conditional distributions of the dependent variable, given various combinations of values of the independent variables, are all equal. For internal estimation, an additional assumption is that the conditional distributions for the dependent variable follow the normal probability distribution.

Deviations Taken from Actual Means

The work involved in finding these regression equations is reduced by taking deviations from the means of the variables under consideration. The regression equation for three variables then becomes:

$$x_1 = b_{12.3} + b_{13.2}\, x_3$$

where $\quad x_1 = (X_1 - \overline{X}_1),\quad x_2 = (X_2 - \overline{X}_2),\quad x_3 = (X_3 - \overline{X}_3)$

The value of $b_{12.3}$ and $b_{13.2}$ can be obtained by solving simultaneously the following two normal equations:

$$\sum x_1 x_2 = b_{12.3} \sum x_2^{\,2} + b_{13.2} \sum x_2 x_3$$

$$\sum x_1 x_2 = b_{12.3} \sum x_2 x_3 + b_{13.2} \sum x_3$$

The value of $b_{12.3}$ and $b_{13.2}$ can also be obtained as follows:

$$b_{12.3} = r_{12.3} \times \frac{\sigma_{1.23}}{\sigma_{2.13}}$$

$$b13.2 = r_{13.2} \times \frac{\sigma_{13.2}}{\sigma_{3.12}}$$

The regression equation of X_1 on X_2 and X_3 can be expressed as follows:

$$(x_1 - \overline{X}_1) = \left(\frac{r_{12} - r_{13}\, r_{23}}{1 - r^2_{23}}\right)\left(\frac{S_1}{S_2}\right)(x_2 - \overline{X}_2)$$

$$+\left(\frac{r_{13} - r_{12}\, r_{23}}{1 - r^2_{23}}\right)\left(\frac{S_1}{S_3}\right)(x_3 - \overline{X}_3)$$

The regression equation of X_3 on X_2 and X_1 can be written as follows:

$$\left(x_3 - \overline{X}_3\right) = \left(\frac{r_{23} - r_{13}\, r_{12}}{1 - r^2_{12}}\right)\left(\frac{S_3}{S_2}\right)\left(x_2 - \overline{X}_2\right)$$

$$+\left(\frac{r_{13} - r_{23}\, r_{12}}{1 - r^2_{12}}\right)\left(\frac{S_3}{S_1}\right)\left(x_1 - \overline{x}_1\right)$$

This method of obtaining regression equation is much simpler as compared to one where simultaneously several normal equations are to be solved. For calculating regression equation for three variables when the above procedure is used, we need the following:

$\overline{X}_1$ $\overline{X}_2$ $\overline{X}_3$

S_1 S_2 S_3

r_{12} r_{13} r_{23}

Other Equations of Multiple Linear Regression

In the case of two variables, there were two equations of regression - one of them indicating regressions of Y on X and the other, that of X on Y. When there are three variables, there will be three equations of regression, one indicating the regression of X_1 on X_2 and X_3, the other indicating the regression of X_2 on X_1 and X_3 and third indicating the regression of X_3 on X_1 and X_2. The first of these has been given earlier. If X_2 and X_3 were to be treated as dependent variables, the regression equation will respectively be:

$$X_2 = a_{2.13} + b_{21.3}\, X_1 + b_{23.1}\, X_3 \qquad \text{...(ii)}$$

$$X_3 = a_{3.12} + b_{31.2}\, X_1 + b_{32.1}\, X_2 \qquad \text{...(iii)}$$

The normal equations for fitting equation (ii) will be:

$$\sum X_2 = N\, a_{2.13} + b_{21.3} \sum X_1 + b_{23.1} \sum X_3$$

$$\sum X_1X_2 = a_{2.13} \sum X_1 + b_{21.3} \sum X_1^2 + b_{23.1} \sum X_1X_2$$

$$\sum X_2X_3 = a_{2.13} \sum X_3 + b_{21.3} \sum X_1X_3 + b_{23.1} \sum X_3^2$$

In case we want to fit equation (iii) the normal equations will be:

$$\sum X_3 = N\, a_{3.12} + b_{31.2} \sum X_1 + b_{32.1} \sum X_2$$

$$\sum X_1X_3 = a_{3.12} \sum X_2 + b_{31.2} \sum X_1^2 + b_{32.1} \sum X_1X_2$$

$$\sum X_2X_3 = a_{3.12} \sum X_3 + b_{31.2} \sum X_1X_2 + b_{32.1} \sum X_2^2$$

Generalizations for More than Three Variables

In case of four variables, the linear regression equation of X_1 on X_2, X_3 and X_4 can be written as

$$X_1 = a_{1.234} + b_{12.34} X_2 + b_{13.24} X_3 + b_{14.23} X_4$$

It represents a hyperplane in four dimensional space. On formal multiplication of both sides of the above equation by X_1, X_2, X_3 and X_4 successively and then summing on both sides, we obtain the normal equations for determination of $a_{1.234}$, $b_{13.24}$, $b_{12.34}$ and $b_{14.23}$ which when substituted to above equation, gives the least square regression equation of X_1 on X_2, X_3 and X_4.

While drawing statistical inference in multiple regression, it should be noted that the regression coefficients for highly intercorrelated independent variables tend to be unreliable. This is for the reason that when independent variables are high intercorrelated, it is extremely difficult to separate out the individual influences of each variable. There is a great deal of concern in fields such as econometrics and applied statistics with this problem of intercorrelation among dependent variables, often referred to as multi-collinearity. As suggested by Morris Hamburg, one of the simplest solutions to the problem of two highly correlated independent variables is merely to discard one of the variables.

Use of Computers in Multiple Regression and Correlation Analysis

The application of multiple regression and correlation analysis requires extensive and highly precise computations. As the number of variables increases, the computations become more and more difficult and time consuming. The computers are being used widely in the application of these techniques. A large number of computer installations have one or more multiple regression and correlation programmes in the programme library that are available to users. In fact, it is becoming increasingly unnecessary nowadays to carry out regression analysis by hand.

The availability of these programmes enables many analysts to obtain the desired regression and correlation result without the analyst having to spend time writing a computer programme. The suitability of a given library programme for use in a particular problem depends upon the input requirements, operating procedure and results computed by the programme. Many library programmes are sufficiently general and comprehensive to fulfil the requirements of a wide variety of users.

It may be pointed out that though with the use of computers it is possible to test and include large number of independent variables in a regression analysis, good judgment and knowledge of the logical relationships involved must always be used as a guide to deciding which variables to include in the construction of a regression equation.

Example 7:

Given the following, determine the regression equation of

(i) X_1 on X_2 and X_3

(ii) X_2 on X_1 and X_3

$r_{12} = 0.8 r_{13} = 0.6;\ r_{23} = 0.5$

$\sigma_1 = 10\ \sigma_2 = 8;\ \sigma_3 = 5$

Solution:

(i) Regression equation of X_1 on X_2 and X_3 is given by

$$X_1 = a_{1.23} + b_{12.3} X_2 + b_{13.2} X_3$$

It the variates X_1, X_2 and X_3 are measured as deviations from their respective means, 'a' will be zero. The values of $b_{12.3}$ and $b_{13.2}$ can be calculate from the data given above but not for 'a'. So let us assume X_1, X_2 and X3 represent deviations from means. So the regression equation of X_1 on X_2 and X_3 is

$$X_1 = b_{12.3} X_2 + b_{13.2} X_3$$

$$b_{12.3} = \frac{\sigma_1}{\sigma_2} \times \frac{r_{12} - r_{12} r_{23}}{1 - r^2_{23}}$$

$$= \frac{10}{8} \times \frac{0.8 - (0.6)(0.5)}{1 - (0.5)^2} = 0.833$$

$$b_{13.2} = \frac{\sigma_1}{\sigma_2} \times \frac{r_{13} - r_{12}\, r_{23}}{1 - r^2_{23}}$$

$$= \frac{10}{5} \times \frac{0.6 - (0.8)(0.5)}{1 - (0.5)^2} = 0.533$$

$\therefore$ Required regression equation is

$$X_1 = 0.833\ X_2 + 0.533\ X_3$$

(ii) Regression equation of X_2 on X_1 and X_3

$$X_2 = b_{12.3} X_1 + b_{23.1} X_3$$

$$b_{12.3} = \frac{\sigma_2}{\sigma_1} \times \frac{r_{12} - r_{23}\, r_{13}}{1 - r^2_{13}}$$

$$= \frac{8}{10} \times \frac{0.8 - (0.5)(0.6)}{1 - (0.6)^2}$$

$$= \frac{8}{10} \times \frac{0.8 - 0.3}{1 - .36}$$

$$= \frac{8}{10} \times \frac{0.5}{0.64} = 0.625$$

$$b_{23.1} = \frac{\sigma_2}{\sigma_3} \times \frac{r_{23} - r_{12}\, r_{13}}{r^2_{13}}$$

$$= \frac{8}{5} \times \frac{0.5 - (0.8)(0.6)}{1 - (0.6)^4}$$

$$= \frac{8}{5} \times \frac{0.02}{0.64} = 0.05$$

Thus, $X_2 = 0.625\ X_1 + 0.05\ X_3$.

RELIABILITY OF ESTIMATES

The problem of determining the accuracy of ^stimates from the multiple regression is basically the same as for estimates from a simple regression equation. Since the correlation is seldom perfect, estimates made from the regression equation will deviate from the correct value of the dependent variable. If an estimate is to be fo maximum usefulness, it is necessary to have some indication of its precision. Just as with the simple regression equation, the measure of reliability is an average of these deviations of the actual value of non-dependent variable from the estimate from the regression equation or, in other words, the standard error estimate.

The standard error of estimate of X_1 on X_2 and X_3 is defined as

$$S_{1.23} = \sqrt{\frac{\sum (X_1 - X_{last})^2}{N - 3}}$$

$S_{1.23}$ represents standard error estimate of X_1 on X_2 and X_3. X_{last} indicates the estimated value of X_1 as calculated from the regression equations.

In terms of the correlation coefficients r_{12}, r_{13} and r_{23}, the standard error of estimate can also be computed from the result:

$$S_{1.23} = S_1 \sqrt{\frac{1 - r^2_{12} - r^2_{13} - r^2_{23} + 2\, r_{23}\, r_{13}\, r_{23}}{1 - r^2_{23}}}$$

The standard error measures the closeness of estimates derived from the regression equation to actual observed values.

Coefficient of Multiple Determination

The coefficient of multiple determination is analogous to the coefficient of determination in the two-variable case. As explained earlier, the fit of a straight line to the two-variable scatter was measured by the simple coefficient of determination r^2 which was defined as the ratio of the explained sum of squares to the total sum of squares. In the same fashion, we can define coefficient of multiple determination which is denoted by R^2. Symbolically:

$$R^2 = \frac{SSR}{SST} = 1 - \frac{SSE}{SST}$$

Similar to the case of r^2 and r in two variable analysis, R^2 is easier to interpret since R^2 is a percentage figure whereas R is not.

As a ratio of explained variation to the total variation in X^1, R^2 can be interpreted as the proportion of the total variation in the dependent variable that is associated with or explained by the regression of X_1 on X_2 and X_3. We may also think of R^2 as a measure of closeness of fit of the regression plane to the actual points. The closer the value of R^2 to 1, the smaller is the scatter of the points about the regression plane and the better is the fit.

The square root of the coefficient of multiple determination is called the coefficient of multiple correlation, denoted as R. This measure is seldom used in practice.

Example 8:

Find the multiple linear regression equation of X_1 on X_2 and X_3 from the data relating to three variables given below:

X_1	4	6	7	9	13	15
X_2	15	12	8	6	4	3
X_3	30	24	20	14	10	4

Solution:

The regression equation of X_1 on X_2 and X_3 is

$$X_1 = a_{1.23} + b_{12.3} X_2 + b_{13.2} X_3.$$

The value of the constants $a_{1.23}$, $b_{12.3}$ and $b_{13.2}$ are obtained by solving the following three normal equations:

$$\Sigma X_1 = Na_{1.23} + b_{12.3} \Sigma X_2 + b_{13.2} \Sigma X_3$$

$$\Sigma X_1 X_2 = a_{1.23} \Sigma X_2 + b_{12.3} \Sigma X_2^2 + b_{13.2} \Sigma X_2 X_3$$

$$\Sigma X_1 X_3 = a_{1.23} \Sigma X_3 + b_{12.3} \Sigma X_2 X_3 + b_{13.2} \Sigma X_3^2$$

Calculating the required values:

X_1	X_2	X_3	X_1X_2	X_1X_3	X_2X_3	X_2^2	X_3^2	X_1^2
4	15	30	60	120	450	225	900	16
6	12	24	72	144	288	144	576	36
7	8	20	56	140	160	64	400	49
9	6	14	54	126	84	36	196	81
13	4	10	52	130	40	16	100	169
15	3	4	45	60	12	9	16	225
ΣX_1 =54	ΣX_2 =48	ΣX_3 =102	ΣX_1X_2 =339	ΣX_1X_3 =720	ΣX_2X_3, =1,034	ΣX_2^2 =494	ΣX_3^2 =2,188	ΣX_1^2 =576

Substituting the values in the normal equations:

$$6\ a_{1.23} + 48\ b_{12.3} + 102\ b_{13.2} = 54 \quad \text{...(i)}$$

$$48\ a_{1.23} + 494\ b_{12.3} + 103\ b_{13.2} = 339 \quad \text{...(ii)}$$

$$102\ a_{1.23} + 1034\ b_{12.3} + 2188\ b_{13.2} = 720 \quad \text{...(iii)}$$

Multiplying Eq. (i) by 8, we get

$$48a_{1.23} + 384\ b_{12.3} + 816\ b_{13.2} = 432 \quad \text{...(iv)}$$

Subtracting Eqn. (ii) from (iv), we get

$$110\ b_{12.3} + 218\ b_{13.2} = 93 \quad \text{...(v)}$$

Multiplying Eqn. (i) by 17, we get

$$102a_{1.23} + 816\ b_{12.3} + 1734\ b_{13.2} = 918 \quad \text{...(vi)}$$

Subtracting Eqn. (iii) from Eqn. (vi), we get

$$218\ b_{12.3} + 454\ b_{13.2} = -\ 198 \quad \text{...(vii)}$$

Multiplying Eq. (v) by 109, we obtain

$$11990\ b_{12.3} + 23762\ b_{13.2} = 10137 \quad \text{...(viii)}$$

Multiplying Eqn. (vii) by 55, we get

$$11990\ b_{12.3} + 23762\ b_{13.2} = -\ 10890 \quad \text{...(ix)}$$

Subtracting Eqn. (viii) from Eqn. (ix), we get

$$1208\ b_{13.2} = -\ 753$$

$$\text{or } b_{13.2} = \frac{-753}{1208} = -\ 0.623$$

Substituting the value of $b_{13.2}$ in Eqn. (v), we get

$$110\ b_{12.3} = 218(-\ 0.623) = -93$$

$$110\ b_{12.3} = 135.814 - 93$$

$$b_{12.3} = \frac{42814}{110} = +\ 0.389$$

Substituting the values of $b_{13.2}$ in Eqn. (i), we get

$$6a_{1.23} + 48\ (0.389) + 102\ (-\ 0.623) = 54$$

$$6a_{1.23} = 54 + 63.564 - 18.672 = 98.874$$

$$a_{1.23} = 16.479$$

Thus, the required regression equation is:

$$X_1 = 16.479 + 0.389\ X_2 - 0.623\ X_3$$

Example 9:

An instructor of mathematics wishes to determine the relationship of grades on a final examination of grades on two quizes given during the semester. Costing X_1, X_2 and X_3 the grades of a student on the first quiz and final examination respectively, he made the following computations for a total of 120 students:

$\overline{X}_1 = 6.8$ $\quad \overline{X}_2 = 0.7$ $\quad \overline{X}_3 = 74$

$S_1 = 1.0$ $\quad S_2 = 0.80$ $\quad S_3 = 9.0$

$r_{12} = 0.60$ $\quad r_{13} = 0.70$ $\quad r_{23} = 0.65$

(i) Find the least square regression equation of X_3 on X_1 and X_2.

(ii) Estimate the final grades of two students who scored respectively 9 and 7, 4 and 5 on the two quizes.

Solution:

The regression equation of X3 and X2 on X1 can be written as:

$$\left(X_3 - \overline{X}_3\right) = \left(\frac{r_{23} - r_{12}\, r_{13}}{1 - r^2_{12}}\right)\left(\frac{S_3}{S_2}\right)\left(X_2 - \overline{X}_2\right) + \left(\frac{r_{13} - r_{23}\, r_{12}}{1 - r^2_{12}}\right)\left(\frac{S_3}{S_1}\right)\left(X_1 - \overline{X}_1\right)$$

Substituting the given values:

$$(X_3 - 74) = \left(\frac{0.65 - (0.7 \times 0.6)}{1 - (0.6)^2}\right)\left(\frac{9}{0.8}\right)(X_2 - 7) + \left(\frac{0.7 - (0.65 \times 0.6)}{1 - (0.6)^2}\right)\left(\frac{9}{1}\right)(X_1 - 6.8)$$

$$(X_3 - 74) = \left(\frac{0.65 - 0.42}{0.64}\right)\left(\frac{9}{0.8}\right)(X_2 - 7) + \left(\frac{0.7 - 0.39}{0.64}\right)(9)(X_1 - 6.8)$$

$$(X_3 - 74) = 4.04\ (X_2 - 7) + 4.36\ (X_1 - 6.8)$$

$$(X_3 - 74) = 4.04\ X_2 - 28.28 + 4.36\ X_1 - 29.65$$

$$X_3 = 16.07 + 4.36\ X_1 + 4.04\ X_2$$

Final grades of students who scored 9 and 7 marks:

$X_1 = 9,\ X_2 = 7$

$$X_3 = 16.07 + 4.36\,(9) + 4.04\,(7)$$
$$= 16.04 + 17.44 + 32.32 = 65.8$$

Example 10:

(a) *If $r_{12} = 0.9$, $r_{13} = 0.75$, $r_{23} = 0.7$, find $R_{1.23}$.*

Solution:

$$R1.23 = \frac{\sqrt{r_{12}^2 + r_{13}^2 - 2r_{12}\,r_{13}\,r_{23}}}{1 - r_{23}^2}$$

Substituting the given values

$$r_{1.23} = \frac{\sqrt{(0.9)^2 + (0.75)^2 - 2 \times 0.9 \times 0.75 \times 0.7}}{1 - (0.7)^2}$$

$$= \frac{\sqrt{0.81 + 0.5625 - 0.945}}{0.51} = \frac{\sqrt{0.4275}}{0.51} = \sqrt{0.838} = 0.916$$

(b) *Given that $r_{12} = 0.77$, $r_{13} = 0.72$, $r_{23} = 0.52$; calculate $R_{1.23}$.*

Solution:

$$R_{1.23} = \frac{\sqrt{r_{12}^2 + r_{13}^2 - 2\,r_{12}\,r_{13}\,r_{23}}}{1 - r_{23}^2}$$

$r_{12} = 0.77$, $r_{13} = 0.72$, $r_{23} = 0.52$

Substituting the values

$$R_{1.23} = \frac{\sqrt{(.77)^2 + (.72)^2 - 2(.77)(.72)(.52)}}{1 - (.52)^3}$$

$$= \frac{\sqrt{.5929 + .5184 - .577}}{1 - .2704} = \frac{\sqrt{.5343}}{.7296} = 0.856$$

Example 11:

Given the following information:

$r_{12} = 0.20$, $r_{13} = 0.40$, $r_{23} = 0.50$

$r_{14} = 0.40$, $r_{24} = 0.30$, $r_{34} = 0.1$

Find $r_{41.23}$.

Solution:

$$r_{41.23} = \frac{r_{41.3} - r_{12.3}\,r_{24.3}}{\sqrt{1 - r^2_{12.3}}\,\sqrt{1 - r^2_{24.3}}}$$

We have to find first $r_{41.3}$, $r_{12.3}$ and $r_{24.3}$

$$r_{41.3} = \frac{r_{14} - r_{13}\,r_{34}}{\sqrt{1-r^2_{13}}\sqrt{1-r^2_{34}}} = \frac{0.4-(0.4\times -0.1)}{\sqrt{1-(0.4)^2}\sqrt{1-(0.1)^2}}$$

$$= \frac{0.4-0.04}{\sqrt{0.84}\sqrt{0.99}} = 0.482$$

$$r_{12.3} = \frac{r_{12} - r_{13}\,r_{23}}{\sqrt{1-r^2_{13}}\sqrt{1-r^2_{23}}} = \frac{0.02-(0.4\times 0.5)}{\sqrt{1-(0.4)^3}\sqrt{1-(0.5)^2}}$$

$$= \frac{-0.2-0.2}{\sqrt{0.84}\sqrt{0.75}} = 0.504$$

$$r_{24.3} = \frac{r_{24} - r_{34}\,r_{23}}{\sqrt{1-r^2_{24}}\sqrt{1-r^2_{23}}} = \frac{0.3-(-0.1\times 0.5)}{\sqrt{1-(-0.1)^2}\sqrt{1-(0.5)^2}}$$

$$= \frac{0.3+0.05}{\sqrt{0.99}\sqrt{0.75}} = \frac{0.3}{0.862} = 0.406$$

Substituting the values:

$$r_{41.23} = \frac{0.482-(-0.504\times 0.406)}{\sqrt{1-(0.504)^2}\sqrt{1-(0.406)^2}} = 0.871$$

Example 12:

Given $r_{12} = 0.5$, $r_{13} = 0.4$ and $r_{23} = 0.1$, find $r_{12.3}$ and $r_{23.1}$.

Solution:

$$\frac{r_{12} - r_{13}\,r_{23}}{\sqrt{1-r^2_{13}}\sqrt{1-r^2_{23}}}$$

Substituting the values:

$$r_{12.3} = \frac{.5-.4\times.1}{\sqrt{1-(.4)^2}\sqrt{1-(.1)^2}}$$

$$= \frac{.5-.04}{\sqrt{.84}\sqrt{.99}} = \frac{.46}{.912} = 0.504$$

$$r_{23.1} = \frac{r_{23} - r_{12}\,r_{13}}{\sqrt{1-r^2_{12}}\sqrt{1-r^2_{13}}}$$

$$= \frac{.1-.5\times .4}{\sqrt{1-(.5)^2}\sqrt{1-(.4)^2}}$$

$$= \frac{-.1}{\sqrt{.75}\sqrt{.84}} = \frac{-.1}{.794} = -\ 0.126$$

Example 13:

The following constants are obtained from measurements on length in mm. (X_1), volume in c.c. (X_2) and weight in gm. (X_2) of 300 eggs:

$\overline{X}_1 = 55.95 \quad S_1 = 2.26 \quad r_{12} = 0.578$

$\overline{X}_2 = 51.48 \quad S_2 = 4.39 \quad r_{13} = 0.581$

$\overline{X}_3 = 56.03 \quad S_3 = 4.41 \quad r_{23} = 0.974$

Obtain the linear regression equation of egg weight on egg length and egg volume. Hence estimate the weight of an egg whose length is 58 mm. and volume is 52.5 c.c.

Solution:

We have to obtain linear regression equation of egg weight on egg length and egg volume, i.e., X_3 on X_1 and X_2. The regression equation of X_2 on X_3 and X_1 can be written as:

$$X_3 - \overline{X}_3 = \left(\frac{r_{23} - r_{13}\,r_{12}}{1 - r^2_{12}}\right)\left(\frac{S_3}{S_2}\right) X_2 - \overline{X}_2 + \left(\frac{r_{13} - r_{23}\,r_{12}}{1 - r^2_{13}}\right)\left(\frac{S_3}{S_1}\right)(X_1 - \overline{X}_1)$$

Substituting the values

$$X_3 - 56.03 = \left(\frac{0.974 - (0.581 \times 0.578)}{1 - (0.578)^2}\right)\left(\frac{4.41}{4.39}\right)$$

$$(X_2 - 51.48) + \left[\frac{0581 - (0.974 \times 0.578)}{1 - (0.578)^2}\right]\left(\frac{4.41}{2.26}\right) \ (X_2 - 55.95)$$

$$X_3 - 56.03 = \left(\frac{0.974 - 0.335}{1 - 0.334}\right)\left(\frac{4.41}{4.39}\right) \ (X_2 - 51.48)$$

$$+ \left(\frac{0.581 - 0.563}{1 - 0.334}\right)\left(\frac{4.41}{2.26}\right) \ (X_1 - 55.95)$$

$X_3 - 56.03 = 0.964\ (X_2 - 51.48)\ 0.053\ (X_1 - 55.95)$

$X_3 - 56.03 = 0.964\ X_2 - 49.63 + 0.053\ X_1 - 2.97$

$X_3 = 3.43 + 0.053\ X_1 + 0.964\ X_2$

When length, i.e., X_1 is 58 and volume, i.e., X_2 is 52.5, the weight of the egg would be:

$X_3 = 3.43 + 0.053\ (58) + 0.964\ (52.5)$

$= 3.43 + 3.074 + 50.61 = 57.114$ or 57.11 gms.

Example 14:

Suppose, a computer has found, for a given set of values of X_1, X_2 and X_3,

$r_{12} = 0.91$, $r_{13} = 0.33$ *and* $r_{23} = 0.81$

explain whether these computations may be said to be free from errors.

Solution:

For determining whether the given computations are correct or not, we calculate the value of $r_{12.3}$. If the value of $r_{12.3}$ is less than one, the computations can be regarded as free errors.

$$r_{12.3} = \frac{r_{12} - r_{12}\, r_{23}}{\sqrt{1 - r^2_{13}}\sqrt{1 - r^2_{23}}}$$

$$= \frac{.91 - (.33)(.81)}{\sqrt{1 - (.33)^2}\sqrt{1 - (.81)^2}}$$

$$= \frac{.91 - .2673}{\sqrt{1 - .1089}\sqrt{1 - .6561}} = \frac{.6427}{\sqrt{.8911 \times .3439}}$$

$$= \frac{.6427}{.5536} = 1.161$$

Since the value of $r_{12.3}$ cannot exceed one, the computations given in the question are not free from errors.

Example 15:

If $r_{13} = 0.65$, $r_{23} = 0.6$ and $r_{21} = 0.4$, calculate the value of $r_{12.3}$.

Solution:

$$r_{12.3} = \frac{r_{12} - r_{13}\, r_{23}}{\sqrt{1 - r^2_{13}}\sqrt{1 - r^2_{23}}}$$

$r_{12} = 0.65$,

$r_{23} = 0.6$, $r_{13} = 0.4$

Substituting the values

$$r_{12.3} = \frac{0.4 - .65 \times .6}{\sqrt{1 - (.65)^2}\sqrt{1 - (.6)^2}}$$

$$= \frac{.4 - .39}{\sqrt{1 - .4225}\sqrt{1 - .36}} = \frac{0.01}{.8 \times .64} = \frac{0.01}{0.512} = 0.02$$

Example 16:

The table shows the corresponding values of three variables, X_1, X_2 and X_3. Find the least square regression equation of X_3 on X_1 and X_2. Estimate X_2 when X_1 = 10 and X_2 = 6.

$\overline{X}_1$	3	5	6	8	12	14
$\overline{X}_2$	16	10	7	4	3	2
$\overline{X}_3$	90	72	54	42	30	12

Solution:

The regression equation of X_3 on X_2 and X_1 can be written as follows:

$$X_3 - \overline{X}_3 = \left(\frac{r_{23} - r_{13}\, r_{12}}{1 - r^2_{12}}\right)\left(\frac{S_3}{S_1}\right)(X_2 - \overline{X}_2)$$

$$+ \left(\frac{r_{13} - r_{23}\, r_{12}}{1 - r^2_{13}}\right)\left(\frac{S_3}{S_1}\right)(X_1 - \overline{X}_1)$$

Calculating $\overline{X}_1$, $\overline{X}_2$, $\overline{X}_3$, S_1, S_2, S_3, r_{12}, r_{13}, r_{23}

X_1	$(X_1-\overline{X}_1)$	X_1^2	X_2	$(X_2-\overline{X}_2)$ x^2	X_2^2	X_3	$(X_3-\overline{X}_3)$ x_3	X_3^2	$X_1 X_2$	X_1X_3	X_2X_3
3	–5	25	16	+9	81	90	+40	1600	–45	–200	+360
5	–3	9	10	+3	9	72	+22	484	–9	–66	+66
6	–2	4	7	0	0	54	+4	16	0	–8	0
8	0	0	4	–3	9	42	–8	64	0	0	+24
12	+4	16	3	–4	16	30	–20	400	–16	–80	+80
14	+6	36	2	–5	25	12	–38	1444	–30	–228	+190
ΣX_1 =48	ΣX_1 =0	ΣX_1^2 =90	ΣX_2 =42	ΣX_2 =0	ΣX_2^2 =140	ΣX_3 =300	ΣX_3 =0	ΣX_3^2 =4008	ΣX_1X_2 =–100	ΣX_1X_3 =–582	ΣX_2X_3 =720

$$\overline{X}_1 = \frac{48}{6} = 8,\ \overline{X}_2 = \frac{42}{6} = 7,\ \overline{X}_3 = \frac{300}{6} = 50$$

$$S_1 = \frac{\sqrt{\Sigma\left(X_1 - \overline{X}_1\right)^2}}{N} = \frac{\sqrt{90}}{6} = \sqrt{15} = 3.87$$

$$S_2 = \frac{\sqrt{\Sigma\left(X_2 - \overline{X}_2\right)^2}}{N} = \frac{\sqrt{140}}{6} = \sqrt{23.33} = 4.83$$

$$S_3 = \frac{\sqrt{\Sigma\left(X_3 - \overline{X}_3\right)^2}}{N} = \frac{\sqrt{4008}}{6} = \sqrt{668} = 25.85$$

$$r_{12} = \frac{\sum X_1 X_2}{\sqrt{\sum X_1^2 \times \sum X_2^2}} = \frac{-100}{\sqrt{90 \times 140}} = -0.891$$

$$r_{13} = \frac{\sum X_1 X_3}{\sqrt{\sum X_1^2 \times \sum X_3^2}} = \frac{-582}{\sqrt{90 \times 4008}} = 0.969$$

$$r_{23} = \frac{\sum X_2 X_3}{\sqrt{\sum X_2^2 \times \sum X_3^2}} = \frac{720}{\sqrt{140 \times 4008}} = 0.961$$

$$X_3 - 50 = \left[\frac{0.961-(-0.969\times-0.891)}{1-(-0.9)^2}\right]\left(\frac{25.85}{4.83}\right)(X_2 - 7)$$

$$+\left[\frac{0.969-(-0.961\times-0.891)}{1-(-0.9)^2}\right]\left(\frac{25.85}{3.87}\right)(x_1 - 8)$$

$x_3 - 50 = 2.546\ (x_2 - 7) - 3.664\ (x_1 - 8)$

$x_3 - 50 = 2.546\ x_2 - 17.822 - 3.664\ x_1 + 29.312$

$x_3 = 2.546\ x_2 - 3.664\ x_1 + 61.49$

$x_1 = 10$ and $x_2 = 6$, X_3 will be:

when $x_3 = 15.276 - 36.64 + 61.49 = 40.126$ or 40.

Example 17:

In a trivariate distribution:

$\sigma_1 = 3,\ \sigma_2 = 4,\ \sigma_3 = 5$

$r_{23} = 0.4,\ r_{31} = 0.6,\ r_{13} = 0.7$

Determine the regression equation of X_1 on X_2 and X_3 if the variates are measured from their means.

Solution:

When the variates are measured from means, the regression equation of X_1, X_2 and X_3 is given by:

$$X_1 = b_{12.3} X_2 + b_{13.2} X_3$$

$$b_{12.3} = \frac{\sigma_1}{\sigma_2} \times \frac{r_{12} - r_{13} r_{23}}{1 - r^2_{23}}$$

Substituting the given values

$$= \frac{3}{4} \times \frac{.7 \times .6 \times .4}{1-(.4)^2}$$

$$= \frac{3}{4} \times \frac{.46}{.84} = 0.411$$

$$b_{13.2} = \frac{\sigma_1}{\sigma_3} \times \frac{r_{13} - r_1 r_{23}}{1 - r^2{}_{23}}$$

$$= \frac{3}{5} \times \frac{.6 - .7 \times .4}{1 - (.4)^2}$$

$$= \frac{3}{5} \times \frac{.32}{.84} = 0.229$$

Hence the required equation is

$$X_1 = 0.41\ X_2 + 0.229\ X_3$$

Example 18:

The simple correlation coəfficients between temperature (X_1), com yield (X_2) and rainfall (X_3) are r_{12} = 0.59, r_{13} = 0.46 and r_{23} = 0.77. Calculate partial correlation coefficient $r_{12.3}$ and multiple correlation coefficient $R_{1.23}$.

Solution:

$$r_{12.3} = \frac{r_{12} - r_{13} r_{23}}{\sqrt{1 - r^2{}_{13}} \sqrt{1 - r^2{}_{23}}}$$

$$r_{12} = .59,\ r_{13} = .46,\ r_{23} = .77$$

$$r_{12.3} = \frac{.59 - .46 \times .77}{\sqrt{1 - (46)^2} \sqrt{1 - (.77)^2}}$$

$$= \frac{.59 - .3542}{\sqrt{1 - .2116} \sqrt{1 - .5529}} = \frac{.2358}{.5665} = 0.416$$

$$R_{1.32} = \frac{\sqrt{r^2{}_{12} + r^2{}_{13} - 2 r_{12} r_{13} r_{23}}}{1 - r^2{}_{23}}$$

$$= \frac{\sqrt{(.59)^2 + (.46)^2 - 2(.59 \times .46 \times .77)}}{1 - (.77)^2}$$

$$= \frac{\sqrt{.3481 + .2116 - .418}}{.4071}$$

$$= \frac{\sqrt{.5597 - .418}}{.4071} = \frac{\sqrt{.1417}}{.4071} = .589$$

Example 19:

From the following information, calculate $R_{1.23}$

$r_{12} = 0.8,\ r_{13} = 0.5,\ r_{23} = 0.3$

Solution:

$$R_{1.23} = \frac{\sqrt{r^2_{12} + r^2_{13} - 2\,r_{12}\,r_{13}\,r_{23}}}{1 - r^2_{23}}$$

$$= \frac{\sqrt{(.8)^2 + (.5)^2 - 2(.8)(.5)(.3)}}{1 - (.3)^2}$$

$$= \frac{\sqrt{.64 + .25 - .24}}{.91} = 0.845$$

Example 20:

Explain multiple and partial correlation in a trivariate distribution $r_{12} = 0.863$, $r_{12} = 0.648$ *and* $r_{23} = 0.709$.

Find $r_{12.3}$ *and* $R_{13.2}$.

Solution:

$$r_{12.3} = \frac{\sqrt{r_{12} - r_{13}\,r_{23}}}{\sqrt{1 - r^2_{13}}\sqrt{1 - r^2_{23}}}$$

$r_{12} = .863,\ r_{13} = .648,$

$r_{23} = .709$

$$r_{12.3} = \frac{.863 - .648 \times .709}{\sqrt{1 - (.648)^2}\sqrt{1 - (.709)^2}}$$

$$= \frac{.863 - .4594}{\sqrt{.580 \times .497}} = \frac{.4036}{\sqrt{.2883}} = \frac{.4036}{.537} = 0.752$$

$$R_{1.23} = \frac{\sqrt{r^2_{12} + r^2_{13} - 2r_{12}\,r_{13}\,r_{23}}}{1 - r^2_{23}}$$

$$= \frac{\sqrt{(.863)^2 + (.648)^2 - 2(.863)(.648)(.789)}}{1 - (.709)}$$

$$= \frac{\sqrt{.745 + .42 - .793}}{.467} = 0.687$$

Example 21:

If r_{12} = 0.8, r_{13} = 0.4 and r_{23} = 0.56, find the value of $r_{23.2}$, $r_{13.2}$ and $r_{23.1}$.

Solution:

$$r_{12.3} = \frac{r_{12} - r_{13}\, r_{23}}{\sqrt{1-r^2_{23}}\sqrt{1-r^2_{23}}}$$

Substituting te given values

$$r_{12.3} = \frac{.8-.4\times.56}{\sqrt{1-(.4)^2}\sqrt{1-(.56)^2}}$$

$$= \frac{.8-.224}{\sqrt{1-.16}\sqrt{1-.316}} = \frac{.576}{\sqrt{.84\times.6864}}$$

$$= \frac{.576}{.759} = 0.759$$

$$r_{13.2} = \frac{r_{13} - r_{12}\, r_{23}}{\sqrt{1-r^2_{12}}\sqrt{1-r^2_{23}}}$$

$$= \frac{.4-.8\times.56}{\sqrt{1-(.8)^2}\sqrt{1-(.56)^2}}$$

$$= \frac{.4-.448}{\sqrt{1-.64}\sqrt{1-3136}} = \frac{-.048}{0.6\times0.828}$$

$$= \frac{-.048}{.497} = -\ 0.097$$

$$r_{23.1} = \frac{r_{23} - r_{12}\, r_{13}}{\sqrt{1-r^2_{12}}\sqrt{1-r^2_{13}}}$$

$$= \frac{.56-.8\times.4}{\sqrt{1-(.8)^2}\sqrt{1-(.4)^2}}$$

$$= \frac{.56-.32}{\sqrt{.36\times.84}} = \frac{.24}{.55} = 0.436$$

Example 22:

In a trivariate distribution, it is found that r_{13} = 0.6, r_{23} = 0.5, r_{12} = 0.8. Find the value of $r_{13.2}$.

Solution:

$$r_{13.2} = \frac{r_{13} - r_{12}\, r_{23}}{\sqrt{1-r^2_{13}}\sqrt{1-r^2_{23}}}$$

$$r_{13} = .6,\ r_{12} = .8,\ r_{23} = .5$$

$$r_{13.2} = \frac{.6-(.8)(.5)}{\sqrt{1-(.8)^2}\sqrt{1-(.5)^2}}$$

$$= \frac{.6-.4}{\sqrt{.36}\sqrt{.75}} = \frac{.2}{.5196} = 0.385$$

Example 23:

Is it possible to get the following fro

m a set of experimental data?

$r_{12} = 0.6,\ r_{23} = 0.8,\ r_{31} = -0.5$

Solution:

$$r_{12.3} = \frac{r_{12} - r_{13}\, r_{23}}{\sqrt{1-r^2_{13}}\sqrt{1-r^2_{23}}}$$

$$= \frac{.6-(-.5)\times(.8)}{\sqrt{1-(.5)^2}\sqrt{1-(.8)^2}}$$

$$= \frac{.6+.4}{\sqrt{.75\times.36}} = \frac{1}{.52} = 1.92$$

The value of $r_{12.3}$ cannot be greater than one. Since, in the given case $r_{12.3}$ has exceeded one, there is some inconsistency in the given data.

Example 24:

If $r_{12} = 0.60,\ r_{13} = 0.70,\ r_{23} = 0.65$ *and* $S_1 = 1.0$, *find* $S_{1.23}$, $R_{1.23}$ *and* $r_{12.3}$.

Solution:

$$S_{1.23} = S_1 \frac{\sqrt{1-r^2_{12}-r^2_{13}-r^2_{23}+2r_{12}\, r_{13}\, r_{23}}}{1-r^2_{23}}$$

$$= 1.0\frac{\sqrt{1-(.6)^2-(.7)^2-(.65)^2+2\times.6\times.7\times.65}}{1-(.65)^2}$$

$$= 1.0\frac{\sqrt{1-0.36-0.49-.423+0.546}}{0.577}$$

$$= 1.0\frac{\sqrt{0.273}}{0.577} = 0.688$$

$$R_{1.23} = \frac{\sqrt{r^2_{12}+r^2_{13}-2r_{12}\,r_{13}\,r_{23}}}{1-r^2_{23}}$$

$$= \frac{\sqrt{(.6)^2+(.7)^2-2(.6)(.7)(.65)}}{1-(.65)^2} = \frac{\sqrt{.304}}{.5775} = 0.726$$

$$r_{12.3} = \frac{r_{12}-r_{13}\,r_{23}}{\sqrt{1-r^2_{13}}\sqrt{1-r^2_{23}}}$$

$$= \frac{.6-.7\times.65}{\sqrt{1-(.7)^2}\sqrt{1-(.65)^2}}$$

$$= \frac{.6-.455}{\sqrt{.51\times5775}} = \frac{.145}{.543} = 0.267$$

Example 25:

The simple correlation coefficients between profits (X_1), sales (X_2) and advertising expenditure (X_3) of a factory are r_{12} = 0.69, r_{13} = 0.45 and r_{23} = 0.58. Find the partial correlation coefficients $r_{12.3}$ and $r_{13.3}$ and interpret them.

Solution:

$$R_{12.3} = \frac{r_{12}-r_{13}\,r_{23}}{\sqrt{1-r^2_{12}}\sqrt{1-r^2_{23}}}$$

$$= \frac{.69-.45\times.58}{\sqrt{1-(.45)^2}\sqrt{1-(.58)^2}}$$

$$= \frac{.69-.261}{\sqrt{1-2022}\sqrt{1-.3364}}$$

$$= \frac{.429}{\sqrt{.7975\times6636}} = \frac{.429}{.7275} = 0.598$$

$$r_{13.2} = \frac{r_{13} - r_{12}\, r_{23}}{\sqrt{1-r^2_{12}}\,\sqrt{1-r^2_{23}}}$$

$$= \frac{.45 - .69 \times .58}{\sqrt{1-(.69)^2}\,\sqrt{1-(.58)^2}}$$

$$= \frac{.45 - .4002}{\sqrt{1-.4761}\,\sqrt{1-.3364}} = \frac{.0498}{\sqrt{.5239 \times .6636}} = 0.085$$

Example 26:

If $r_{12} = 0.80$, $r_{13} = -0.56$ *and* $r_{23} = 0.40$, *then obtain* $r_{12.3}$ *and* $R_{1.23}$.

Solution:

$$r_{12.3} = \frac{r_{12} - r_{13}\, r_{23}}{\sqrt{1-r^2_{13}}\,\sqrt{1-r^2_{23}}}$$

$$r_{12} = 0.8,\ r_{13} = -0.56,\ r_{23} = -0.4$$

$$r_{12.3} = \frac{.8 - (-.56)(-.4)}{\sqrt{1-(-.56)^2}\,\sqrt{1-(-.4)^2}}$$

$$= \frac{.8 - .224}{\sqrt{.6864}\,\sqrt{.84}} = \frac{.576}{.759} = 0.759$$

$$R_{1.23} = \frac{\sqrt{r^2_{12} + r^2_{13} - 2\, r_{12}\, r_{13}\, r_{23}}}{1 - r^2_{23}}$$

$$= \frac{\sqrt{(.8)^2 + (-.56)^2 - 2(.8)(-.56)(-.4)}}{1-(-.4)^2}$$

$$= \frac{\sqrt{.64 + .3136 - .3584}}{1 - .16} = \frac{\sqrt{.5952}}{.84} = 0.842$$

Example 27:

If $r_{12} = 0.5$, $r_{31} = 0.3$, $r_{23} = 0.45$, *find* $R_{3.12}$.

Solution:

$$R_{3.12} = \frac{\sqrt{r^2_{13} + r^2_{23} - 2\, r_{12}\, r_{13}\, r_{23}}}{1 - r^2_{13}}$$

$$= \frac{\sqrt{(0.3)^2 + (0.45)^2 - 2(0.5 \times 0.3 \times 0.45)}}{1-(0.5)^2}$$

$$= \frac{\sqrt{0.09 + 0.2025 - 2(0.0675)}}{1-0.25}$$

$$= \sqrt{0.57} = 0.755$$

Example 28:

If $r_{12} = 0.77$, $r_{13} = 0.72$ and $r_{23} = 0.52$, find the partial correlation coefficient $r_{12.3}$ and multiple correlation coefficient $R_{1.23}$.

Solution:

$$r_{12.3} = \frac{r_{12} - r_{13}\, r_{23}}{\sqrt{1-r^2_{13}}\sqrt{1-r^2_{23}}}$$

$$r_{12} = 0.77,\ r_{13} = 0.72,\ r_{23} = 0.52$$

$$r_{12.3} = \frac{.77 - .72 \times .52}{\sqrt{1-(.72)^2}\sqrt{1-(0.52)^2}}$$

$$= \frac{.77-.37}{\sqrt{1-.5184}\sqrt{1-.2704}}$$

$$= \frac{.4}{\sqrt{.4816 \times .7296}} = \frac{.4}{.593} = 0.674$$

$$R_{1.23} = \frac{\sqrt{r^2_{12} + r^2_{13} - 2\, r_{12}\, r_{13}\, r_{23}}}{1-r^2_{23}}$$

$$= \frac{\sqrt{(.77)^2 + (.72)^2 - 2(.77)(.72)(.52)}}{1-(.52)^2}$$

$$= \frac{\sqrt{.5929 + .5184 - .5766}}{1-.2704}$$

$$= \frac{\sqrt{.5347}}{.7296} = 0.856.$$

LIST OF FORMULAS

Partial Correlation Coefficients $r_{12.3} = \dfrac{r_{12} - r_{13}\, r_{23}}{\sqrt{1 - r_{13}^2}\,\sqrt{1 - r_{23}^2}}$

$$r_{13.2} = \frac{r_{13} - r_{12}\, r_{23}}{\sqrt{1 - r_{12}^2}\,\sqrt{1 - r_{23}^2}}$$

$$r_{23.1} = \frac{r_{23} - r_{12}\, r_{13}}{\sqrt{1 - r_{12}^2}\,\sqrt{1 - r_{13}^2}}$$

Multiple Correlation Coefficient $R_{1.23} = \dfrac{\sqrt{r_{12}^2 - r_{13}^2 - 2r_{13}\, r_{12}\, r_{23}}}{1 - r_{23}^2}$

Multiple Regression of X_1 on X_2 and X_3 $X_{1.23} = a^{1.23} + b_{12.3}\, X^2 + b_{13.2}\, X^3$ $b_{12.3} = \dfrac{\sigma_1}{\sigma_2} \times \dfrac{r_{12} - r_{13}\, r_{23}}{1 - r^2{}_{23}}$

Partial Regression Coefficient $b_{13.2} = \dfrac{\sigma_1}{\sigma_2} \times \dfrac{r_{13} - r_{12}\, r_{23}}{1 - r^2{}_{23}}$

EXERCISES

1. The coefficient of multiple correlation between X_2 on the one hand and X_1 and X_3 on the other is denoted by
   (i) $r_{12.3}$ (ii) $r_{13.2}$ (iii) $R_{1.23}$ (iv) $R_{2.13}$ (v) $R_{3.12}$
2. In case of three variables, the regression equation X_1 on X_2 and X_3 has the form :
   (i) $X_1 = a_{1.23} + b_{12.3}\, X_2 + b_{13.2}\, X_3$
   (ii) $X_1 = a_{2.13} - b_{12.3}X_2 + b_{12.3}X_3$
   (iii) $X_1 = a_{1.22} + b_{12.3}X_2 + b_{13.2}X_3$
   (iv) $X_2 = a_{1.23} + b_{12.3}X_2 - b_{13.2}X_1$
   (v) none of these.
   (g) $b_{12.3}$ is called the partial regression coefficient of :
   (i) X_2 on X_3, keeping X_1 constant.
   (ii) X_1 on X_3, keeping X_3 constant.
   (iii) X_1 on X_3, keeping X_2 constant.
   (iv) none of these.
   [**Ans.** (a) (ii), (b) (iii), (c) (i), (d) (v), (e) (iv), (f) (i), (g) (ii)]
3. (a) What is meant by multi-collinearity?
   (b) Explain the concept of multiple regression and cite an example of its utility in the practical field.

4. (a) Distinguish between simple, multiple and partial correlations.
5. Fill in the blanks :
   (a) Partial correlation coefficients such as $r_{12.3}$ and $r_{13.2}$ are referred to as ....
   (b) Second order partial correlation coefficients can be obtained from .... coefficients.
   (c) By squaring $R_{1.23}$ we obtain the coefficient of ....
   (d) The coefficient of multiple correlation lies between .... and ....
   (e) The dependent variable is always denoted by ....
   (a) First order coefficients,
   (b) first order,
   (c) multiple determination,
   (d) 0, 1, (e) X_1
6. (a) The following information about a trivariate population is known:
   $\sigma_1 = 3$ $\sigma_2 = 4$ and $\sigma_3 = 5$
   $r_{23} = 0.40$ $r_{31} = 0.60$ and $r_{12} = 0.7$
   Determine the regression equation of X_1 on X_2 and X_3.
   (b) If $r_{12} = 0.9$, $r_{13} = 0.75$, $r_{23} = 0.7$, find $R_{1.23}$.
7. Find the multiple regression equation of X_1 and X_3 from the data relating to three variables given below :

X_1	X_2	X_3	X_1	X_2	X_3
4	15	30	9	6	14
6	12	24	13	4	10
7	8	20	15	3	4

   Also predict the value of X_1 when $X_2 = 10$ and $X_3 = 22$.
8. The table shows the corresponding values of three variables X_1, X_2 and X_3 :

X_1 :	3	5	6	8	12	14
X_2 :	16	10	7	4	3	2
X_3 :	90	72	54	42	30	12

   Find the least squares regression equation of X_3 on X_1 and X_2. Estimate X_3 when $X_1 = 10$ and $X_2 = 6$.
9. Find the multiple linear regression of X_1 on X_2 and X_3 from the data relating to three variables given below :

X_1 :	11	17	26	28	31	35	41	49	63	69
X_2 :	2	4	6	5	8	7	10	11	13	14
X_3 :	2	3	4	5	6	7	8	10	11	13

10. Estimate the relationship between the use of inputs and labour on productivity from the following data :

Productivity :	15	18	16	20	24	27
Input :	5	8	7	6	10	9
Labour :	40	45	50	55	60	50

   (i) What will be the productivity if inputs will be 12 and labour 65?

   (ii) Compute the coefficient of multiple determination.

11. Tick the correct answer :

   (a) In multiple correlation analysis, there are at least
      (i) 2 variables,
      (ii) 3 variables,
      (iii) 5 variables,
      (iv) 10 variables,
      (v) none of these.

   (b) $r_{13.2}$ means are coefficient of partial correlation between
      (i) X_1 and X_2, keeping the effect of X_3 constant,
      (ii) X_2 and X_3, keeping the effect of X_1 constant,
      (iii) X_1 and X_2, keeping the effect of X_2 constant,
      (iv) X_1, X_2 and X_3.

   (c) In case of 3 variables, X_1, X_2 and X_3, there are
      (i) 3 partial correlation coefficients,
      (ii) 2 partial correlation coefficients,
      (iii) 4 partial correlation coefficients,
      (iv) at least 10.

   (d) $r_{12.4}$ is calculated as :

(i) $\dfrac{r_{12} - r_{13}\ r_{24}}{\sqrt{\left(1 + r_{13}^2\right)\left(1 - r_{24}^2\right)}}$ (ii) $\dfrac{r_{13} - r_{12}\ r_{24}}{\sqrt{\left(1 + r_{12}^2\right)\left(1 - r_{24}^2\right)}}$

(iii) $\dfrac{r_{12} + r_{13}\ r_{24}}{\sqrt{\left(1 + r_{13}^2\right)\left(1 - r_{24}^2\right)}}$ (v) $\dfrac{r_{13} - r_{14}\ r_{24}}{\sqrt{\left(1 - r_{14}^2\right)\left(1 - r_{24}^2\right)}}$

12. What is multiple linear regression? What is its importance in economic analysis?

13. When is multiple regression needed? Explain with the help of an example.

8

Sampling and Designs

SAMPLE METHOD OF CENSUS

Under the census or complete enumeration survey method, data are collected for each and every unit (person, household, field, shop, factory, etc., as the case may be) of the population or universe which is the complete set of items which are of interest in any particular situation.

(i) Data are obtained from each and every unit of the population.

(ii) The results obtained are likely to be more representative, accurate and reliable.

(iii) It is an appropriate method of obtaining information on rare events such as areas under some crops and yield thereof, the number of persons of certain age groups, their distribution by sex, educational level of people, etc. This is the reason why throughout the world the population data are obtained by conducting a census generally every 10 years by the census method.

(iv) Data of a complete enumeration census can be widely exploited as a basis for various surveys.

Sampling is simply the process of learning about the population on the basis of a sample drawn from it. Thus, in the sampling technique instead of every unit of the universe only a part of the universe is studied and the conclusions are drawn on that basis for the entire universe. A sample, is a subset of population units. The process of sampling involves three elements:

(a) Selecting the sample,

(b) collecting the information, and

(c) Marking an inference about the population.

The three elements cannot generally be considered in isolation from one another. Sample selection, data collection, and estimation are all interwoven and each has an impact on the others. Sampling is not haphazard selection; it embodies definite rules for selecting, the sample. But having followed a set of rules for sample selection, we cannot consider the estimation process independent of it; estimation is guided by the manner in which the sample has been selected.

THEORETICAL BASIS OF SAMPLING

On the basis of sample study we can predict and generalise the behaviour of mass phenomena. This is possible because there is no statistical population whose elements would vary from each other without limit. For example, wheat varies to limited extent in colour, protein content, length, weight etc., but it can always be identified as wheat. Similarly apples of the same tree may vary in size, colour, taste, weight, etc. but they can always be identified as apples. Thus we find that although diversity is a universal quality of mass data, every population has characteristic properties with limited variation. This makes possible to select a relatively small unbiased random sample that can portray fairly well the trail of the population.

There are two important laws on which the theory of sampling is based:

1. Law of 'Statistical Regularity', and
2. Law of 'Inertia of Large Numbers'.

Law of Statistical Regularity

This Law is derived from the mathematical theory of probability. In the words of King, "The low of statistical regularity lays down that a moderately large number of items chosen at random from a large group are almost sure on the average to possess the characteristics of the large group." In other words, this law points out that if a sample is taken at random from a population, it is likely to possess almost the same characteristics as that of the population. This law directs our attention to one very important point, that is, the desirability of choosing the sample at random.

Law of Inertia of Large Numbers

This law is a corollary of the law of statistical regularity. It is of great significance in the theory of sampling. It states that, others things being equal, larger the size of the sample, more accurate the results are likely to be. This is because large numbers are more stable as compared to small ones. The difference in the aggregate result is likely to be insignificant, when the number in the sample is large, because when large numbers are considered the variation in the component part tend to balance each other and, therefore, the variation in the aggregate is insignificant.

Essentials of Sampling

If the sample results are to have any worthwhile meaning, it is necessary that a sample possesses the following essentials:

(i) *Representativeness* : A sample should be so selected that it truly represents the universe otherwise the results obtained may be misleading. To ensure representativeness the random method of selection should be used.

(ii) *Adequacy:* The size of sample should be adequate otherwise it may not represent the characteristics of the universe.

(iii) *Independence*: All items of the sample should be selected independently of one another and all items of the universe should have the same chance of being selected in the sample. By independence of selection we mean that the selection of a particular item in one draw has no influence on the probabilities of selection in any other draw.

(iv) *Homogeneity*: When we talk of homogeneity we mean that there is no basic difference in the nature of units of the universe and that of the sample. If two samples from the same universe are taken, they should give more or less the same result .

METHODS OF SAMPLING

The various methods of sampling or different sampling designs can be grouped under two broad heads-random sampling and non-random sampling. Random sampling is also referred to as probability sampling since if the sampling process is random the laws of probability can be applied. It may be noted that the term random sample is not used to describe the data in the sample but the process employed to select the sample. Randomness is thus a property of the sampling procedure instead of an individual sample. As such randomness can enter a processed sampling in a number of ways and hence random samples may be of many kinds.

Advantages of Probability Sampling

The following are the basic advantages of probability sampling :

(1) Probability sampling does not depend upon the existence of detailed information about the universe for its effectiveness.

(2) Probability sampling provides estimates which are essentially unbiased and have measurable precision.

(3) It is possible to evaluation the relative efficiency of various sample designs only when probability sampling is used.

Limitations of Probability Sampling

Despite the great advantages of probability sampling techniques mentioned above it has certain limitations because of which non-probability sampling is quite often used in practice. These limitations are:

(1) Probability sampling requires a very high level of skill and experience for its use.

(2) It requires a lot of time to plan and execute a probability sample.

(3) The costs involved in probability sampling are larger as compared to non-probability sampling.

Non-random sampling is a process of sample selection without the use of randomization. In other words, a non-random sample is selected on a basis other than the probability considerations such as convenience, judgement, etc.

The most important difference between random and non-random sampling is that whereas the pattern of sampling variability can be ascertained in case of random sampling, in non-random sampling, there is no way knowing the pattern of variability in the process.

In the following few pages some of the important sampling methods that are popularly used in practice are discussed.

A. Random Sampling Methods:

(a) Simple or unrestricted random sampling: and

(b) Restricted random sampling:

(i) Stratified sampling,

(ii) Systematic sampling,

(iii) Cluster sampling,

B. Non-Random Sampling Methods:

(i) Judgment sampling;

(ii) Convenience sampling; and

(iii) Quota sampling.

A brief description of these methods is given below.

RANDOM SAMPLING METHODS

(a) Simple or Unrestricted Random Sampling

Simple random sampling refers to that sampling technique in which each and every unit of the population has an equal opportunity of being selected in the sample. In simple random sampling which items get selected in the sample is just a matter of chance-personal bias of the investigator does not influence the selection. It should be noted that the word *'random'* does not mean *'haphazard'* or *'hit-or-miss'-it* rather means that the selection process is such that chance only determines which items are included in the sample. As pointed out by Chou, when a sample of size n is drawn from a population with N elements, the sample is a *'simple random sample'* if any of the following is true. And, if any of the following is true, so are the other two:

(1) All n items of the sample are selected independently of one another and all N items in the population have the same chance of being

included in the sample. By independence of selection we mean that the selection of a particular item in one draw has no influence on the probabilities of selection in any other draw.

(2) At each selection, all remaining items in the population have the same chance of being drawn. If sampling is made with replacement i.e., when each unit drawn from the population is returned prior to drawing the next unit, each item has a probability of 1/N of being drawn at each selection. If sampling is without replacement, i.e., when each unit drawn from the population is not returned prior to drawing the next unit, the probability of selection of each item remaining in the population at the first draw is 1/N, at the second draw 1/(N–1), at the third draw is 1/(N–2), and so on. It should be noted that sampling with replacement has very limited and special uses in statistics-we are mostly concerned with sampling without replacement.

(3) All the possible samples of a given size n are equally likely to be selected.

To ensure randomness of selection one may adopt either the Lottery Method or consult table of random numbers.

Lottery Method: This is a very popular method of taking a random sample. Under this method, all items of the universe are numbered or named on separate slips of paper of identical size and shape. These slips are then folded and mixed up in a container or drum. A blindfold selection is then made of the number of slips required to constitute the desired sample size. The selection of items thus depends entirely on chance. The methods would be quite clear with the help of an example. If we want to take a sample of 10 persons out of a population of 100, the procedure is to write the name of all the 100 persons on separate slips of paper, fold these slips, mix them thoroughly and then make a blindfold selection of 10 slips.

The random numbers are generally obtained by some mechanism which, when repeated a large number of times, ensures approximately equal frequencies for the numbers from 0 to 9 and also proper frequencies for various combinations of numbers (such as 00, 01,...99 ; 000,001,...999; etc.) that could be expected in a random sequence of the digits 0 to 9.

Several standard tables of random numbers are available, among which the following may be specially mentioned, as they have been tested extensively for randomness:

(1) Tippett's (1927) random number tables consisting of 41,600 random digits grouped into 10,400 sets of four-digited random numbers;

(2) Fisher and Yates (1938) table of random numbers with 15,000 random digits arranged into 1,500 sets of ten-digited random numbers.

(3) Kendall and B.B. Smith (1939) table of random numbers leaving 10,00,000 random digits grouped into 25,000 sets of four-digited random numbers;

(4) Rand corporation (1955) table of random numbers consisting of 1,00,000 random digits grouped into 2,00,000 sets of five-digited random numbers; and

(5) C.R. Rao, Mitra and Matthai (1966) table of random numbers with 20,000 random digits grouped into 5,000 sets of four-digited random numbers.

Merits

1. The analyst can easily assess the accuracy of this estimate because sampling errors follow the principle of chance. The theory of random sampling is further developed than that of any other type of sampling which enables the analyst to provide the most reliable information at the lest cost.
2. As compared to judgement sampling, a random sample represents the universe in a better way. As the size of the sample increases, it becomes increasingly representative of the population.
3. Since the selection of items in the sample depends entirely on chance there is no possibility of personal bias affecting the results.

Limitations

1. Random sampling may produce the most non-random-looking results. For example, thirteen cards from a well-shuffled pack of playing cards may consist of one suit. But the probability of this type of occurrence is very-very low.
2. From the point of view of field survey it has been claimed that case selected by random sampling tend to be too widely dispersed geographically and that the time and cost of collecting data become too large.
3. The size of the sample required to ensure statistical reliability is usually larger under random sampling than stratified sampling.
4. The use of simple random sampling necessitates a completely catalogued universe from which draw the sample. But it is often difficult for the investigator to have up-to-date lists of all he items of the population to be sampled. This restricts the use of this method in economic and business date where very we have to employ restricted random sampling designs.

(b) Restricted Random Sampling

(i) Stratified Sampling

Stratified random sampling or simply stratified sampling is one of the random methods which, by using the available information concerning the population, attempts to design a more efficient sample than obtained by the simple random procedure.

When this method of sampling is adopted, the population is divided into different groups or classes called stratas and a sample is drawn from each stratum at random. For example, if we are interested in studying the consumption pattern of the people of Delhi, the city of Delhi may be divided into various parts (such as zones or wards) and from each part a sample may be taken at random. Before deciding on stratification we must have knowledge of the traits of the population. Such knowledge may be based upon expert judgment, past data, preliminary observation from pilot studies, etc.

The purpose of stratification is to increase the efficiency of sampling by dividing a heterogeneous universe in such a way that (i) there is as great homogeneity as possible within each stratum and (ii) as marked a difference as possible between the strata.

Example:

You are given the following data of the number of lecturers, readers and professors in a University:

Length of service	Lecturers	Readers	Professors	Total
Less than 5 Yrs.	2,000	250	50	2,300
5-10 Yrs.	3,000	220	80	3,300
10-15 Yrs.	1,500	170	30	1,700
More than 15 Yrs.	880	80	40	1,000
Total	7,380	720	200	8,300

Work out how many lecturers, readers and professors would be selected from each category if,

(i) we follow stratified proportionate sampling method and take 10% of the universe equivalent to the sample size,

(ii) if the size of the sample is 10% of the universe but the lecturers, readers and professors are to be in the ratio of 5:3:2 and weightage to the length of service is to be in the ratio of 4:3:2:1.

Solution :

(i) The sample size is 10% of the universe hence 830 persons would be selected in the sample. Since 12 strata are formed and we want to follow proportionate stratified sampling method. We will take 10% from each stratum. The number of Persons selected shall be as follows:

Length of service	Lecturers	Readers	Professors	Total
Less than 5 yrs.	200	25	5	230
5-10 yrs.	300	22	8	330
10-15 yrs.	150	17	3	170
Above 15 yrs.	88	8	4	100
Total	738	72	20	830

(ii) In the second case also the size of sample is 830 but the lecturers, readers and professors are to be in the ratio of 5:3:2 of the sample, i.e., we take $\frac{830 \times 5}{10} = 415$ lecturers, $\frac{830 \times 3}{10} = 249$ readers, and $\frac{830 \times 2}{10} = 166$ professors. Since the weightage to length of service is 4 : 3 : 2 : 1 the number selected from each category shall be :

Length of service	Lecturers	Readers	Professors	Total
Less than 5 yrs.	$\frac{415 \times 4}{10} = 166$	$\frac{249 \times 4}{10} = 99.6$ or 100	$\frac{166 \times 4}{10} = 66.4$ or 66	332
5 – 10	$\frac{415 \times 3}{10} = 124.5$ or 124	$\frac{249 \times 3}{10} = 74.7$ or 74	$\frac{166 \times 3}{10} = 49.80$ or 50	248
10 – 15	$\frac{415 \times 2}{10} = 83$	$\frac{249 \times 2}{10} = 49.8$ or 50	$\frac{166 \times 2}{10} = 33.2$ or 33	166
Above 15	$\frac{415 \times 1}{10} = 41.5$ or 42	$\frac{249 \times 1}{10} = 24.9$ or 25	$\frac{166 \times 1}{10} = 16.6$ or 17	84
Total	415	249	166	830

Merits

1. Greater geographical concentration. As compared with random sample, stratified samples can be more concentrated geographically i.e., the units from the different strata may be selected in such a way the all of them are localised in one geographical area. This would greatly reduce the time and expenses of interviewing.
2. Greater accuracy. Stratified sampling ensures greater accuracy. The accuracy is maximum if each stratum is so formed that it consists of uniform or homogeneous items.
3. More representative. Since the population is first divided into various strata and then a sample is drawn from each stratum there is little possibility of any essential group of the population being completely excluded. A more representative sample is thus secured. C.J. Grohmann has rightly pointed out that this type of sampling balances the uncertainity of random sampling against the bias of deliberate selection."

Limitations

1. The items from each stratum should be selected at random. But this may be difficult to achieve in the absence of skilled sampling supervisors and a random selection within each stratum may not be ensured.
2. Utmost care must be exercised in dividing the population into various stratas. Each stratum must contain, as possible, homogeneous items as otherwise the results may not be reliable. If proper stratification of the population is not done the sample may have the effect of bias.

(ii) Systematic Sampling

A systematic sample is formed by selecting one unit at random and then selecting additional units at evently-spaced intervals until the sample has been formed. ; This method is popularly used in those cases where a complete list of the population from which sample is to be drawn is available. The list may be prepared in alphabetical, geographical, numerical or some other. The items are serially numbered. The first item is selected at random generally by following the Lottery method. Subsequent items are selected are selected by taking every kth item from the list where 'k' refers to the sampling interval or sampling ratio i.e., the ratio of population size of the sample. Symbolically,

$$k = \frac{N}{n}$$

where k = Sampling interval, N = Universe size and n = Sample size.

While calculating k, it is possible that we get a fractional value. In such a case we should use approximation procedure i.e., if the fraction is less than

0.5 it should be omitted and if it is more than 0.5 it should be taken as 1. If it is exactly 0.5 it should be omitted if the number is even and should be taken as 1, if the number is odd. This is based on the principle that the number after approximation should preferably be even. For example if the number of students are respectively 1020, 1150 and 1100 and we want to take a sample of 200, k shall be:

(i) $8\,k = \frac{1020}{200} = 5.1$ or 5

(ii) $82\,k = \frac{1150}{200} = 5.75$ or 6

(iii) $2\,k = \frac{1100}{200} = 5.5$ or 6s

Example:

In a class there are 96 students with Roll Nos. from 1 to 96. It is desired to take a sample of 10 students. Use the systematic sampling method to

Solution :

$$k = \frac{N}{n} = \frac{96}{10} = 9.6 \text{ or } 10$$

From 1 to 96 Roll Nos. The first student between 1 and k i.e., 1 and 10 will be selected at random and then we will go on taking every kth student. Suppose the first student comes out to be 4th. The sample would them consist of the following Roll Nos.

4, 14, 24, 434, 44, 54, 64, 74, 84, 94,

Systematic sampling is relatively a simple technique and may be more efficient than simple random sampling provided the lists are arranged wholly at random. However, it is rarely that this requirement is fulfilled. The nearest approach to randomness is provided by alphabetical lists such as are found in telephone directory, although even these may have certain non-random characteristics.

Merits

The systematic sampling design is simple and convenient to adopt. The time and work involved in sampling by this method are relatively smaller. The result obtained are also found to be generally satisfactory provided care is taken to see that there are no periodic features associated with the sampling interval. If populations are sufficiently large systematic sampling can often be expected to yield results similar to those obtained by proportional stratified sampling.

Limitation

The main limitation of the method is that it becomes less representative if we are dealing with populations having hidden *periodicities*. Also if the population is ordered in a systematic way with respect to the characteristics the investigator is interested in, then it is possible that only certain types of items will be included in the population, or at least more of certain types than others. For instance, in a study of workers' wages the list may be such that every tenth worker on the list gets wages above Rs. 150 per month.

(iii) Multi-stage or Sampling Cluster Sampling

As the name implies this method refers to a sampling procedure which is carried out in several stages. The material is regarded as made up for a number of second stage sampling units, each of which is made of a number of second stage units, etc. At first, the first stage units are sampled by some suitable method, such as simple random sampling, Then, a sample of second stage units is selected from each of the selected first stage units, again by some suitable method which may be the same as or different from the method employed for the first stage units. Further stages may be added as required. The producer may be illustrated as follows:

Suppose we want to take a sample of 5,000 households from the Stage of U.P. At the first stage, the Stage may be divided into a number of districts and a few districts selected at random. At the second stage, each district may be sub-divided into a number of villages and a sample of villages may be taken at random. At the third stage, a number of households may be selected from each of the villages selected at the second stage. To take another example suppose in a particular survey, we wish to take a sample of 10,000 students from Delhi University. We may take colleges primary units-at the first stage, then draw departments as the second stage, and choose students at the third and last stage.

Merits

Multi-stage sampling introduces flexibility in the sampling method which is lacking in the other methods. It enables existing divisions and sub-divisions of the population to be used as units at various stages, and permits the field work to be concentrated and yet large area to be covered, Another advantage of the method is that sub-division into second stage units (i.e., the construction of the second stage frame) need be carried out for only those first stage units which are included in the sample. It is, therefore, particularly valuable in surveys of underdeveloped area where no frame is generally sufficiently detailed and accurate for sub-division of the material into reasonably small sampling units.

Limitations

However, a multi-stage sample is in general less accurate than a sample containing the same number of final stage units which have been selected by some suitable single stage process.

We have discussed above the various random procedures as independent design. In practice we often combine two or more of these methods into a single design.

B. Non-Random Sampling Methods

(i) Judgment Sampling

In judgment sampling the choice of sampling items depends exclusively on the discretion of the investigator. In other words, the investigator exercises his judgment in the choice and includes those items in the sample which he thinks are most typical of the universe with regard to the characteristics under investigation. For example, if sample of ten students is to be selected from a class of sixty for analysing the spending habits of students, the investigator would select 10 students who, in his opinion, are representative of the class.

Merits

Though the principles of sampling theory are not applicable to judgement sampling, the method is often used in solving many types of economic and business problems. The use of judgment sampling is justified under a variety of circumstance:

(i) In solving everyday business problems and making public policy decisions, executives and public officials are often pressed for time and cannot wait for probably sample designs. Judgment sampling is then the only practical method to arrive at solutions to their urgent problems.

(ii) When we want to study some unknown traits of a population, some of whose characteristics are known, we may then stratify the population according to these known properties and select sampling units from each stratum on the basis of judgment. This method is used to obtain a more representative sample.

(iii) when only a small number of sampling units is in the universe, simple random selection may miss the more important elements, whereas judgment selection would certainly include them in the sample.

Limitations

This method, though simple, is not scientific because the population units to be sampled may be affected by the personal prejudicer or bias of the investigator. Thus, judgement sampling involves the risk that the investigator may establish foregone conclusions by including those items in the sample which conform to his preconceived notions. For example, if an investigator holds the view that the wages of workers in a certain establishment are very low, and if he adopts the judgment sampling method, he may include only those workers in the sample whose wages are low and thereby establish his point of view which may be far from the truth. Since an element of subjectiveness is possible, this method cannot be recommended for general use.

(ii) Convenience Sampling

A convenience sample is obtained by selecting 'convenient' population units.

The method of convenience sampling is also called the chunk. A chunk refers to that fraction of the population being investigated which is selected neither my probability nor by judgment but by convenience. A sample obtained from readily available lists such as automobile registrations, telephone directories, etc. is a convenience sample and not a random sample even if the sample is drawn at random from the lists. If a person is to submit a project report on labour-management relations in textile industry and he takes a textile mill close to his office and interview some people over there, he is following convenience sampling method. Convenance samples are prove to bias by their very nature-selecting population elements which are convenient to choose almost always makes them special or different from the best of the elements in the population is some way.

(iii) Quota Sampling

Quota sampling is a type of judgment sampling. In a quota sample, quotas are set up according to some specified characteristics such as so many in each of several income groups, so many in each age, so many with certain political or religious affiliations, and so on. Each interviewer is then told to interview a certain number of persons which constitutes his quota. Within the quotas, the selection of sample items depends on personal judgment.

Quota sampling is often used in public opinion studies. It occasionally provides satisfactory results if the interviewers are carefully trained and if they follow their instructions closely.

Selection of Appropriate Method of Sampling

Having discussed the various methods of sampling, the question now arises as to which method to adopt in a particular situation. It should be noted that one method can be regarded as best under all circumstances- each method has its own speciality. A number of factors such as nature of the problem, size of universe, size of the sample, availability of finance time, etc. would influence the selection of a particular method of sampling.

SIZE OF SAMPLE

An important decision that has to be taken adopting a sampling technique is about the size of the sample. Size of sample means the number of sampling units selected from the population for investigations. Different opinions have been expressed by experts on this point.

The following factors should be considered while deciding the sample size:

(i) *Nature of Study:* For an intensive and continuous study a small sample may be suitable. But for studies which are not likely to be repeated and are quite extensive in nature, it may be necessary to take a large sample size.

(ii) *Homogeneity or Heterogeneity of the Universe* : If the universe consists of homogeneous units a small sample may be inevitable.

(iii) *The degree of accuracy or precision desired :* The greater the degree of accuracy desired the larger should be the sample always ensure greater accuracy. If a sample is selected by experts by following scientific method, it may ensure better results even when it is small compared to a situation in which a large sample size is selected by inexperienced people.

(iv) *The resources available :* If the resources available are vast a larger sample size could be taken. However, in most cases resources constitute a big constraint on sample size.

(v) *The size of the universe:* The larger the size of the universe, the bigger should be the sample size.

(iv) *Nature of respondents:* Where it is expected that a large number of respondents, will not cooperate and send back the questionnaires, a larger sample should be selected.

The above factors have to be properly weighed before arriving at the sample size. However, the selection of optimum sample size is not that simple as it might seem to be. If the sample is used which is larger than necessary, resources are wasted, if the sample is smaller than required, the objectives of the analysis man not be achieved.

(vii) *Methods of Sampling adopted :* The size of sample is also influenced by the type of sampling plan adopted. For example, if the sample is a simple random sample it may necessitate a bigger sample size. However, in a properly drawn stratified sampling plan, even a small sample may give better results.

Mathematical Formula for Determining the Sample Size

A number of formulae have been devised for determining the sample size depending upon the availability of information. A few formulae are given below :

$$n = \left(\frac{Z\sigma}{d}\right)^2$$

n = Sample size

Z = value at a specified level of confidence or desired degree of precision.

s = Standard deviation of the population

d = difference between population mean and sample mean.

The steps in computing the sample from the above formula are:

(i) Select the desired degree of precision, i.e., specified level or confidence and designate it as small 'z' (at 1% level of significance or 99% confidence level the valde of 'z' is 2.576, and at 5% level of significance of 95% confidence level 1.96).

(ii) Multiply the 'z' selected in step 1 by the standard deviation of the universe which may be assumed.

(iii) Divide the product of the preceding step by the standard error of mean or difference between population and sample mean. Square the resultant quotient. The result is the size of sample required.

MERITS AND LIMITATIONS OF SAMPLING

Merits

The sampling technique has the following merits over the complete enumeration survey:

1. *More reliable results* : Although the sampling technique involves certain inaccuracies owing to sampling errors, the result obtained is generally more reliable than that obtained from a complete count. There are several reasons for it. First, it is always possible to determine the extent of sampling errors. Secondly, other types of errors to which a survey is subject, such as inaccuracy of information, incompleteness of returns, etc., are likely to be more serious in a complete census than in a sample survey. This is because more effective precautions can be taken in a sample survey to ensure that the information is accurate and complete. For these reasons not only may the total error be expected to the smaller in a sample survey but sample results can also be used with a greater degree of confidence because of our knowledge of the probable size of error. Thirdly, it is possible to avail of the services of experts and to impart thorough training to the investigators in a sample survey which further reduces the possibility of errors. Follow-up work can also be undertaken much more effectively in the sampling method. Indeed, even a

complete census can only be tested for accuracy by some type of sampling check.

2. *Less time consuming* : Since the sample is a study of a part of population considerable time and labour are saved when a sample survey is carried out. Time saved not only in collecting data but also in processing it. For these reasons a sample provides more timely data in practice that a census.
3. *Less cost* : Although the amount of effort and expense involved in collecting information is always greater per unit of the sample than a complete census, the total financial burden of a sample survey is generally less than that of a complete census. This is because of the fact that in sampling, we study only a part of population and the total expenses of collecting data is less than that required when the census method is adopted. This is a great advantage particularly in an underdeveloped economy where much of the information would be difficult to collect by the census method for lack of adequate resources.
4. *More detailed information* : Since the sampling technique saves time and money, it is possible to collect more detailed information in a sample survey. For example, if the population consists of 1,000 persons in a survey of the consumption pattern of the people, the two alternative techniques available are as follows:
   (a) We may collect the necessary data from each one of the 1,000 people through a questionnaire containing, say, 10 questions (census method), or
   (b) We may take a sample of 100 persons (i.e., 10% of population and prepare a questionnaire containing as many as 10 questions. The expenses involved in the latter case would almost be the same as in the former but it will enable nine time more information to be obtained.
5. The sample method is often used to judge the accuracy of the information obtained on a census basis. For example, in the population census which is conducted very often 10 years in our country the field officers employ the sample method to determine the accuracy of information obtained by the enumerators on the census basis.
6. *Sampling Method is the only method that can be used in certain cases.* There are some cases in which the census method is inapplicable and the only practicable means is provided by the sample method. For example, if one is interested in testing the breaking strength of chalks manufactured in a factory under the census method all the chalks would be broken in the process of testing. Hence, census method is impracticable and resort must be had to the sample method.

Similarly if the producer wants to find out whether the tensile strength of a lot of steel wires meets the specified standard. He must resort to sample method because census would mean complete destruction of all the wires. Also if the population under investigation is infinite, sampling is the only possible solution.

Limitations

Despite the various advantages of sampling, it is not altogether free from limitations. Some of the difficulties involved in sampling are stated below:

1. If the information is required for each and every unit in the domain of study, a complete enumeration survey is necessary.
2. At time the sampling plan may be so complicated that it requires more time, labour and money than a complete count. This is so if the size of the sample is a large proportion of the total population and if complicated weighed procedures are used. With each additional complication in the survey, the chance of errors multiply and greater care has to be taken which, in turn, means more time and labour.
3. Sampling generally requires the services of experts, if only for consultation purposes. In the absence of qualified and experienced persons, the information obtained from sample surveys cannot be relied upon. In India, shortage of experts in the sampling field is a serious hurdle in the way of reliable statistics.
4. A sample survey must be carefully planed and executed otherwise the results obtained may be inaccurate and misleading. Of course, even for a complete count care must be taken but serious errors may arise in sampling, if the sampling procedure is not perfect.

SAMPLING AND NON-SAMPLING ERRORS

To appreciate the need for sample surveys, it is necessary to understand clearly the role of sampling and non-sampling errors in complete enumeration and sample surveys. The error arising due to drawing inferences about the population on the basis of few observations (sampling) is termed *sampling error*. Clearly the sampling error in this sense is nonexistent in complete enumeration survey, since the whole population is surveyed. However, the error mainly arising at the stages of ascertainment and processing of data, which are termed non-sampling errors, are common both in complete enumeration and sample surveys.

I. Sampling Errors

Even if utmost care has been taken in selecting a sample. The results derived from a sample study may not be exactly equal to the true value in the population. The reason is that estimate is based on a part and not on the whole and samples are seldom, if ever, perfect miniature of the population. Hence sampling gives rise to certain errors known as sampling errors (or sampling fluctuations). These errors would not be present in a complete enumeration survey. However, these errors can be controlled. The modern sampling theory helps in designing the survey in such a manner that the sampling errors can be made small.

Sampling errors are of two types-biased and unbiased:

(1) *Biased errors.* These errors arise from any bias in selection, estimation, etc.

(2) *Unbiased errors.* These errors arise due to chance differences, between the members of population included in the sample and those not included. An error in statistics is the difference between the value of a statistic and that of the corresponding parameter.

Thus the total sampling error is made up of error due to bias, if any, and the random sampling error. The essence of bias is that it forms a constant component of error that does not decrease in a large population as the number in the sample increases. Such error is, therefore, also known as *cumulative or non-compensating error.* The random sampling error, on the other hand, decreases on an average as the size of the sample increases. Such error is, therefore, also known as non-cumulative or compensating error.

Causes of Bias

Bias may arise due to:

(i) faulty process of selection;

(ii) faulty work during the collection of information; and

(iii) faulty methods of analysis.

(i) Faulty Selection : Faulty selection of the sample may give rise to bias in a number of ways, such as:

(1) An appeal to the vanity of the person questioned may give rise to yet another kind of bias. For example, the question 'Are you a good student?' Is such that most of the students would succumb to vanity and answer 'Yes.'

(2) *Non-response.* If all the items to be included in the sample are not covered, there will be bias even though no substitution has been attempted. This fault particularly occurs in mailed questionnaires, which are incompletely returned. Moreover, the information supplied by the informants may also be biased.

(3) *Substitution.* Substitution of an item in place of one chosen in a random sample sometimes leads to bias. Thus if it was decided to interview every 50th householder in the street, it would be inappropriate to interview the 51st or any other number in his

place as the characteristics possessed by them may differ from those who were originally to be included in the sample.

(4) *Conscious or unconscious bias in the selection of a 'random' sample.* The randomness of selection may not really exist, even though the investigator claims that he has a random sample if he allows his desire to obtain a certain result to influence his selection.

(5) Deliberate selection of a 'representative' sample.

(ii) Bias Due to Faulty Collection of Data : Any consistent error in measurement will give rise to bias whether the measurements are carried out on a sample or on all the units of the population. The danger of error is, however, likely to be greater in sampling work, since the units measured are often smaller. Bias may arise due to improper formulation of the decision, problem wrongly defining the population, specifying the wrong decision, securing an inadequate frame, and so on. Biased observations may result from a poorly designed questionnaire, an ill-trained interviewer, failure of a respondent's memory, etc. Bias in the flow of data may be due to unorganised collection procedure, faulty editing or coding of responses.

(iii) Bias in Analysis : In addition to bias which arises from faulty process of selection and faulty collection of information, faulty methods of analysis may also introduce bias. Such bias can be avoided by adopting the proper methods of analysis.

Avoidance of Bias

If possibilities of bias exist, fully objective conclusions cannot be drawn. The first essential of any sampling or census procedure must, therefore, be the elimination of all sources of bias. The simplest and the only certain way of avoiding bias in the selection process is for the sample to be drawn either entirely at random, or at random, subject to restrictions which, while improving the accuracy, are of such a nature that they do not introduce bias in the results. In certain cases, systematic selection may also be permissible.

Method of Reducing Sampling Errors

Once the absence of bias has been ensured, attention should be given to the random sampling errors. Such errors must be reduced to the minimum so as to attain the desired accuracy.

Apart from reducing errors of bias, the simplest way of increasing the accuracy of a sample is to increase its size. The sampling error usually decreases with increase in sampling size (number units selected in the samples) and in act in many situations the decrease is inversely proportional to the square-root of the sample size as can be seen from the diagram below.

From this diagram it is clear that though the reduction in sampling error is substantial for initial increases in sample size, it becomes marginal after a certain stage.

Non Sampling Errors

When a complete enumeration of units in the universe is made one would expect that it would give rise to data free from errors. However, in practice it is not so.

Non-sampling errors can occur at every stage of planning and execution of the census or survey. Such errors can arise due to a number of causes such as defective methods of data collection and tabulation, faulty definition, incomplete coverage of the population or sample, etc. More specifically, non-sampling errors, may arise from one or more of the following factors:

1. Errors committed during presentation and printing of tabulated results.
2. Errors in data proceeding operations such as coding, punching, verification, etc.
3. Errors due to non-response, i.e., incomplete coverage in respect of units.
4. Lack of adequate inspection and supervision of primary staff.
5. Lack of trained and experienced investigators.
6. Inaccurate or inappropriate methods of interview, observation or measurement with inadequate or ambiguous schedules, definitions or instructions.
7. Inappropriate statistical unit.
8. Data specification being inadequate and inconsistent with respect to the objectives of the census or survey.

These sources are not exhaustive, but are given to indicate some of the possible sources of error. In a sample survey, non-sampling errors may also arise due to defective frame and faulty selection of sampling units.

Control of Non-sampling Errors

In some situations the non-sampling errors may be large and deserve greater attention than the sampling errors. While in general, sampling errors decreases with increase in sample size, non-sampling errors tend to increase with the sample size. In the case of complete enumeration non-sampling error and in the case of sample surveys both sampling and non-sampling errors require to be controlled and reduced to a level at which their presence does not vitiate the use of final results.

9

Tests of Significance

INTRODUCTION

A hypothesis my be defined as an assumption or a statment about the population parameter. It is based on the logical arguments and procedure associated with the problem. By significance of statistics we mean non-chance difference between obtained scores on the basis of sample and scores based on some hypothesis. If observed difference is significant then we say that observed difference is not influenced by chance defying null hypothesis On the other hand if observed difference is not significant then one can say that observed difference is obtained by chance.' Here we shall deal only one method of test of significance known as "standard error" and will learn about the use of students' t-est and chi-square (χ^2) test.

It referes to the process of selecting and using a sample statistic to draw inference about a population parameter based on a subset of it–the sample drawn from the population. Statistical inference treats two different classes of problems :

1. *Hypothesis testing,* i.e., to test some hypothesis about parent population from which the sample is drawn.
2. *Estimation, i.e.,* to use the 'statistics' obtained from the sample as estimate of the unknown 'parameter' of the population from which the is drawn.

Hypothesis Testing

Hypothesis testing begins with an assumption, called a *Hypothesis,* that we make about a population parameter. A hypothesis is a supposition made as a basis for reasoning. According to Prof. Morris Hamburg. "A hypothesis in statistics is simply a quantitative statement about a population." Palmer O Johnson has beautifully described hypothesis as "islands in the uncharted seas of thought to be used as bases for consolidation and recuperation as we advance into the unknown."

Procedure of Testing Hypothesis

The procedure of testing hypothesis is briefly described below :

(1) **Set up a Hypothesis :** The first thing in hypothesis testing is to set up a hypothesis about a population parameter. Then we collect sample data, produce sample statistics, and use this information to decide how likely it is that our hypothesized population parameter is correct. Say, we assume a certain value for a population mean. To test the validity of our assumption, we gather sample data and determine the difference between the hypothesized value and the actual value of the sample mean. Then we judge whether the difference is significant. The smaller the difference, the greater the likelihood that our hypothesized value for the mean is correct. The larger the difference, the smaller the likelihood.

The conventional approach to hypothesis testing is not to construct a single hypothesis about the population parameter, but rather to set up two different hypotheses. These hypotheses must be so constructed that if one hypothesis is accepted, the other is rejected and *vice versa.*

The two hypothesis in a statistical test are normally referred to as:

(i) Null hypothesis, and

(ii) Alternative hypothesis

The null hypothesis is a very useful tool in testing the significance of difference. In its simplest form the hypothesis asserts that there is no real difference in the sample and the population in the particular matter under consideration (hence the word "null" which means invalid, void, or amounting to nothing) and that the difference found is accidental and unimportant arising out of fluctuations of sampling. The null hypothesis is akin to the legal principle that a man is innocent until he is proved guilty. It constitutes a challenge; and the function of the experiment is to give the facts a chance to refute (or fail to refute) this challenge. For example, if we want to find out whether extra coaching has benefited the students or not, we shall set up a null hypothesis that "extra coaching has not benefited the students". Similarly, if we want to find out whether a particular drug is effective in curing malaria we will take the null hypothesis that "the drug is not effective in curing malaria". The rejection of the null hypothesis indicates that the differences have statistical significance and the acceptance of the null hypothesis indicates that the differences are due to chance. Since many practical problems aim at establishment of statistical significance of differences, rejection of the null hypothesis may thus indicate success in statistical project.

As against the null hypothesis, the alternative hypothesis specifies those values that the researcher believers to hold true, and of course, he hopes that the sample data lead to acceptance of this hypothesis as true. The alternative hypothesis may embrace the whole range of values rather than single point. Now a days, it is usually accepted common practice not to associate any special meaning to the null or alternative hypothesis but merely to let these terms represent to different assumptions about the population parameter. However, for statistical convenience it will make a difference as to which hypothesis is called the null hypothesis and which is called the alternative.

The null and alternative hypotheses are distinguished by the use of two different symbols, H_0 representing the null hypothesis and H_a the alternative hypothesis. Thus a psychologist who wishes to test whether or not a certain class of people have mean I.Q. higher than 100 might establish the following null and alternative hypotheses:

$H_0 : \mu = 100$ (null hypothesis)

$H_a : \mu \neq 100$ (alternative hypothesis)

Or, if he is interested in testing the differences between the mean I.Q. of two groups, this psychologist may like to establish the null hypothesis that the two groups have equal means ($\mu_1 - \mu_2 = 0$) and the alternative hypothesis that their means are not equal ($\mu_1 - \mu_2 \neq 0$)

$H_0 : \mu_1 - \mu_2 = 0$ (null hypothesis)

$H_a : \mu_1 - \mu_2 \neq 0$ (alternative hypothesis)

(2) **Set yo a Suitable Significance Level :** Having set up the hypothesis, the next step is to test the validity of H_0 against that of H_a at as certain level of significance. The confidence with which an experimenter rejects–or retains–a null hypothesis depends upon the significance level adopted. The significance level is customarily expressed as a percentage, such as 5 per cent, is the probability of rejecting the null hypothesis if it is true. When the hypothesis in question is accepted at the 5 per cent level, the statistician is running the risk that, in the long run, he will be making the wrong decision about 5 per cent of the time. By rejecting the hypothesis at the same level the runs the risk of rejecting a true hypothesis in 5 out of every 100 occasions. By testing at the 1 per cent level he seeks to reduce the chance of making a false judgment but some element of risk remains (1 out of 100 occasions) that he will make the wrong decision, i.e., he may accept where he ought to have rejected or *vice versa.*

(3) **Setting a Test Criterion :** The third step in hypotheses testing procedure is to construct a test criterion. This involves selecting an appropriate probability distribution for the particular test, that is, a probability distribution which can properly be applied. Some probability distributions that are commonly used in testing procedures are t, F and χ^2. Test criteria must employ an appropriate probability distribution; for example, if only small

sample information is available, the use of the normal distribution would be inappropriate.

(4) **Doing Computations :** Having taken the first three steps, we have completely designed a statistical test. We now proceed to the fourth step–performance of various computations–from a random sample of size n, necessary for the test. These calculations include the testing statistic and the standard error of the testing statistic.

(5) **Making Decisions :** Finally, as a fifth step, we may draw statistical conclusions and take decisions. A statistical conclusion or statistical decision is a decision either to reject or to accept the null hypothesis. The decision will depend on whether the computed value of the test criterion falls in the region of rejection or the region of acceptance. If the hypothesis is being tested at 5 per cent level and the observed set of results has probabilities less than 5 per cent, we consider significant. In other words, we think that the sample result is so rare that it cannot be explained by chance variation alone. We then decide to reject H_0 and state: "the null hypothesis is false", or "the sample observations are not consistent with the null hypothesis" (the rejection of H_0 automatically leads to acceptance of H_a).

On the other hand, if at 5 per cent level of significance the observed set of results has probability more than 5 per cent we give reason that the difference between the sample result and the hypothetical parameter can be explained by chance variations and, therefore, is not significant statistically. Consequently, we decide not to reject H_0 and state: "The sample observations are not inconsistent with the null hypothesis." If the probability is about 5 per cent, the wisest course may be to resolve judgment and draw another sample, if possible.

The reader might have noted above that the rejection statement is much stronger than the acceptance statement. In other words, if the null hypothesis is not rejected, the statistician does not then categorically conclude that the hypothesis is true. The difference in attitudes arises essentially from the fact that, in logic, it is always easier to prove something false than to prove it true.

It should be clearly noted that the practical "managerial decision" is outside the responsibility of the statistician. He does not make the decision; he purely provides information on the basis of which the businessman or administrator can be assisted in making his decisions.

Two Types of Errors in Testing of Hypothesis

When a statistical hypothesis is tested there are four possibilities:

1. The hypothesis is true but our test rejects. (Type I error)
2. The hypothesis is false but our test accepts it. (Type II error)

3. The hypothesis is true but our test accepts it. (Correct decision)
4. The hypothesis is false but our test rejects it. (Correct decision)

Obviously, the first two possibility lead to errors.

In a statistical hypothesis testing experiment, a Type I error is committed by rejecting the null hypothesis when it is true. The probability of committing a Type I error is denoted by a (pronounced as alpha), where

α = Prob. (Type I error)

= Prob. (Rejecting H_0/H_a is true)

On the other hand, a Type II error is committed by not rejecting (i.e., accepting) the null hypothesis when it is false. The probability of committing a Type II error is denoted by β (pronounced as beta), where

β = Probability (Type II error)

= Probability (Not rejecting or accepting H_0/H_a is false)

Measuring the Power of a Hypothesis Test

It is quite important to know how well a hypothesis test is working. The measure of how well the test is working, is called the *power of the test.*

In hypothesis testing α and β (the probabilities of type I and type II errors) should both be small. Type I error occurs when we reject a null hypothesis that is true and α (the significance level of the test) is the probability of making a type I error. Once a significance level is decided nothing can be done about α. Type II error occurs when we accept a null hypothesis that is false, the probability of type II error is β. The smaller the β, the better it is. Alternatively $(1 - \beta)$, i.e., the probability of rejecting a null hypothesis when it is false, should be as large as possible.

Since rejecting a null hypothesis when it is false is exactly what a good test ought to do, a high value of $1 - \beta$ (something near 1) means the test is working quite well (it is rejecting the null hypothesis when it is false). A low value of $1 - \beta$ (something near 0) means that the test is working very poorly (it is not rejecting the null hypothesis when it is false). $(1 - \beta)$ is the measure of how well the test is working and is called power of the test. If we plot the values of $(1 - \beta)$ for each value of μ for which the alternative hypothesis is true, the resulting curve is known as a *power curve.*

STANDARD ERROR AND SAMPLING DISTRIBUTION

Before discussing the various types of tests of hypothesis let us acquaint ourselves with the concept of standard error which is of fundamental importance in testing hypothesis.

The standard deviation of the sampling distribution is called the standard error. It is so called because it measures the sampling variability due to chance or random forces. Hence to clarify the term standard error it is necessary to describe a sampling distribution. If we select a number of independent random samples of a definite size from a given population and calculate some statistic (lie the mean, standard deviation, etc.) from each sample, we shall get a series of values of these statistics or functions. These values obtained from the different samples can be put in the form of a statistic is called the *sampling distribution* or the *probability distribution* of that statistic. Thus if we draw 100 random samples from a given population and calculate their means, we shall get a series of 100 means which would form a frequency distribution. This distribution will be known as the sampling distribution of the means.

An explanation of sampling distribution would be incomplete without describing the universe distribution, the sample distribution and showing the relationship of these two with the sampling distribution.

Universe Distribution

Such a distribution emerges when each and every item of the universe is studied, and we have full knowledge of its mean and standard deviation. The mean of the universe which is also called the true mean is symbolized by μ (the lower-case m in Greek and called mu) and its standard deviation by σ (lower case sigma). Greek letters are used for these measures to emphasize their difference from corresponding measures taken from a sample. Measures characterizing a universe, such as μ and σ, are called parameters.

The Sample Distribution

If instead of all the items of the universe we study only a small part of it i.e., take a sample, we will arrive at a sample distribution. The symbols $\overline{X}$ and S are used to designate the mean and standard deviation of the sample distribution. A measure characterizing a sample such as $\overline{X}$ or S, is called a statistic. It may be noted that several sample distributions are possible from a given universe.

The sampling distribution of a statistic reveals some important features:

(1) First, a sampling distribution is generated from a population distribution, known or assumed.

(2) Secondly, the same population may generate an infinite number of sampling distributions for the statistic, each for special sample size n.

(3) Finally, a population may generate distributions for two or more different statistics.

For example, the sampling distributions of mean has the following important properties :

(1) The sampling distribution of means is normally distributed. Since the sampling distribution of the mean is normally distributed, we can

develop a method in which we can use the mean of our sample to estimate the population mean. For example, if we compute an interval of $\overline{X} \pm 1.96\, \sigma\sqrt{n}$, the interval will include 95 per cent of all the sample means.

(2) The arithmetic mean of the sampling distribution is the same as the mean of the universe from which samples were taken. For this reason the mean of the sampling distribution may be denoted by the same symbol as that for the mean of the universe involved, namely μ.

(3) The sampling distribution of mean has a standard deviation (a standard error) equal to the population standard deviation divided by the square root of the sample size, i.e., $\frac{\sigma}{\sqrt{n}}$.

Utility of Concept of Standard Error

The concept of standard error is of great significance in statistical work because of the following reasons :

(1) Standard error provides an idea about the unreliability of a sample. The greater the standard error, the greater is the departure of actual frequencies from the expected ones and hence, the greater the unreliability of the sample. The reciprocal of S.E., i.e., $\frac{1}{\text{S.E.}}$, is a measure of reliability (or precision) of the sample. The reliability or precision of an observed proportion varies as the square root of the number of items in the sample. In other words, if we want to double the precision (which is the same thing as reducing the standard error to one-half) the number of observations should be increased four times.

(2) It is used as an instrument in testing a given hypothesis. The hypotheses are generally tested at 5 per cent level of significance. If the difference between observed and expected means is more than 1.90 standard error (S.E.), we say that the result of the experiment does not support the hypothesis at 5 per cent level or, in other words, the difference is regarded as significant, i.e., it could not have arisen due to fluctuations of sampling. On the other hand, if the difference between observed and expected results is less than 1.96 S.E., it is not regarded as significant, i.e., it could have arisen due to fluctuations of simple sampling, i.e., we say that the result of the experiment does not provide any evidence against the hypothesis. If the difference is more than 2.58 S.E., it is considered to be significant at 1 per cent level. In practice, quite often a hypothesis is accepted if the difference is less than 3 S.E., because the probability of a difference greater than 3 S.E., arising by chance, is only about 3 in thousand (0.27%)

as 99.73 per cent items are covered between mean ± 3s on either side of the mean. However, it must be emphasised that the use of 3s rule is justified only if n is large. Some people apply a criterion of 2 S.E. in order to determine whether or not the difference could have arisen due to fluctuations of sampling. However, instead of 3 S.E. or 2 S.E., it is suggested that we should use either 5 per cent level or 1 per cent level of significance. (In practice 5% level is more popular).

(3) With the help of S.E. we can determine the limits within which the parameter values are expected to lie. This is made possible because four large samples, sampling distributions tend to approximate a normal distribution. In a normal distribution 68.27 per cent of the samples will have their mean values (or any other constant) within a range of the population mean ± 1 standard deviation or standard error as it is alternatively called. Similarly a range of mean ± 2 S.E. will give 95.45 per cent values and mean ± 3 S.E. will give 99.73 per cent values. Thus a range of ± 3 S.E. should be taken as the determining limit outside which the value of the parameter probability does not fall. The chance of a value lying outside ± 3 S.E. limits is only 0.27% (i.e., approximately 3 in 1,000).

ESTIMATION

When data are collected by sampling from a population, the most important objective of statistical analysis is to draw inferences or generalisations about that population from the information embodied in the sample. Statistical estimation, or briefly estimation, is concerned with the methods by which population characteristics are estimated from sample information. It may be pointed out that the true value of a parameter is an unknown constant that can be correctly ascertained only by an exhaustive study of the population. However, it is ordinarily too expensive or it is infeasible to enumerate complete populations to obtain the required information. In case of finite populations, the cost of complete censuses may be prohibitive and in case of infinite population, complete enumerations are impossible. A realistic objective may be to obtain a guess or estimate of the unknown true value or an interval of plausible values from the sample data and also to determine the accuracy of the procedure. Statistical estimation procedures provide us with the means of obtaining estimates of population parameters with desired degrees of precision. With respect to estimating a parameter*. The following two types of estimates are possible:

1. Point estimates, and
2. Interval estimates.

Point Estimates

A point estimate is a single number which is used as an estimate of the unknown population parameter. The procedure in point estimation is to select a random sample of n observations, x_1, x_2...x_3 from a population f(x, θ^*) and then to use some preconceived method to arrive from these observations at a number say $\hat{\theta}$ (read theta hat) which we accept as an estimator of θ. The estimator θ is a single point on the real number scale and thus the name point estimation. $\hat{\theta}$ depends on the random variables that generate the sample and hence, it too is a random variable with its own sampling distribution.

Interval Estimates

As distinguished from a point estimate which provides one single value of the parameter, an interval estimate of a population parameter is a statement of two values between which it is estimated that the parameter lies. An interval estimate would always be specified by two values, i.e., the lower one and the upper one. In more technical terms, interval estimation refers to the estimation of a parameter by a random interval, called the *Confidence interval,* whose end points L and U with L < U, are functions of the observed random variables such that the probability that the inequality $L < \theta < U$ is satisfied in terms of pre-determined number, 1 – a, L and U are called the confidence limits and are the random end points of interval estimate. Since in an interval estimate, we determine an interval of plausible values, hence the name interval estimation. Thus, on the basis of sample study, if we estimate the average income of the people living in a village as Rs. 875 it will be a point estimate. On the other hand, if we say that the average income could lie between Rs. 800 and Rs. 950, it will be an interval estimate.

Properties of a Good Estimator

A distinction is made between an *estimate* and an *estimator*. The numerical value of the sample mean is said to be an estimate of the population mean figure. On the other hand, the statistical measure used, that is, the method of estimation, is referred to as an *estimator.* For example, the sample mean, $\overline{X}$ is an estimator of the population mean.

A good estimator, as common sense dictates, is close to the parameter being estimated. Its quality is to be evaluated in terms of the following properties :

(i) *Unbiasedness.* An estimator is said to be unbiased if its expected value is identical with population parameter being estimated. That is if $\hat{\theta}$ is an unbiased estimate of θ, then we must have $E(\hat{\theta}) = \theta$. Many estimators are "Asymptomatically unbiased" in the sense that the biases reduce to practically insignificant values zero when n becomes sufficiently large. The estimator S^2 is an example.

It should be noted that bias in estimation is not necessarily undesirable. It may turn out to be an asset in some situations. For example, it may happen that an unbiased estimator is less desirable than a biased estimator if the former has a greater variability than the latter and, as a consequence, the expected value of the latter is closer than that of the former to the parameter being estimated.

(ii) *Consitency*. If an estimator, say $\hat{\theta}$, approaches the parameter q closer and closer as the sample size n increases. $\hat{\theta}$ is said to be a consistent estimator of θ. Stating somewhat more regorously, the estimator $\hat{\theta}$ is said to be a consistent estimator of θ if, as n approaches infinity, the probability approaches I that $\hat{\theta}$ will differ from the parameter θ by not more than an arbitrary small constant.

The sample mean is an unbiased estimator of μ no matter what form the population distribution assumes, while the sample median is an unbiased estimate of m only if the population distribution is symmetrical.. The sample mean is better than the sample median as an estimate of μ in terms of both unbiasedness and consistency.

In case of large samples consistency is a desirable property for an estimator to possess. However in small samples, consistency is of little importance unless the limit of probability defining consistency is reached even with a relatively small size of the simple.

(iii) *Efficiency*. The concept of efficiency refers to the sampling variability of an estimator. If two competing estimators are both unbiased, the one with the smaller variance (for a given sample size) is said to be relatively more efficient. Stated in a somewhat different language, estimator $\hat{\theta}_1$ is said to be more efficient than another estimator $\hat{\theta}_2$ for θ if the variance of the first is less than the variance of the second. The smaller the variance of the estimator, the more concentrated is the distribution of the estimator around the parameter being estimated and, therefore, the better this estimator is.

(iv) *Sufficiency*. An estimator is said to be sufficient if it conveys as much information as is possible about the parameter which is contained in the sample. The significance of sufficiency lies in the fact that if a sufficient estimator exists, it is absolutely unnecessary to consider a sample can furnish with respect to the estimation of a parameter is being utilised.

Many methods have been devised for estimating parameters that may provide estimators satisfying these properties. The two important methods are the least square method and the method of maximum likelihood.

Having discussed the above concepts let us now discuss the various situations where we have to apply different tests of significance. For the sake of convenience and clarity these situations may be summed up under the following three heads :

- Test of Significance for Attributes.
- Tests of Significance for Variables (Large Samples).
- Tests of Significance for Variables (Small Samples).

The various tests of significance for attributes are discussed under the following heads :

(i) Tests for Number of Successes,

(ii) Tests for Proportion of Successes,

(iii) Tests for Difference between Properties.

Tests for Number of Successes : The sampling distribution of the number of successes follows a binomial probability distribution. Hence its standard error is given by the formula :

$$\text{S. E. of no. of successes} = \sqrt{npq}$$

where n = size of sample

p = probability of success in each trial

q = (1 – p), i.e., probability of failure.

Example 1:

In a hospital 480 female and 520 male babies were born in a week. Do these figures confirm the hypothesis that males and females are born in equal number?

Solution:

Let us take the hypothesis that male and female babies are born in equal number, i.e., $p = \frac{1}{2}$

$$\text{S.E.} = \sqrt{npq} = \sqrt{1000 \times \frac{1}{2} \times \frac{1}{2}} = 15.81$$

Difference between observed and expected number of female babies = 520 – 500 = 20

$$\frac{\text{Difference}}{\text{S.E.}} = \frac{20}{15.81} = 1.265.$$

Since the difference is less than 1.96 S.E. (5% level) it can be concluded that the male and female babies are born in equal number.

Example 2:

A person throws 10 dice 500 times and obtains 2560 times, 4, 5, or 6. Can this be attributed to fluctuations of sampling?

Solution:

Let us take the hypothesis that there is no significant difference in the sample results and the expected results.

$$\text{S.E.} = \sqrt{npq}$$

$$n = 5000,\ p = \frac{1}{2},\ \{q = \frac{1}{2}$$

$$\text{S.E.} = \sqrt{500 \times \frac{1}{2} \times \frac{1}{2}} = 35.36$$

$$\frac{\text{Difference}}{\text{S.E.}} = \frac{2560 - 2500}{35.36} = 1.7$$

Since the difference is less than 1.96 S.E. at 5% level of significance the hypothesis holds true. Hence the difference could be attributed to fluctuations of sampling.

Example 3:

(a) A coin was tossed 400 times and the head turned up 216 times. Test the hypothesis that the coin is unbiased.

Solution:

Let us take the hypothesis that the coin is unbiased. On the basis of this hypothesis the probability of getting head or tail would be equal, i.e., 1/2 * Hence in 400 throws of a coin we should expect 200 heads and 200 tails.

Observed number of heads = 216

Difference between observed number of heads and expected number of heads = 216 – 200 = 16

$$\text{S.E. of no. of heads} = \sqrt{npq}$$

$$n = 400,\ p = q = \frac{1}{2}$$

$$\therefore \text{S.E.} = \sqrt{400 \times \frac{1}{2} \times \frac{1}{2}} = 10$$

$$\frac{\text{Difference}}{\text{S.E.}} = \frac{16}{10} = 1.6$$

The difference between observed and expected number of heads is less than 1.96 S.E. (5% level of significance). Hence our hypothesis is true and we conclude that the coin is unbiased.

(b) In 324 throws of a xis-faced dice, odd points appeared 180 times. Would you say that the dice is fair at 5 per cent level of significance?

Solution:

Let us take the hypothesis that the dice is fair. In a fair dice we would expect 162 odd points in 324 throws

$$\text{S.E.} = \sqrt{npq} = \sqrt{324 \times \frac{1}{2} \times \frac{1}{2}} = 9$$

$$\frac{\text{Diff.}}{\text{S.E.}} = \frac{180-162}{9} = 2$$

Since the difference is more than 1.96 at 5 per cent level of significance, the hypothesis rejected. Hence we cannot say that the dice is fair at 5 per cent level of significance.

Example 4:

In a sample of 500 people from a village in Rajasthan, 280 are found to be rice eaters and the rest wheat eaters. Can we assume that both the food articles are equally popular?

Solution:

Let us take the hypothesis that the food articles are equally popular. The expected frequency for wheat eaters and rice eaters would be $\frac{500}{2} = 250$ each.

$$\text{S.E.} = \sqrt{npq} = \sqrt{500 \times \frac{1}{2} \times \frac{1}{2}} = 11.18$$

Difference between observed and expected number of rice eaters = 280 – 250 = 30

$$\frac{\text{Diff.}}{\text{S.E.}} = \frac{30}{11.18} = 2.68\,.$$

Since the difference is more than 2.58 S.E. (1% level), it could not have arisen due to fluctuations of sampling. Hence we cannot assume that both the food articles are equally popular.

Test for Proportion of Successes : Instead of recording the number of successes in each sample, we might record the proportion successes, that is $\frac{1}{n}$th of the number of successes in each sample. As this would amount to dividing all figures of the record by n, the mean proportion of successes must be p, and the standard deviation of the proportion of successes $\sqrt{\frac{pq}{n}}$. Thus we have the following formula:

$$S.E._p = \sqrt{\frac{pq}{n}}.$$

Example 5:

A wholesaler in apples claims that only 4% of the apples supplied by him are defective. A random sample of 600 apples contained 36 defective apples. Test the claim of the wholesaler.

Solution:

The wholesaler claims that only 4% of the apples are defective. Hence the 95% confidence limits are given by $\overline{X} \pm 1.96$ S.E.

$$S.E._p = \sqrt{\frac{pq}{n}} = \sqrt{\frac{.96 \times .04}{600}} = 0.008$$

$$\begin{aligned} \text{95\% confidence limits} &= \overline{X} \pm 1.96 \text{ S.E.} \\ &= .96 \pm 1.96\ (0.008) \\ &= .96 \pm 0.0157 \\ &= 0.9443 \text{ to } 0.9757. \end{aligned}$$

Out of 600 apples, the good apples may vary between $.9943 \times 600 = 566.50$ and $0.9757 \times 600 = 585.42$ i.e., 576 and 585.

The number of defectives is thus expected to lie between 15 and 33. Since the actual number is 36, the wholesaler's claim that only 4% of his apples are defective cannot be accepted.

Test for Difference between Properties : If two samples are drawn from different populations, we may be interested in finding out whether the difference between the proportion of successes is significant or not. In such a case we take the hypothesis that the difference between p_1, i.e., the proportion of successes in one sample, and p_2, i.e., the proportion of successes in another sample, is due to fluctuations of random sampling.

The standard error of the difference between proportions is calculated by applying the following formula :

$$S.E.\ (p_1 - p_2) = \sqrt{pq\left(\frac{1}{n_1} + \frac{1}{n_2}\right)}$$

where p = the pooled estimate of the actual proportion in the population. The value of p is obtained as follows :

$$p = \frac{n_1 p_1 + n_2 p_2}{n_1 + n_2} \text{ or } p = \frac{x_1 + x_2}{n_1 + n_2}$$

where x_1 and x_2 stand for the number of occurrences in the two samples of sizes n_1 and n_2 respectively.

If $\frac{p_1 - p_2}{S.E.}$ is less than 1.96 S.E. (5% level of significance), the difference is regarded as due to random sampling variation, i.e., as not significant.

The following illustrations shall explain the method.

Example 6:

In a random sample of 1,000 persons from town A, 400 are found to be consumers of wheat. In a sample of 800 from town B, 400 are found to be consumers of wheat. Do these data reveal a significant difference between town A and town B, so far as the proportion of wheat consumers is concerned?

Solution:

Let us set up the hypothesis that the two towns do not differ so far as proportion of what consumers is concerned, i.e., $H_0 : p_1 = p_2$ against $H_a : p_1 \neq p_2$.

Computing the standard error of the difference of proportions :

$$S.E. (p_1 - p_2) = \sqrt{pq\left(\frac{1}{n_1} + \frac{1}{n_2}\right)}$$

$$n_1 = 100, \; p_1 = \frac{400}{1000} = 0.4; \; n_2 = 800; \; p_2 = \frac{400}{800} = 0.5$$

$$p = \frac{(1000 \times 0.4) + (800 \times 0.5)}{1000 + 800} = \frac{400 + 400}{1800}$$

$$\text{or simply } p = \frac{x_1 + x_2}{n_1 + n_2} = \frac{400 + 400}{1000 + 800} = \frac{4}{9}$$

$$q = 1 - \frac{4}{9} = \frac{5}{9}$$

$$S.E. (p_1 - p_2) = \sqrt{\frac{4}{9} \times \frac{5}{9}\left(\frac{1}{1000} + \frac{1}{800}\right)} = \sqrt{\frac{20}{21} \times \frac{9}{4000}} = 0.024$$

$$p_1 - p_2 = 0.4 - 0.5 = -0.1*$$

$$\frac{\text{Difference}}{S.E.} = \frac{0.1}{0.024} = 4.17$$

Since the difference is more than 2.58 S.E. (1% level of significance). It could not have arisen due to fluctuations of sampling. Hence the data reveal a significant difference between town A and town B so far as the proportion of wheat consumers is concerned.

Example 7:

In a random sample of 1000 persons from U.P. 510 were found to be consumers of cigarettes. In another sample of 800 persons from Rajasthan, 480 were found to be consumers of cigarettes. Discuss the data reveal a significant difference between U.P. and Rajasthan so far as the proportion of consumers of cigarettes in concerned.

Solution:

Let us take the hypothesis that there is no significant difference between UP and Rajasthan so far as the proportion of consumers of cigarettes is concerned.

$$S.E.(p_1 - p_2) = \sqrt{pq\left(\frac{1}{n_1} + \frac{1}{n_2}\right)}$$

$$p_1 = \frac{510}{1000} = 0.51$$

$$p_2 = \frac{480}{800} = 0.60$$

$$p = \frac{x_1 + x_2}{n_1 + n_2} = \frac{510 + 480}{1000 + 800} - 0.55$$

$$q = 1 - p = 1 - .55 = .45$$

$$S.E. = \sqrt{.55 + .45\left(\frac{1}{1000} + \frac{1}{800}\right)} = 0.024$$

$$\frac{\text{Diff.}}{\text{S.E.}} = \frac{|.51 - .6|}{.024} = 3.75$$

Since the difference is more than 2.58 S.E. at 1% level of significance the hypothesis is rejected. Hence the data reveal a significant difference between UP and Rajasthan so far as proportion of consumers of cigarettes is concerned.

At times we may be interested in comparing the proportion of persons possessing an attribute in a sample with the proportion given by the population. In such a case the following formula is applicable :

$$S.E.\ (p_1 - p_2) = \sqrt{p_0 q_0 \times \frac{n_2}{n_1\left(n_1 + n_2\right)}}$$

where p_0 = Population proportion

$q_0 = 1 - p_0$

n_1 = Number of observations in the sample

$n_1 + n_2$ = Size of population

n_2 = (Size of population – n_1).

Example 8:

500 apples are taken at random from a large basket and 50 are found to be bad. Estimate the proportion of bad apples in the basket and assign limits within which the percentage most probably lies.

Solution:

The proportion of bad apples in the given sample = $\frac{50}{500} = 0.1$.

Hence p = 0.1 and q = 0.9

$$\text{S.E.} = \sqrt{\frac{pq}{n}} = \sqrt{\frac{0.1 \times 0.9}{500}} = \sqrt{\frac{0.09}{500}} = 0.013$$

The limits within which percentage of bad apples lies

$$= \left[p \pm 3\sqrt{\frac{pq}{n}}\right] \times 100$$

$= 0.1 \pm (0.013)] \times 100$

$= 0.1 \pm (0.039)] \times 100$

$= 6.1$ and 13.9

Thus the percentage of bad apples in the consignment almost certainly lies between 56.1 and 13.9.

Example 9:

In a simple random sample of 600 men taken from a big city 400 are found to be smokers. In another simple random sample of 900 men taken from another city 450 are smokers. Do the data indicate that there is a significant difference in the habit of smoking in the two cities?

Solution:

Let us take the hypothesis that there is no significant difference in the habit of smoking in the two cities.

$p_1 - p_2 = 0.4 - 0.5 = -0.1$. However, the conclusion remains the same irrespective of the fact whether the difference is positive or negative.

S.E. $(p_1 - p_2)$ =

$$p_1 = \frac{400}{600} = 0.667;\ p_2 = \frac{400}{900} = 0.5$$

$$p = \frac{x_1 + x_2}{n_1 + n_2},\ q = 1 - p$$

$$n_1 = 600,\ n_2 = 900,\ x_1 = 400,\ x_2 = 450$$

$$p = \frac{400 + 450}{600 + 900} = \frac{850}{1500} = \frac{17}{30}$$

$$q = 1 - \frac{17}{30} = \frac{13}{30}$$

$$\text{S.E. } (p_1 - p_2) = \sqrt{\frac{17}{30} \times \frac{13}{30}\left(\frac{1}{600} + \frac{1}{900}\right)}$$

$$= \sqrt{\frac{17}{30} \times \frac{13}{30} \times \frac{1500}{600 \times 900}} = 0.026$$

Since the difference is more than 2.58 S.E. at 1% level of significance, the hypothesis is rejected. Hence there is a significant difference in the habit of smoking of the two cities.

Example 10:

A machine produced 20 defective articles in a batch of 400. After overhauling it produced 10 defectives in a batch of 300. Has the machine improved?

Solution:

Let us take the hypothesis that the machine has not improved after overhauling. Applying test of difference of proportions.

$$p_1 = \frac{20}{400} = 0.050$$

$$p_2 = \frac{10}{300} = 0.033$$

Pooled estimate of actual proportion in the population.

$$p = \frac{x_1 + x_2}{n_1 + n_2} = \frac{20 + 10}{400 + 300} = 0.043$$

$$q = 1 - p = 1 - 0.43 = 0.957$$

$$\text{S.E. } (p_1 - p_2) = \sqrt{pq\left(\frac{1}{n_1} + \frac{1}{n_2}\right)}$$

$$= \sqrt{.043 \times .957\left(\frac{1}{400} + \frac{1}{300}\right)} = 0.0155$$

$$\frac{\text{Diff.}}{\text{S.E.}} = \frac{p_1 - p_2}{.0155} = \frac{0.05 - 0.033}{0.0155} = 1.1$$

Since the difference is less than 1.96 S.E. (at 5% level) our hypothesis is true. Machine has not improved after overhanding.

Example 11:

Before an increase in excise duty on tea 400 people of a sample of 500 persons were found to be tea drinkers. After an increase in the duty, 400 persons were known to be tea drinkers in a sample of 600 people. Do you think that there has been a significant decrease in the consumption of tea after the increase in the excise duty.

Solution:

Let us take the hypothesis that there is no significant decrease in the consumption of tea after increase in the excise duty.

Computing the standard error of the difference of proportions

$$\text{S.E. } (P_1 - p_2) = \sqrt{pq\left(\frac{1}{n_1} + \frac{1}{n_2}\right)}$$

$$p_1 = \frac{400}{500} = 0.8;\ p_2 = \frac{400}{600} = 0.667$$

$$p = \frac{400 + 400}{500 + 600} = \frac{8}{11};\ q = 1 - \frac{8}{11} = \frac{3}{11}$$

$$\text{S.E. } (p_1 - p_2) = \sqrt{\frac{8}{11} \times \frac{3}{11}\left(\frac{1}{500} + \frac{1}{600}\right)} = 0.027$$

$$\frac{p_1 - p_2}{\text{S.E.}} = \frac{0.8 - 0.667}{0.027} = 4.93.$$

Since the difference is more than 2.58 S.E. (1% level of significance) it could not have arisen due to fluctuations of sampling. Hence, there is significant decrease in the consumption of tea after the increase in excise duty.

Example 12:

There are 1,000 students in a college. Out of 20,000 in the whole university, in a study 200 were found smokers in the college and 1,000 in the whole university. Is there a significant difference between the proportion of smokers in the college and university?

Solution:

Let us take the hypothesis that there is no significant difference in the proportion of smokers in the college and university. Proportion of smokers in the college.

Proportion of smokers in the university

$$= \frac{1000}{20000} = 0.05$$

Difference between the two proportions

$$= 0.2 - 0.05 = 0.15$$

p_0, i.e., proportion of smokers in the university = 0.05

$q_0 = 1 - p_0 = 0.95,\ n_1 = 1000,\ n_1 + n_2 = 20000$

So $\quad n_2 = 20000 - 100 = 19000$

$$\text{S.E. of difference} = \sqrt{p_0 q_0 \times \frac{n_2}{n_1(n_1 + n_2)}}$$

$$= \sqrt{0.05 \times 0.95 \left(\frac{19000}{1000 + 20000}\right)} = \sqrt{0.000015} = 0.0067$$

$$\frac{\text{Difference}}{\text{S.E.}} = \frac{0.15}{0.0067} = 22.39\ .$$

Since the difference is more than 2.58 S.E. (1% level of significance) it could not have arisen due to fluctuations of sampling. Hence there is a significant difference between the proportion of smokers in the college and university–the proportion being significantly higher in the college.

There are three main objects in studying problems relating to sampling of variables :

(1) To compare observation with expectation and to see how far the deviation of one from the other can be attributed to fluctuations of sampling;

(2) To estimate from samples some characteristic of the parent population, such as the mean of a variable; and

(3) To gauge the reliability of our estimates.

Difference between Small and Large Samples

In this Bection we shall be studying problem relating to large samples only. Though it is difficult to draw a clear-cut line of demarcation between large and small samples, it is normally agreed amongst statisticians that a sample is to be recorded as large only if its size exceeds 30. The tests of different from the ones used for small samples for the reason that the assumptions that we make in case of large samples do not hold good for small samples. The assumptions made while dealing with problems relating to large samples are :

(i) The random sampling distribution of a statistic is approximately normal, and

(ii) Values given by the samples are sufficiently close to the population value and can be used in its place for calculating the standard error of the estimate.

While testing the significance of a statistic in case of large samples, the concept of standard error discussed earlier is used. The following is a list of the formulae for obtaining standard error for different statistics:

(1) Standard Error of Mean*

(i) When standard deviation of the population is known

$$\text{S.E. } \overline{X} = \frac{\sigma_p}{\sqrt{n}}$$

where S.E. $\overline{X}$ referes to the standard error of the mean

s_p = Standard deviation of the population

n = Number of observations in the sample.

(ii) When standard deviation of population is not known, we have to use standard deviation of the sample in calculating standard error of mean. Consequently, the formula for calculating standard error is

$$\text{S.E. } \overline{X} = \frac{\sigma\,(\text{sample})}{\sqrt{n}}$$

where s denotes standard deviation of the sample.

It should be noted that if standard deviations of both sample as well as population are available then we should prefer standard deviation of the population for calculating standard error of mean.

Feudalize limits of population mean :

95% fiducial limits of population mean are

$$\overline{X} \pm 1.96\,\frac{\sigma}{\sqrt{n}}$$

99% fiducial limits of population mean are

$$\overline{X} \pm 2.58\,\frac{\sigma}{\sqrt{n}}$$

(2) S.E. of Median or $S.E._{Med} = 1.25331 \dfrac{\sigma}{\sqrt{n}}$

(3) S.E. of Quartile or $S.E._{a} = 1.36263 \dfrac{\sigma}{\sqrt{n}}$

(4) S.E. of Quartile Deviation of $S.E._{QD} = 0.78672 \dfrac{\sigma}{\sqrt{n}}$

(5) S.E. of Mean Deviation or $S.E._{MD} = 0.6028 \dfrac{\sigma}{\sqrt{n}}$

(6) S.E. of Standard Deviation or $S.E._{\sigma} = \dfrac{\sigma}{\sqrt{2n}}$

* The standard error of the mean measures only sampling errors. Sampling errors are errors involved in estimating a population parameter from a sample instead of including all the essential information in the population.

(7) S.E. of Variance or $S.E._{\sigma}{}^{2} = \sigma^2 \sqrt{\dfrac{2}{n}}$

(8) S.E. of Coefficient of Correlation or $S.E._{r} = \dfrac{1-r^2}{\sqrt{n}}$

(9) S.E. of Regression Coefficient of Y on X, i.e., S.E. b_{yx}

$$= \frac{\sigma_x \sqrt{1-r^2}}{\sigma_x \sqrt{n}}$$

(10) S.E. of Regression Coefficient of X on Y, i.e., S.E. b_{XY}

$$= \frac{\sigma_x \sqrt{1-r^2}}{\sigma_y \sqrt{n}}$$

(11) S.E. of Regression Estimate of Y on X or $S.E._{xy} = \sigma_x \sqrt{1-r^2}$

(12) S.E. of Regression Estimate of X on Y or $S.E._{yx} = \sigma_y \sqrt{1-r^2}$

(13) S.E. of Coefficient of Association, i.e., S.E.Q.

$$= \frac{1-Q^2}{2} \sqrt{\frac{1}{(AB)} + \frac{1}{(A\beta)} + \frac{1}{(\alpha B)} + \frac{1}{(\alpha\beta)}}$$

(14) S.E. of Rank Correlation Coefficient, i.e., $S.E._{m}$

$$= \frac{1}{\sqrt{n-1}}$$

The following examples will illustrate how standard error of some of the statistics is calculated :

Example 13:

If it costs a rupee to draw one number of a sample, how much would it cost in sampling from a universe with mean 100 and standard deviation 10 to take sufficient number as to ensure that the mean of a sample would in a 5 per cent probability be within 0.01 per cent of the true? Find the extra cost necessary to double this precision.

Solution:

$$\overline{X} = 100\%$$

Difference between universe mean and sample mean = 0.01 per cent

$$\text{S.E. } \overline{X} = \frac{\sigma}{\sqrt{n}}$$

For 95% confidence, difference between sample mean and population mean should be equal to 1.96 S.E., i.e.,

$$1.96 \times \frac{10}{\sqrt{n}}$$

$$\therefore \quad \frac{1.96 \times 10}{\sqrt{n}} = 0.01 \text{ or } \sqrt{n} \times 0.01 = 19.6$$

$$\sqrt{n} = \frac{19.6}{0.01} = 1960$$

or $\qquad n = (1960)^2 = 38{,}41{,}600$

Double the precision would imply that the difference between sample mean and population mean should be = 0.005 only.

$$\frac{1.96 \times 10}{\sqrt{n}} = 0.005 \text{ or } \sqrt{n} = \frac{19.6}{0.005} = 3920$$

$\therefore n = (3920)^2 = 1{,}53{,}66{,}400$

$\therefore$ Extra cost = 1,53,66,400 – 38,41,600

= Rs. 1,15,24,800

Hence in order to double the precision the extra cost required is Rs. 1,15,24,800.

Example 14:

A sample of 100 tyres is taken from a lot. The mean life of tyres is found to be 39,350 kms. with a standard deviation of 3,260. Could the sample come from a population with mean life of 40,000 kms? Establish 99% confidence limits within which the mean life of tyres is expected to lie.

Solution:

Let us take the hypothesis that there is no significant difference between the sample mean and the hypothetical population mean.

$$\text{S.E.}\quad \overline{x} = \frac{\sigma}{\sqrt{n}} = \frac{3260}{\sqrt{100}} = 326$$

$$\frac{\text{Diff.}}{\text{S.E.}} = \frac{40{,}000 - 39{,}350}{326} = \frac{650}{326} = 1.994\,.$$

Since the difference is less than 2.58 S.E. (at 1% level of significance) the hypothesis is accepted and the difference could have arisen due to fluctuations of sampling.

99% confidence limits $\overline{X} \pm 2.58$ (S.E.)

$$39350 + 2.58\ (326)$$

$$39350 \pm 841.08$$

$$38508.92 - 40191.08$$

Hence the mean life of tyres is expected to lie between 38,509 and 40191.

Example 15:

Find the regression coefficient of Bombay prices over Calcutta prices from the following data and also calculate its standard error:

	Bombay Rs.	*Calcutta Rs.*
Average price per quintal of wheat	*120*	*130*
Standard deviation	*4*	*5*
Coefficient of correlation	*= 0.6*	
n	*= 100*	

Solution:

Regression coefficient of Bombay prices (X) over Calcutta prices (Y)

$$b_{xy} = r\frac{\sigma_x}{\sigma_y} = 0.6\,\frac{4}{5} = 0.48$$

The standard error of regression coefficient is

$$\text{S.E.}_{b_{xy}} = \frac{\sigma_x\sqrt{1-r^2}}{\sigma_y\sqrt{n}} = \frac{4\sqrt{1-(0.6)^2}}{5\sqrt{100}} = \frac{4 \times 0.8}{50} = 0.064$$

Thus, the regression coefficient of Bombay prices over Calcutta prices is 0.48 and its standard error is 0.064.

Two-tailed Test for Difference Between the Means of Two Samples

(1) If two random samples with $\overline{X}_1$, σ_1, n_1 and $\overline{X}_2$, σ_2, n_2 respectively are drawn from different populations, then the S.E. of the difference between the mean is given by the formula :

$$= \sqrt{\frac{\sigma_1^2}{n_1} + \frac{\sigma_2^2}{n_2}}$$

and where σ_1 and σ_2 are unknown.

S.E. of the difference between means

$$= \sqrt{\frac{S_1^2}{n_1} + \frac{S_2^2}{n_2}}$$

where S_1 and S_2 represent standard deviations of the two samples.

The null hypothesis to be tested is that there is no significant difference in the means of the two samples, i.e.,

$H_0 : \mu_1 = \mu_2 \leftarrow$ null hypothesis, there is no difference

$H_1 : \mu_1 \neq \mu_2 \leftarrow$ alternative hypothesis, a difference exists.

Example 16:

An auto company decided to introduce a new six cylinder car whose mean petrol consumption is claimed to be lower than that of the existing auto engine. It was found that the mean petrol consumption for the 50 cars was 10 km per litre with a standard deviation of 3.5 km per litre. Test for the company at 5% level of significance, whether the claim the new car petrol consumption is 9.5 km per litre on the average is acceptable.

Solution:

Let us take the hypothesis that there is no significant difference between the sample average and the company's claim, i.e.,

$$H_0 = \overline{X} - m' = 0$$

$$\text{S.E. } \overline{X} = \frac{\sigma}{\sqrt{n}} = \frac{3.5}{\sqrt{50}} = \frac{3.5}{7.07} = 0.495$$

$$\frac{\text{Difference}}{\text{S.E.}\overline{X}} = \frac{10 - 9.5}{0.495} = 1.01$$

Since the difference is less than 1.96 S.E. at 5% level of significance, the hypothesis is accepted. Hence the company's claim that the new car petrol consumption is 9.5 km. per litre is acceptable.

Example 17:

Two samples of 100 electric bulbs each has a means 1500 and 1550, Standard deviation 50 and 60. Can it be concluded that two brands differ significantly at 1% level of significance in equality.

Solution:

Let us take the hypothesis that there is no significant difference in the mean life of the two makes of bulbs.

$$\text{S.E. } \left(\overline{X} - \overline{X}_2\right) = \sqrt{\frac{\sigma_1^2}{n_1} + \frac{\sigma_2^2}{n_2}}$$

$$= \sqrt{\frac{(50)^2}{100} + \frac{(60)^2}{100}} = \sqrt{25 + 36} = 7.81$$

$$\frac{\text{Difference}}{\text{S.E.}} = \frac{1550 - 1500}{7.81} = 6.4$$

Since the difference is more than 2.58 S.E. (1% level of significance), the hypothesis is rejected. Hence there is a significant difference in the mean life of the two brands of bulbs.

Example 18:

Intelligence test on two groups of boys and girls gave the following results :

	Mean	*S.D.*	*N*
Girls	*75*	*15*	*150*
Boys	*70*	*20*	*250*

Is there a significant difference in the mean scores obtained by boys and girls?

Solution:

Let us take the hypothesis that there is no significant difference in the mean scores obtained by boys and girls.

$$\text{S.E. } \left(\overline{X} - \overline{X}_2\right) = \sqrt{\frac{\sigma_1^2}{n_1} + \frac{\sigma_2^2}{n_2}}$$

$\sigma_1 = 15$, $\sigma_2 = 20$, $n_1 = 150$, $n_2 = 250$.

Substituting the values

$$\text{S.E., } \left(\overline{X} - \overline{X}_2\right) = \sqrt{\frac{(15)^2}{150} + \frac{(20)^2}{250}} = \sqrt{1.5 + 1.6} = 1.761$$

$$\frac{\text{Difference}}{\text{S.E.}} = \frac{75-70}{1.781} = 2.84$$

Since the difference is more than 2.58 S.E. (1% level of significance), the hypothesis is rejected. There seems to be a significant difference in the mean scores obtained by boys and girls.

(b) A man buys 50 electric bulbs of 'Philips' and 50 electric bulbs of 'HMT'. He finds that 'Philips' bulbs give an average life of 15000 hours with a standard deviation of 60 hours and 'HMT' bulbs gave an average life of 1512 hours with a standard deviation of 80 hours. Is there a significant difference in the mean life of the two makes of bulbs?

Solution:

Let us set up the hypothesis that there is no significant difference in the mean life of the two makes of bulbs. Calculating standard error of difference of means

$$\text{S.E. } \left(\overline{X} - \overline{X}_2\right) = \sqrt{\frac{\sigma_1^2}{n_1} + \frac{\sigma_2^2}{n_2}}$$

$\sigma_1 = 60,\ n_1 = 50,\ s_2 = 80,\ n_2 = 50$

$$\text{S.E. } \left(\overline{X}_1 - \overline{X}_2\right) = \sqrt{\frac{(60)^2}{50} + \frac{(80)^2}{50}} = \sqrt{\frac{3600 + 6400}{50}}$$

$$= \sqrt{200} = 14.14$$

Observed difference between the two means = 512 – 1500 = 12

$$\frac{\text{Difference}}{\text{S.E.}} = \frac{1}{14.14} = 0.849\,.$$

Since the difference is less than 2.58 S.E. (1% level of significance), it could have arisen due to fluctuations of sampling. Hence the difference in the mean life of the two makes is not significant.

In case of two large random samples, each drawn from a normally distributed population, the S.E. of the difference between the standard deviations is given by :

$$\text{S.E.}_{(\sigma_1-\sigma_2)} = \sqrt{\frac{\sigma_1^2}{2n_1} + \frac{\sigma_2^2}{2n_2}}$$

Where population standard deviations are not known

$$\text{S.E.}_{(S_1-S_2)} = \sqrt{\frac{S_1^2}{2n_1} + \frac{S_2^2}{2n_2}}\,.$$

Example 19:

The mean produce of wheat of a sample of 100 fields is 200 lb. per acre with a standard deviation of 10 lb. Another sample of 150 fields gives the mean at 220 with a standard deviation of lb. Assuming the standard deviation of the mean field at 1 lb. of the universe, find at 1% level if the two results are consistent.

Solution:

$$S.E._{(\sigma_1-\sigma_2)} = \sqrt{\frac{\sigma^2}{2}\left(\frac{1}{n_1}+\frac{1}{n_2}\right)} = \sqrt{\frac{(11)^2}{2}\left(\frac{1}{100}+\frac{1}{150}\right)}$$

$$= \sqrt{\frac{60.5}{100}+\frac{60.5}{150}} = \sqrt{0.605 + .403} = 1.004$$

Difference between two standard deviations = (12 – 10) = 2 lb.

$$\frac{\text{Difference}}{\text{S.E.}} = \frac{2}{1.004} = 1.992\,.$$

Since the difference is less than 2.58 S.E. (1% level of significance) the two results may be regarded as consistent.

$$S.E.\bar{x} = \frac{13.11}{\sqrt{100}} = \frac{13.11}{10} = 1.311\,.$$

Example 20:

The mean height obtained from a random sample of size 100 is 64 inches. The standard deviation of the distribution of height of the population is known to be 3 inches. Test the statement that the mean height of the population is 67 inches at 5% level of significance. Also set up 99% limits of the mean height of the population.

Solution:

Let us take the hypothesis that there is no significant difference between the sample mean and the population mean.

$$S.E.\bar{x} = \frac{\sigma}{\sqrt{n}} = \frac{3}{\sqrt{100}} = 0.3$$

$$\frac{\text{Diff.}}{\text{S.E.}} = \frac{67-64}{0.3} = 10$$

Since the difference is more than 1.96 S.E. (5% level) the hypothesis is rejected. Hence the mean height of the population could not be 67 inches.

99% probable limits of the mean height of the population

$= \overline{X} \pm 2.58$ S.E.

$= 64 \pm 2.58\ (.3)$

$= 64 \pm 0.774 = 63.2$ to 64.8.

Example 21:

To study the correlation between the stature of father and the stature of son, a sample of 1,600 is taken from the universe of fathers and sone. The sample study gives the correlation between the two to be 0.80. Within what limits does it hold true for the universe?

Solution:

The standard error of the correlation coefficient between the stature of the father and son is

$$\text{S.E.}_{\cdot r} = \frac{1-r^2}{\sqrt{n}}$$

$r = 0.8$, $n = 1{,}600$

$$\text{S.E.}_{\cdot r} = \frac{1-(0.8)^2}{\sqrt{1600}} = \frac{1-0.64}{40} = \frac{0.36}{40} = 0.009\ .$$

If the sampling was simple random sampling the correlation in the universe cannot deviate from the correlation in the sample by more than thrice the standard error, i.e., .027. Hence correlation in the universe most probably lies between r ± 3 S.E. or S ± 0.027 or 0.773 and 0.827.

Example 22:

From the following data compute the value of Yule's Coefficient of Association and find out its standard error. Also determine 95 per cent fiducial limits of the true values in the population :

Married and failures = *100*

Married and non-failures = *50*

Unmarried and failures = *20*

Unmarried and non-failures = *80*

Solution :

Let A denote married and B failures

∴ α will denote unmarried and β non-failures. The given information in terms of these symbols is

$(AB) = 100$, $(A\beta) = 50$, $(\alpha\beta) = 20$, $(\alpha\beta) = 80$

$$Q = \frac{(AB)\,(\alpha\beta) - (A\beta)\,(\alpha B)}{(AB)\,(\alpha\beta) + (A\beta)\,(\alpha B)}$$

$$= \frac{(100 \times 80) - (50 \times 20)}{(100 \times 80) + (50 \times 20)} = \frac{7{,}000}{9{,}000} = +\ 0.778$$

$$\text{S.E.} = \frac{1-Q^2}{2}\sqrt{\frac{1}{(AB)} + \frac{1}{(A\beta)} + \frac{1}{(\alpha B)} + \frac{1}{(\alpha\beta)}}$$

$$= \left\{\frac{1-(0.778)^2}{2}\right\}\sqrt{\frac{1}{100} + \frac{1}{50} + \frac{1}{20} + \frac{1}{80}}$$

$$= \left\{\frac{1-0.605}{2}\right\}\sqrt{0.01 + 0.02 + 0.05 + 0.0125}$$

$$= 0.1975\sqrt{0.0925}$$

$$= 0.1975 \times 0.304 = 0.06.$$

95% fiducial limits :

Q ± 1.96 S.E. will give the required limits

0.778 ± 1.96 (0.06)

0.778 ± 0.1176 = 0.66 and 0.896.

Example 23:

The mean life time of a sample of 400 fluorescent light bulbs produced by a company is found to be 1,570 hours with a standard deviation of 150 hours. To the hypothesis that the mean life time of the bulbs produced by the company is 1600 hours against the alternative hypothesis that it is greater than 1,600 hours at 1% level of significance.

Solution:

Let us take the hypothesis that there is no significant difference between the sample mean and hypothetical population mean i.e. $X_1 - \mu = 0$

$$\text{S.E.}\bar{x} = \frac{\sigma}{\sqrt{n}} = \frac{150}{\sqrt{400}} = \frac{150}{20} = 7.5$$

$$\frac{\text{Diff.}}{\text{S.E.}} = \frac{[1570 - 1600]}{7.5} = 4$$

The difference is more than 2.58 S.E. at 1% level of significance. The hypothesis is rejected. Hence there is significant difference between sample mean and hypothetical population mean.

Example 24:

(a) A simple sample of the height of 6,400 Englishmen has a mean of 67.85 inches and a standard deviation of 2.56 inches while a simple sample

of heights of 1600 Austrians has a mean of 68.55 inches and standard deviation of 2.52 inches. Do the data indicate that the Austrians are on the average taller than the Englishmen? Give reasons for your answer.

Solution:

Let us take the hypothesis that there is no significant difference in the mean height of *Englishmen and Austrians.*

$$S.E.\left(\overline{X}_1 - \overline{X}_2\right) = \sqrt{\frac{\sigma_1^2}{n_1} + \frac{\sigma_2^2}{n_2}}$$

$\sigma_1 = 2.56,\ n_1 = 6400,\ \sigma_2 = 2.52,\ n_2 = 1600$

$$S.E. = \sqrt{\frac{(2.56)^2}{6400} + \frac{(2.52)^2}{1600}} = \sqrt{.001 + .004} = .0707$$

$$\frac{\text{Diff.}}{\text{S.E.}} = \frac{|67.85 - 68.55|}{.0707} = \frac{0.7}{.0707} = 9.9$$

Since the difference is more than S.E. (at 1% level of significance), the hypothesis is rejected. Hence the data indicates that the Austrians are on the average taller than the Englishmen.

(b) In a survey buying habits, 400 women shoppers are chosen at random in super market A located in a certain section of Bombay city. Their average monthly food expenditure is Rs. 250 with a standard deviation of Rs. 40. For 400 women shoppers chosen at random in super market B in another section of the city, the average monthly food expenditure is Rs. 220 with a standard deviation of Rs. 55. Test at 1% level of significance whether the average food expenditure of the two populations of shoppers from which the samples were obtained are equal.

Solution:

Let us take the hypothesis that there is no difference in the average food expenditure of the two populations of shoppers.

$$S.E.\left(\overline{X}_1 - \overline{X}_2\right) = \sqrt{\frac{\sigma_1^2}{n_1} + \frac{\sigma_2^2}{n_2}}$$

$n = 400,\ \overline{X}_1 = 250,\ \sigma_1 = 40;\ n_1 = 400,\ \overline{X}_2 = 220,\ \sigma_2 = 55$

Substituting the values

$$S.E.\left(\overline{X}_1 - \overline{X}_2\right) = \sqrt{\frac{(40)^2}{400} + \frac{(55)^2}{400}} = \sqrt{\frac{1600}{400} + \frac{3025}{400}} = 3.4$$

Difference of Means = $\left(\overline{X}_1 - \overline{X}_2\right)$ = 250 – 220 = 30

$$\frac{\text{Difference}}{\text{S.E.}} = \frac{30}{3.4} = 8.82.$$

Since the difference is more than 2.58 S.E. (1% level) the hypothesis is rejected. Hence the average food expenditures of the populations of shoppers are not equal.

Example 25:

For a given group of adults, the coefficient of correlation between height and weight is 0.7, standard deviations of height and weight are 2 inches and 10 1b. respectively and the means of height and weight for the entire group are 70 inches and 130 lb. respectively. Find out the best estimate of the weight of an individual who is 65 inches tall. Assign limits to this estimate in which in all probability his actual weight would be lying.

Solution:

The regression of weight (Y) on height (X) is

$$Y - Y - \overline{Y} = r\frac{\sigma_y}{\sigma_x}\left(X - \overline{X}\right)$$

$$\overline{Y} = 130\ \overline{X} = 70,\ s_y = 10,\ s_x = 2,\ r = 0.7$$

$$Y - 130 = 0.7\ \frac{10}{2}(X - 70)$$

$$Y - 130 = 3.5X - 245 \text{ or}$$

$$Y = 3.5X - 115$$

when $X = 65$, Y will be

$$Y = 3.5\ (65) - 115 = 112.5 \text{ lb.}$$

The standard error of estimate of Y on X or $S.E._{yx}$ is :

$$S.E._{yx} = \sigma_y\sqrt{1 - r^2}$$

$$= 10\sqrt{1 - (0.7)^2} = 10\sqrt{0.51} = 10 \times 0.714 = 7.14 \text{ lb.}$$

Thus the best estimate of the weight of an individual who is 65 inches tall is 112.5 lb. and this estimate cannot deviate from the actual weight by more than three times the standard error. Hence in all probability his actual weight would be lying between 112.5 ± 35 (1.428) or 112.5 ± 4.284, i.e., 108.2 lbs and 116.8 lbs.

Interested not so much in the value of the correlation in the parent population, but more generally whether this value could have arisen from an uncorrelated population, i.e. whether it is significant of correlation in the parent population.

The Assumption of Normality

While dealing with small samples also, an assumption is made that the parent population is normal, unless otherwise stated. Strictly speaking, therefore, our results will be true only for the normal population. However, as pointed out earlier, the assumption of normality is not very much warranted in case of small samples. Experiments have, therefore, been made to ascertain whether the results are true for other types of population. Theoretical work confirms that the results remain true for populations which do not deviate markedly from normality. However, if there is any good reason to suspect that the parent population is markedly skew, i.e., U, or J-shaped, the methods given below cannot be applied with much confidence.

Since in many of the problems it becomes necessary to take a small size sample, considerable attention has been paid in developing suitable tests for dealing with problems of small samples. The greatest contribution to the theory of small samples is that of Sir William Gusset and R.A. Fasher. Sir William Gusset published his discovery in 1905 under the pen name 'Student'. He gave a test popularly known as 't-test' and Fisher gave another test known as 'z-test'. These tests are based on 't'-distribution and 'z'-distribution.

Student's t-Distribution

Theoretical work on t-distribution was done by W.S. Gusset (1876 – 1937) in the early 1900. Gusset was employed by the Guinness & Son, a Dublin bravery, Ireland, which did not permit employees to publish research findings under their own names. So Gusset adopted the pen name "student" and published his findings under this name. Thereafter, the t-distribution is commonly called *Student's t-distribution* or simply *Student's distribution.*

The t-distribution is used when sample size is 30 or less and the population standard deviation is unknown.

The "t-statistic" is defined as :

$$t = \frac{\overline{X} - \mu}{S} \times \sqrt{n}$$

where $S = \sqrt{\frac{\Sigma\left(X - \overline{X}\right)^2}{n - 1}}$

The t-distribution has been derived mathematically under the assumption of a normally distributed population.

It has the following form

$$f(t) = C\left(1+\frac{t^2}{v}\right)^{-(v+1)/2}$$

$$\text{where } t = \frac{(\overline{X}-\mu)}{S}\sqrt{n}$$

C = a constant required to make the area under the curve equal to unity v = n − 1, the number of degrees of freedom.

Properties of t-Distribution

(1) The variance of the t-distribution is greater than one, but approaches one as the number of degrees of freedom and, therefore, the sample size becomes large. Thus the variance of the t-distribution approaches the variance of the standard normal distribution as the sample size increases. It can be demonstrated that from an infinite number of degrees of freedom ($v - \infty$), the t-distribution and normal distribution are exactly equal. Hence there is a widely practised rule of thumb that samples of size $n > 30$ may be considered be used as an approximation to t-distribution, where the latter is the theoretically correct functional form.

(2) Like the standard normal distribution, the t-distribution is symmetrical and has a mean zero.

(3) The constant c is actually a function of v (pronounced as nu), so that for a particular value of v, the distribution of f(t) is completely specified. Thus f(t) is a family of functions, one for each value of v.

(4) The variable t-distribution ranges from minus infinity to plus infinity.

The t-Table: The t-table given at the end is the probability integral of t-distribution. It gives, over a range of values of v, the probabilities of exceeding by chance value of t at different levels of significance. The t-distribution has a different value for each degree of freedom and when degrees of freedom are infinitely large, the t-distribution is equivalent to normal distribution and the probabilities shown in the normal distribution tables are applicable.

Application of the t-Distribution

The following are some of the examples to illustrate the way in which the 'Student' distribution is generally used to test the significance of the various results obtained from small samples.

(1) To Test the Significance of the Mean of a Random Sample : In determining whether the mean of a sample drawn from a normal population deviates significantly from a stated value (the hypothetical value of the populations mean), when variance of the population is unknown we calculate the statistic :

$$t = \frac{\left(\overline{X} - \mu\right)\sqrt{n}}{S}$$

where $\overline{X}$ = the mean of the sample

μ = the actual or hypothetical mean of the population

n = the sample size

S = the standard deviation of the sample

$$X = \sqrt{\frac{\Sigma\left(X - \overline{X}\right)^2}{n-1}} \text{ or } S = \sqrt{\frac{\Sigma d^2 - n\left(\overline{d}\right)^2}{n-1}}$$

$$= \sqrt{\frac{1}{(n-1)}\left[\Sigma d^2 - \frac{(\Sigma d)^2}{n}\right]}$$

where d = deviation from the assumed mean.

If the calculated value of [t] exceeds $t_{0.05,}$ we say that the difference between $\overline{X}$ and m is significant at 5% level, if it exceeds $t_{0.01,}$ the difference is said to be significant at 1% level. If $|t| < t_{0.05,}$ we conclude that the difference between $\overline{X}$ and m is not significant and hence the sample might have been drawn from a population with mean = μ.

Fiducial Limits of Population Mean. Assuming that the sample is a random sample from a normal population of unknown mean the 95% fiducial limits of the population mean (μ) are :

$$\overline{X} \pm \frac{S}{\sqrt{n}} t_{0.05}$$

and 99% limits are

$$\overline{X} \pm \frac{S}{\sqrt{n}} t_{0.05}$$

The following examples will illustrate this test :

Example 26:

In a sample of 1000 the mean is 17.5 and the s.d. 2.5. In another sample of 800 the mean is 18 and s.d. 2.7. Assuming that the samples are independent discuss whether the two samples can have come from a population which have the same s.d.

Solution :

Let us take the hypothesis that there is no significant difference in the standard deviations of the two samples

$$S.E._{(\sigma_1-\sigma_2)} = \sqrt{\frac{\sigma_1^2}{2n_1} + \frac{\sigma_2^2}{2n_2}}$$

$\sigma_1 = 2.5, n_1 = 1000, \sigma_2 = 2.7, n_2 = 800$

$$S.E._{(\sigma_1-\sigma_2)} = \sqrt{\frac{(2.5)^2}{2000} + \frac{(2.7)^2}{1600}} = \sqrt{\frac{6.25}{2000} + \frac{7.29}{1600}}$$

$$= \sqrt{0.00312 + 0.00456} = 0.0876$$

$$\frac{\text{Diff.}}{\text{S.E.}} = \frac{2.7 - 2.5}{0.0876} = \frac{.2}{.0876} = 2.283$$

Since the difference is more than 1.96 S.E. at 5% level of significance the hypothesis rejected. Hence the two samples have not come from a population which has the same standard deviation.

Example 27:

The life time of electric bulbs for a random sample of 10 from a large consignment gave the following data :

Item :	*1*	*2*	*3*	*4*	*5*	*6*	*7*	*8*	*9*	*10*
Life in '000 hours:	*4.2*	*4.6*	*4.1*	*5.2*	*3.8*	*3.9*	*4.3*	*4.4*	*5.6*	

Can we accept the hypothesis that the average life time of bulbs is 4,000 hours.

Solution:

Let us take the hypothesis that there is no significant difference in the sample mean and the hypothetical population mean. Applying the t-test.

$$t = \frac{\overline{X} - \mu}{S}\sqrt{n}$$

Calculation of $\overline{X}$ and S

X	$(X - \overline{X})$	$(X - \overline{X})^2$
4.2	–.02	0.04
4.6	+ 0.2	0.04
3.9	– 0.5	0.25
4.1	– 0.3	0.09
5.2	+ 0.8	0.64
3.8	– 0.6	0.36
3.9	– 0.5	0.25
4.3	– 0.1	0.01
4.4	0	0
5.6	+ 1.2	1.44
$\Sigma X = 44$		$\Sigma (X - \overline{X})^2 = 3.12$

$$\Sigma = \sqrt{\frac{\Sigma\left(X-\overline{X}\right)^2}{n-1}} = \sqrt{\frac{3.12}{9}} = 0.589$$

$$t = \frac{4.4-4}{0.589}\sqrt{10} = \frac{0.4 \times 3.162}{0.589} = 2.148$$

$v = n - 1 = 10 - 1 = 9$. For $v = 9$, $t_{0.05} = 2.262$

The calculated value of t is less than the table value. The hypothesis is accepted. The average life time of the bulbs could be 4000 hours.

Example 28:

Prices of shares of a company on different days in a month were found to be :

66, 65, 70, 69, 71, 70, 63, 64 and 68. Discuss whether the price of the shares be 65 (the table value of t = 2.262).

Solution:

Let us take the hypothesis that there is no difference between the mean price of the shares and the hypothetical mean. Applying t-test,

$$t = \frac{\overline{X}-\mu}{S}\sqrt{n}$$

Calculating Mean and Standard Deviation

X	**(X – 67)** **d**	**d²**
66	–1	1
65	–2	4
69	+2	4
70	+3	9
69	+2	4
71	+4	16
70	+3	9
63	–4	16
64	–3	9
68	+1	1
$\Sigma X = 675$	$\Sigma d = 5$	$\Sigma d^2 = 73$

$$\overline{X} = \frac{675}{10} = 67.5;\ \overline{d} = \frac{5}{10} = 0.5$$

$$S = \sqrt{\frac{\Sigma d^2 - n\left(\overline{d}\right)^2}{n-1}} = \sqrt{\frac{73-10\,(.5)^2}{9}} = 2.799$$

$$= \frac{67.5 - 65}{2.799} \times \sqrt{10} = 2.82$$

$v = 10 - 1 = 9$. For $v = 9$, $t_{0.05} = 2.262$.

The calculated value of t is greater than the table value. Our hypothesis does not hold good. Hence the mean price of the shares could not be Rs. 65.

Example 29:

Two laboratories A and B carry out independent estimates of fat content in ice-cream made by a firm. A sample is taken from each, halved, and the separated halves sent to the two laboratories. That fat content obtained by the laboratories is recorded below :

Bath No.	*1*	*2*	*3*	*4*	*5*	*6*	*7*	*8*	*9*	*10*
Lab. A	*7*	*8*	*7*	*3*	*8*	*6*	*9*	*4*	*7*	*8*
Lab. B	*9*	*8*	*8*	*4*	*7*	*7*	*9*	*6*	*6*	*6*

Is there a significant difference between the mean fat content obtained by the two laboratories A and B?

You may use the following extracts from t table in answering the question:

Degrees of freedom	*6*	*7*	*8*	*9*	*10*	*16*	*18*	*20*
5% value of t	*1.45*	*2.36*	*2.31*	*2.26*	*2.23*	*2.12*	*2.10*	*2.09*

Solution :

Let us take the hypothesis that there is no significant difference between the mean fat content obtained by the two laboratories, A and B, Applying t-test:

$$t = \frac{\overline{X}_1 - \overline{X}_2}{S} \times \sqrt{\frac{n_1 n_2}{n_1 + n_2}}$$

Since the actual means are in fractions, we shall take deviations from the assumed means.

	Lab. A			**Lab. B**		
Batch No.	X_1	$(X_1 - A_1)$ $A_1 = 16.$	$(X_1 - A_1)^2$	X_2	$(X_2 - A_2)$ $A_2 \neq 7$	$(X_2 - A_2)^2$
1	7	+1	1	9	+2	4
2	8	+2	4	8	+1	1
3	7	+1	1	8	+1	1
4	3	–3	9	4	–3	9
5	8	+2	4	7	0	0
6	6	0	0	7	0	0

7	9	+3	9	9	+2	4
8	4	–2	4	6	–1	1
9	7	+1	1	6	–1	1
10	8	+2	4	6	–1	1
	$\Sigma X_1=67$	$\Sigma(X_1-A_1) = 7$	$\Sigma(X_1-A_1)^2 = 37$	$\Sigma X_2 = 70$	$\Sigma(X_2-A_2) = 0$	$\Sigma(X_2-A_2)^2 = 22$

$$X_1 = \frac{67}{10} = 6.7; \overline{X}_2 = 7.0$$

$$S = \sqrt{\frac{\Sigma(X_1 - A_1)^2 + \Sigma(X_2 - A_2)^2 - n_1(\overline{X}_1 - A_1)^2 - n_2(\overline{X}_2 - A_2)^2}{n_1 + n_2 - 2}}$$

$$= \sqrt{\frac{37+22-10(6.7-6)^2 - 10(7-7)^2}{10+10-2}} = \sqrt{\frac{59.49}{18}} = \sqrt{\frac{54.1}{18}} = 1.734$$

$$t = \frac{6.7-7}{1.734}\sqrt{\frac{10 \times 10}{10+10}} = \frac{0.3}{1.734} \times 2.236 = 0.87$$

$v = n_1 + n_2 - 2 = 18$; For $v = 18$, $t_{0.05} = 2.10$.

The calculated value of $|t|$ is less than the table value and hence there is reason to doubt the hypothesis. We, therefore, conclude that the mean fat contents obtained by two laboratories A and B do not differs significantly.

(2) **Testing Difference Between Means of Two Samples (Independent Samples) :** Given two independent random samples of size n_1 and n_2 with means $\overline{X}_1$ and $\overline{X}_2$ and standard deviations S_1 and S_2 we may be interested in testing the hypothesis that the samples come from the same normal population. To carry out the test, we calculate the statistic as follows :

$$t = \frac{\overline{X}_1 - \overline{X}_2}{S} \times \sqrt{\frac{n_1 n_2}{n_1 + n_2}}$$

where $\overline{X}_1$ = mean of the first sample

$\overline{X}_2$ = mean of the second sample

n_1 = number of observations in the first sample

n_2 = number of observations in the second sample

S = combined standard deviation.

The value of S is calculated by the following formula

$$S = \sqrt{\frac{\Sigma(X_1 - \overline{X}_1)^2 + \Sigma(X_2 - \overline{X}_2)^2}{n_1 + n_2 - 2}}$$

When the actual means are in fraction the deviations should be taken from assumed means. In such a case the combined standard deviation is obtained by applying the following formula :

$$S = \sqrt{S(X_1 - A_1)^2 + S(X_2 - A_2)^2 - n_1(\overline{X} - A_1)^2 - n_2(\overline{X}_2 - A_2)^2}\ n_1+n_2-2$$

A_1 = Assumed mean of the first sample

A_2 = Assumed mean of the second sample

$\overline{X}_1$ = Actual mean of the first sample

$\overline{X}_2$ = Actual mean of the second sample.

The degrees of freedom = $(n_1 + n_2 - 2)$

When we are given the number of observations and standard deviation of the two samples, the pooled estimate of standard deviation can be obtained can be obtained as follows :

$$S = \sqrt{\frac{(n_1 - 1)\ S_1^2 + (n_2 - 1)\ S_2^2}{n_1 + n_2 - 2}}$$

If the calculated value of t be > $t_{0.05}$ ($t_{0.01}$), the difference between the sample means is said to be significant at 5% (1%) level of significance otherwise the data are said to be consistent with the hypothesis.

Example 30:

To verify whether a course in accounting improved performance, a similar test was given to 12 participants both before and after the course. The original marks recorded in alphabetical order of the participants – were 44, 40, 61, 52, 44, 70, 41, 67, 72, 53, and 72. After the course, the marks were in the same order, 53, 38, 69, 57, 46, 39, 73, 48, 73, 74, 60 and 78. Was the course useful?

Solution:

Let us take the hypothesis that there is no difference in the marks obtained before and after the course, i.e., the course has not been useful.

Applying t-test (difference formula) :

$$t = \frac{\overline{d}\sqrt{n}}{S}$$

Participants	Before (1st Test)	After (2nd Test)	(2nd–1st Test d	d^2
A	44	53	+9	81
B	40	38	–2	4
C	61	69	+8	64
D	52	57	+5	25
E	32	46	+14	196
F	44	39	–5	25
G	70	73	+3	9
H	41	48	+7	49
I	67	73	+6	36
J	72	74	+2	4
K	53	60	+7	49
L	72	78	+6	36
			$\Sigma d = 60$	$\Sigma d^2 = 578$

$$\bar{d} = \frac{\Sigma d}{n} = \frac{60}{12} = 5$$

$$S = \sqrt{\frac{\Sigma d^2 - n(\bar{d})^2}{n-1}} = \sqrt{\frac{578 - 12(5)^2}{12-1}} = \sqrt{\frac{278}{11}} = 5.03$$

$$t = \frac{5 \times \sqrt{12}}{5.03} = \frac{5 \times 3.464}{5.03} = 3.443$$

$v = n - 1 = 12 - 1 = 11$; For $v = 11$, $t_{0.05} = 2.201$

The calculated value of t is greater than the table value. The hypothesis is rejected. Hence the course has been useful.

(3) Testing Difference between Means of Two Samples (Dependent Samples or Matched Paired Observations) : In the previous test it was assumed that the two samples were independent i.e., the values of observations in one sample does not depend on the other. However, there are many situations in which this condition does not hold true, i.e., we have dependent (or paired) samples. Two samples are said to be dependent when the elements in one sample are related to those in the other in any significant or meaningful manner. In fact, the two samples may consist of pairs of observations made on the same object, individual or, more generally, on the same selected population elements. When samples are dependent they comprise the same number of elementary units. We may carry out some experiment, say, to find out the effect of training on some employees, find out the efficacy of a coaching class or determine whether there is a significant difference in the efficacy of two drugs–one made within the country and another imported. Often the use of dependent or paired samples will enable us to perform a more precise analysis, because they will allow us to control the entraneous factors. The t-test based on paired observations is defined by the following formula :

$$t = \frac{\bar{d} - 0}{S} \times \sqrt{n} \text{ or } t = \frac{\bar{d}\sqrt{n}}{S}$$

where $\bar{d}$ = the mean of the differences

S = the standard deviation of the differences

The value of S is calculated as follows :

$$S = \sqrt{\frac{\Sigma\left(d - \bar{d}\right)^2}{n-1}} \text{ or } \sqrt{\frac{\Sigma d^2 - n\left(\bar{d}\right)^2}{n-1}}$$

It should be noted that t is based on n – 1 degrees of freedom.

The following examples will illustrate the application of difference test:

Example 31:

The manufacturer of a certain make of electric bulbs claims that his bulbs have a mean life of 25 months with a standard deviation of 5 months A random sample of 6 such bulbs gave the following values.

Life of months 24, 26, 30, 20, 20 18.

Can you regard the producer's claim to be valid at 1% level of significance? (Given that the table values of the appropriate test statistic at the said level are 4.032, 3.707 and 3.499 for 5, 6 and 7 degrees of freedom respectively).

Solution:

Let us take the hypothesis that there is no significant difference in the mean life of bulbs in the sample and that of the population. Applying t-test:

$$t = \frac{\left(\bar{X} - \mu\right)}{S}\sqrt{n}$$

Calculation of $\bar{X}$ and S

X	$(X - \bar{X})$ x	x^2
24	+1	1
26	+3	9
30	+7	49
20	–3	9
20	–3	9
18	–5	25
ΣX = 138		$\Sigma x^2 = 102$

$$\overline{X} = \frac{\Sigma X}{n} = \frac{138}{6} = 23$$

$$S = \sqrt{\frac{\Sigma x^2}{n-1}} = \sqrt{\frac{102}{5}} = \sqrt{20.4} = 4.517$$

$$= \frac{23-25}{4.517}\sqrt{6} = \frac{2 \times 2.449}{4.517} = 1.084$$

$v = n - 1 = 6 - 1 = 5$. For $v = 5$, $t_{0.01} = 4.032$.

The calculated value of t is less than the table value. The hypothesis is accepted. Hence the producer's claim is not valid at 1% level of significance.

Example 32:

Is a correlation coefficient of 0.5 significant if obtained from a random sample of 11 pairs of values from a normal population? Use t-test.

Solution:

Let the null hypothesis be that there is no correlation or the given correlation coefficient is not significant.

$$t = \frac{r}{\sqrt{1-r^2}} \times \sqrt{n-2}$$

$r = 0.5$, $n = 11$

$$t = \frac{0.5}{\sqrt{1-(0.5)^2}} \times \sqrt{11-2} = \frac{0.5}{\sqrt{1-0.25}} \times = \frac{0.5}{0.866} \times 3 = 1.732$$

$v = 11 - 2 = 9$ For $v = 9$, $t_{0.05} = 2.26$

The calculated value of t is less than the table value and hence the given correlation coefficient is not significant.

Example 33:

A random sample of size 16 has 53 as mean. The sum of the squares of the deviations taken from mean is 135. Can this sample be regarded as taken from the population having 56 as mean? Obtain and 95% and 99% confidence limits of the mean of the population. (for $v = 15$, $t_{0.05} = 2.13$ for $v = 15$ $t_{0.01} = 2.95$)

Solution:

Let us take the hypothesis that there is no significant difference between the sample mean and hypothetical population mean. Applying t test:

$$t = \frac{\overline{X} - \mu}{S}\sqrt{n}$$

$\overline{X} = 53$, $\mu = 56$, $n = 16$, $S(X - \overline{X})^2 = 135$

$$S = \sqrt{\frac{\Sigma(X-\overline{X})}{n-1}} = \sqrt{\frac{135}{15}} = 3$$

$$t = \frac{|53-56|}{3}\sqrt{16} = \frac{3\times 4}{3} = 4$$

$v = 16 - 1 = 15$. For $v = 16$, $t_{0.05} = 2.13$

The calculated value of t is more than the table value. The hypothesis is rejected. Hence the samples has not come from a population having 56 as mean.

95% confidence limits of population mean

$$\overline{X} \pm \frac{S}{\sqrt{n}} t_{0.05}$$

$$= 53 \pm \frac{3}{\sqrt{16}} \times 2.13$$

$$= 53 \pm 1.6 = 51.4 \text{ to } 54.6$$

99% confidence limits of the population mean

$$\overline{X} \pm \frac{S}{\sqrt{n}} t_{0.01}$$

$$= 53 \pm \frac{3}{\sqrt{16}} \times 2.95$$

$$= 53 \pm \frac{3}{4} \times 2.95$$

$$= 53 \pm 2.212 = 50.788 \text{ to } 55.212.$$

Example 34:

From a sample of 19 pairs of observations the correlation is 0.5 and the corresponding population value is 0.3. Is the difference significant?

Solution:

Using Z-transformation.

$$Z = \frac{1}{2}\log_e \frac{1+r}{1-r}$$

$$= 1.1513 \log_{10} \frac{1+r}{1-r} = 1.1513 \log_{10} \frac{1+0.5}{1-0.5}$$

$$= 1.1513 \log 3 = 1.1513 \times 0.4771 = 0.549$$

$$\xi = \log_{10} \frac{1+\rho}{1-\rho} \times 1.1513 = \log_{10} \frac{1+0.3}{1-0.3} \times 1.1513$$

$$= \log 1.857 \times 1.1513 = 0.2688 \times 1.1513 = 0.309$$

$$(Z - \xi) = 0.549 - 0.309 = 0.24$$

$$S.E._z = \frac{1}{\sqrt{n-3}} = \frac{1}{\sqrt{16}} = 0.25$$

$$\frac{\text{Difference}}{\text{S.E.}} = \frac{0.24}{0.25} = 0.96$$

Since the difference between Z and x is less than 2.58 S.E. (1% level) the difference could have arisen due to fluctuations of sampling. Hence the given correlation coefficient is not significantly different from 0.3.

To test the significance of the difference between two independent correlation coefficients. To test the significance of two correlation coefficients derived from two separate samples we have to compare the difference of the two corresponding values of Z with the standard error of that difference– remembering that the standard error of the difference of two statistical quantities is the square root of the sum of their variances. In other words, we have to apply the following formula :

$$Z = \frac{Z_1 - Z_2}{\sqrt{\frac{1}{n_1 - 3} + \frac{1}{n_2 - 3}}}$$

where $$Z_1 = \frac{1}{2}\log_e\left(\frac{1+r_1}{1-r_1}\right) \text{ or } 1.1513 \log_{10}\left(\frac{1+r_1}{1-r_1}\right)$$

and $$Z_2 = \frac{1}{2}\log_e\left(\frac{1+r_2}{1-r_2}\right) \text{ or } 1.1513 \log_{10}\left(\frac{1+r_2}{1-r_3}\right)$$

$$S.E._{(Z_1-Z_2)} = \sqrt{\frac{1}{n_1 - 3} + \frac{1}{n_2 - 3}}$$

If the absolute value of this statistic is greater than 1.96, the difference will be significant at 5% level. The following examples will illustrate the application of this test.

Example 35:

A drug is given to 10 patients, and the increments in their blood pressure were recorded to be 3, 6, –2, 4, –3, 4, 6, 0, 0, 2. Is it reasonable to believe that the drug has no effect on change of blood pressure? (5% value of t for 9 d.f. = 2.26).

Solution:

Let us take the hypothesis that the drug has no effect on change of blood pressure. Applying the difference test :

$$t = \frac{\bar{d}\sqrt{n}}{S}$$

'd	d^2
3	9
6	36
–2	4
4	16
–3	9
4	16
6	36
0	0
0	0
2	4
$\Sigma d = 0$	$\Sigma d^2 = 130$

$$\bar{d} = \frac{\Sigma d}{n} = \frac{20}{10} = 2$$

$$S = \sqrt{\frac{\Sigma d^2 - n(\bar{d})^2}{n-1}} = \sqrt{\frac{130 - 10(2)^2}{10-1}} = 3.162$$

$$t = \frac{2\sqrt{10}}{3.162} = \frac{2 \times 3.162}{3.162} = 2$$

$v = n - 1 = 10 - 1 = 9$; For $v = 9$, $t_{0.05} = 2.26$.

The calculated value of t is less than the table value. The hypothesis is accepted. Hence it is reasonable to believe that the drug has no effect on change of blood pressure.

(4) Testing the Significance of an Observed Correlation Coefficient: Given a random sample from a bivariate normal population. If we are to test the hypothesis that the correlation coefficient of the population is zero, i.e., the variables in the population are uncorrelated, we have to apply the following test :

$$t = \frac{r}{\sqrt{1-r^2}} \times \sqrt{n-2}$$

here t is based on (n – 2) degrees of freedom.

If the calculated value of t exceeds $t_{0.05}$ for (n – 2), d.f., we say that the value of r is significant at 5% level. If $t < t_{0.05}$ the data are consistent with the hypothesis of an uncorrelated population.

The following examples will illustrate the test:

Example 36:

Two types of drugs were used on 5 and 7 patients for reducing their drugs for six month was as follows :

Drug A :	*10*	*12*	*13*	*11*	*14*		
Drug B :	*8*	*9*	*12*	*14*	*15*	*10*	*9*

If the bias correlation due to small is ignored, pooled estimate of the standard deviation can be obtained by :

$$S = \sqrt{\frac{n_1 S_1^2 + n_2 S_2^2}{n_1 + n_2}}$$

Is there a significant difference in the efficacy of the two drugs? If not, which drug should you buy. (For v = 10, $t_{0.05}$ = 2.223).

Solution:

Let us take the hypothesis that there is no significant difference in the efficacy of the two drugs. Applying t-test :

$$t = \frac{\overline{X}_1 - \overline{X}_2}{S} \sqrt{\frac{n_1 n_2}{n_1 + n_2}}$$

X_1	$(X_1 - \overline{X}_1)$	$(X_1 - \overline{X}_1)^2$	X_2	$(X_2 - \overline{X}_2)$	$(X_2 - \overline{X}_2)^2$
10	–2	4	8	–3	9
12	0	0	9	–2	4
13	+1	1	12	+1	1
11	–1	1	14	+3	9
14	+2	4	15	+4	16
			10	–1	1
			9	–2	4
$\Sigma X_1 = 60$		$\Sigma(X_1 + \overline{X}_1)^2 = 10$	$\Sigma X_2 = 77$		$\Sigma(X_2 - \overline{X}_2)^2 = 44$

However, it is advisable to take account of bias.

$$\overline{X}_1 = \frac{\Sigma X_1}{n_1} = \frac{60}{5} = 12; \ \overline{X}_2 = \frac{\Sigma X_2}{n_2} = \frac{77}{7} = 11$$

$$S = \sqrt{\frac{\Sigma(X_1 - \overline{X}_1)^2 + \Sigma(X_2 - \overline{X}_2)^2}{n_1 + n_2 - 2}} = \sqrt{\frac{10 + 44}{5 + 7 - 2}} = \sqrt{\frac{54}{10}} = 2.324$$

$$t = \frac{\overline{X}_1 - \overline{X}_2}{S} \sqrt{\frac{n_1 n_2}{n_1 + n_2}}$$

$$\frac{12 - 11}{2.324} \sqrt{\frac{5 \times 7}{5 + 7}} = \frac{1.708}{2.324} = 0.735$$

$$v = n_1 + n_2 - 2 = 5 + 7 - 2 = 10.$$

For $v = 10,\ t_{0.05} = 2.228.$

The calculated value of t is less than the table value, the hypothesis is accepted. Hence there is no significance in the efficacy of two drugs. Since drug B is indigenous and there is no difference in the efficacy of imported and indigenous drug, we should buy indigenous drug, i.e., B.

$$S = \sqrt{\frac{n_1S_1^2 + n_2S_2^2}{n_1 + n_2}}.$$

Example 37:

Two working designs are under consideration for adoption in a plant. A time and motion study shows that 12 workers using design A have a mean assembly time of 300 seconds with a standard deviation of 12 seconds and that 15 workers using design B have a mean assembly time of 335 seconds with a standard deviation of 15 seconds. Is the difference in the mean assembly time between the two working designs significant at one per cent level of significance? The following table gives some t-values which may be used:

Level of significance	*Degrees of Freedom*		
	25	*26*	*27*
$p_r = 0.05$	*2.06*	*2.06*	*2.05*
$p_r = 0.01$	*2.79*	*2.78*	*2.77*

Solution:

Let us take the hypothesis that the difference between the two designs 'A' and 'B' in respect of mean assembly time is not significant. Applying t-test:

$$t = \frac{\overline{X}_1 - \overline{X}_2}{S}\sqrt{\frac{n_1n_2}{n_1 + n_2}}$$

$$\overline{X}_1 = 300,\ S_1 = 12,\ n_1 = 12$$

$$\overline{X}_2 = 335,\ S_2 = 15,\ n_2 = 15$$

$$S = \sqrt{\frac{(n_1 - 1)\ S_1^2 + (n - 2)\ S_2^2}{n_1 + n_2 - 2}}$$

$$S = \sqrt{\frac{11(12)^2 + 14(15)^2}{12 + 15 - 2}} = \sqrt{\frac{5.103}{25}} = 13.8$$

$$t = \frac{|300 - 335|}{13.8} \times \sqrt{\frac{12 \times 15}{12 + 15}} = \frac{35}{13.8} \times 2.582 = 6.55$$

$$v = n_1 + n_2 - 2 = 12 + 15 - 2 = 25$$

For $\quad v = 25, t_{0.01} = 2.79$

The calculated value of v is greater than the table value. Hence the hypothesis does not hold true. We, therefore, conclude that the difference in the mean assembly time between the two working designs is significant at 1 per cent level.

Cautions while using t-test. While drawing inference on the basis of t-test it should be remembered that the conclusions arrived at on the basis of the 't-test' are justified only if the assumptions upon which the test is based are true. If the actual distribution is not normally distributed then, strictly speaking the t-test is not justified for small samples. If it is not a random sample, then the assumption that the observations are statistically independent is not justified and the conclusions based on the t-test may not be correct. The effect of violating the normality assumption is slight when making inference about means provided that the sampling is fairly large when dealing with small samples. However, it is a good idea to check the normality assumption, if possible. A review of similar samples or related research may provide evidence as to whether or not the population is normally distributed.

Z-test of the Significance of the Correlation Coefficient. Prof. Fisher has given a method of testing the significance of the correlation coefficient in small samples. According to this method the coefficient of correlation is transformed into Z and hence the name *Z-transformation.* The statistic Z given by Prof. Fisher is used to test (i) whether an observed value of r differs significantly from some hypothetical value, or (i) whether two sample values of r differ significantly.

For testing whether r differs significantly from zero, the t-test is preferable.

In order to apply the test we have to calculate Z and ξ by applying Fisher's transformation and then calculate the value of the standard normal variate.

$$\frac{Z - \xi}{1/\sqrt{n-3}}$$

The difference is significant at 5% level.

Here $$Z = \frac{1}{2}\log_e \frac{1+r}{1-r} \text{ or } 1.1513 \log_{10}\left(\frac{1+r}{1-r}\right)$$

$$\xi = \frac{1}{2}\log_e \frac{1+\rho}{1-\rho} \text{ or } 1.1513 \log_{10}\left(\frac{1+\rho}{1-\rho}\right)$$

r refers to the population correlation coefficient.

$$S.E._z = \frac{1}{\sqrt{n-3}}$$

The following examples will illustrate the test :

Example 38:

12 students were given intensive coaching and 5 tests were conducted in a month. The scores of tests 1 and 5 are given below. Does the scores from the 1 to 5 show an improvement?

No. of Students	*1*	*2*	*3*	*4*	*5*	*6*	*7*	*8*	*9*	*10*	*11*
Marks in 1st test	*50*	*42*	*51*	*26*	*35*	*42*	*60*	*41*	*70*	*55*	*62*
Marks in 5th test	*62*	*40*	*61*	*35*	*30*	*52*	*68*	*51*	*84*	*63*	*72*

(the value of 'r' for 11 degrees of freedom at 5% level of significance is 2.20).

Solution:

Let us take the hypothesis that there is no difference in the scores obtained in the first and fifth test. Applying t-test :

$$t = \frac{\bar{d}\sqrt{n}}{S}$$

No. of Students	**Marks in 1st Test**	**Marks in 5th Test**	**(5th – 1st Test) d**	**d^2**
1	50	62	+12	144
2	42	40	–2	4
3	51	61	+10	100
4	26	35	+9	81
5	35	30	–5	25
6	42	52	+10	100
7	60	68	+8	64
8	41	51	+10	100
9	70	84	+14	196
10	55	63	+8	64
11	62	72	+10	100
12	38	50	+12	144
			$\Sigma d = 96$	$\Sigma d^2 = 1122$

$$\bar{d} = \frac{\Sigma d}{n} = \frac{96}{12} = 8$$

$$S = \sqrt{\frac{\Sigma d^2 - n(\bar{d})^2}{n-1}}$$

$$= \sqrt{\frac{1122 - 12(8)^2}{12 - 1}}$$

$= 5.673$

$$t = \frac{8\sqrt{12}}{5.673} = 4.885$$

For $v = 11$, $t_{0.05} = 2.20$.

The calculated value of t is greater than the table value hypothesis stands rejected. Hence, the scores from test 1 to 5 show an improvement.

Example 39:

For a random sample of 10 persons, fed on diet A, the increased weight in pounds in certain period were:

10, 6, 16, 17, 13, 12, 8, 14, 15, 9,

For another random sample of 12 persons, fed on diet B, the increase in the same period were :

7, 13, 22, 15, 12, 14, 18, 8, 21, 23, 10, 17

Test whether the diets A and B differ significantly as regards their effect on increase in weight.

Given the following :

Degrees of freedom:	*19*	*20*	*21*	*22*	*23*
Value of t at 5% level:	*2.09*	*2.09*	*2.08*	*2.07*	*2.07*

Solution:

Let us take the null hypothesis that A and B do not differ significantly weight regard to their effect on increase in weight. Applying t-test :

$$t = \frac{\overline{X}_1 - \overline{X}_2}{S}\sqrt{\frac{n_1 n_2}{n_1 + n_2}}$$

Calculating the required values :

Persons fed on diet A			**Persons fed on diet B**		
Increases in weight	**Deviations from mean 12**	$(X_1 - \overline{X}_1)^2$	**Increase in weight**	**Deviations from actual mean 15**	$(X_2 - \overline{X}_2)^2$
X_1	$(X_1 - \overline{X}_1)$		X_2	$(X_2 - \overline{X}_2)$	
10	–2	4	7	–8	64
6	–6	36	13	–2	4
16	+4	16	22	+7	49
17	+5	25	15	0	0

13	+1	1	12	–3	9
12	0	0	14	–1	1
8	–4	16	18	+3	9
14	+2	4	8	–7	49
15	+3	9	21	+6	36
9	–3	9	23	+8	64
			10	–5	25
			17	+2	4
$\Sigma X_1 = 120$	$\Sigma(X_1-\overline{X}_1) = 0$	$\Sigma(X_1-\overline{X}_1)^2 = 120$	$\Sigma X_2 = 180$	$\Sigma(X_2-\overline{X}_2) = 0$	$\Sigma(X_2-\overline{X}_2)^2 = 314$

Mean increase in weight of 10 persons fed on diet A

$$\overline{X}_1 = \frac{\Sigma X_1}{n_1} = \frac{120}{10} = 12 \text{ pounds}$$

Mean increase in weight of 12 persons fed on diet B

$$\overline{X}_2 = \frac{\Sigma X_2}{n_2} = \frac{180}{12} = 15 \text{ pounds}$$

$$S = \sqrt{\frac{\Sigma(X_1-\overline{X}_1)^2 + \Sigma(X_2-\overline{X}_2)^2}{n_1+n_2-2}} = \sqrt{\frac{120+314}{10+12-2}} = \sqrt{\frac{434}{20}} = 4.66$$

$\overline{X}_1 = 12$, $\overline{X}_2 = 15$, $n_1 = 10$, $n_2 = 12$, $S = 4.66$. Substituting the values in the above formula :

$$t = \frac{12-15}{4.66} \times \sqrt{\frac{10 \times 12}{10+2}} = \frac{3}{4.66} \times 2.34 = 1.51$$

$v = n_1 + n_2 - 2 = 10 + 12 - 2 = 20.$

For v = 20, the table value of t at 5 per cent level is 2.09. The calculated value is less than the table value and hence the experiment provides no evidence against the hypothesis. We, therefore, conclude that diets A and B do not significantly as regards their effect on increase in weight is concerned.

Example 40:

A random sample of 27 pairs of observations from a normal population gives a correlation coefficient of 0.42. Is it likely that the variables in the population are uncorrelated?

Solution:

Let us take the hypothesis that there is no significant difference in the sample correlation and correlation in the population. Applying t-test:

$$t = \frac{r}{\sqrt{1-r^2}} \times \sqrt{n-2}$$

Here $r = 0.42,\ n = 27$

$$t = \frac{.42}{\sqrt{1-(.42)^2}} \times \sqrt{27-2}$$

$$= \frac{.42}{.908} \times 5 = 2.31$$

$$v = n - 2 = 27 - 2 = 25$$

For $v = 25$, $t_{0.05} = 1.708$. The calculated value of t is more than the table value. The hypothesis is rejected. Hence it is likely that the variables in the population are uncorrelated.

Example 41:

How many pairs of observations must be included in a sample in order than an observed correlation coefficient of value 0.42 shall have a calculated value of t greater than 2.72?

Solution:

$$t = \frac{r}{\sqrt{1-r^2}} \times \sqrt{n-2}$$

We are given the value of t and r and we have to find out n.

$$\therefore \quad \frac{0.42}{\sqrt{1-(0.42)^2}} \times \sqrt{n-2} > 2.72$$

or $$\frac{0.42}{0.908} \times \sqrt{n-2} < 2.72$$

or $$0.42 \times \sqrt{n-2} > 2.72 \times 0.908$$

or $$\sqrt{n-2} > \frac{2.72 \times 0.908}{0.42} = 5.88$$

or $$n - 2 > (5.88)^2 = 34.57$$

or $$n - 24.57 + 2 = 36.57 \text{ or } 37$$

Hence we should include 37 observations.

Example 42:

The mean life of a sample of 10 electric light bulbs was found to be 1,456 hours with standard deviation of 423 hours. A second sample of 17 bulbs chosen from a different batch showed a mean life of 1,280 hours with standard deviation of 398 hours. Is there a significant difference between the means of the two batches?

Solution:

Let us take the hypothesis that the means of two batches do not differ significantly. Applying t-test :

$$t = \frac{\overline{X}_1 - \overline{X}_2}{S}\sqrt{\frac{n_1 n_2}{n_1 + n_2}}$$

$$\overline{X}_1 = 1456,\ n_1 = 10,\ S_1 = 423,\ \overline{X}_2 = 1280,\ n_2 = 17,\ S_2 = 398$$

$$S = \sqrt{\frac{(n_1 - 1)\,S_1^2 + (n_2 - 1)\,S_2^2}{n_1 + n_2 - 2}} = \sqrt{\frac{9(423)^2 + 16(398)^2}{10 + 17 - 2}}$$

$$= \sqrt{\frac{1610361 + 2534464}{25}} = \sqrt{165793} = 407.18$$

$$t = \frac{1456 - 1280}{407.18}\sqrt{\frac{10 \times 17}{10 + 17}} = \frac{176}{407.18} \times 2.51 = 1.085$$

$v = 10 + 17 - 2 = 25$; For $v = 25$, $t_{0.05} = 2.06$

The calculated value of t is less than the table value. Hence the hypothesis holds true. We, therefore, conclude that the means of two batches do not differs significantly.

Example 43:

The following table gives the ages in years of 10 husbands and their wives at marriage. Compute the correlation coefficient and test for its significance.

Solution:

We set up the hypothesis that there is no correlation in the population, Applying t-test

$$t = \frac{r}{\sqrt{1 - r^2}} \times \sqrt{n-2} = \frac{0.939}{\sqrt{1-(0.939)^2}} \times \sqrt{10-2}$$

$$= \frac{0.939}{\sqrt{1-0.882}} \times 2.8284 = \frac{0.939}{0.344} \times 2.8284 = 7.72$$

$v = n - 2 = (10 - 2) = 8$

For $v = 8$, $t_{0.05} = 2.31$

The calculated value being much higher than the table value, the correlation is significant.

Example 44:

In a test given to two groups of students, the marks obtained are as follows :

First Group :	*18*	*20*	*36*	*50*	*49*	*36*	*34*	*49*	*41*
Second Group :	*29*	*28*	*26*	*35*	*30*	*44*	*46*		

Examine the significance of difference between the arithmetic mean of the marks secured by the students of the above two groups.

(The value of t at 5% level of significance for v = 14 is 2.14)

Solution:

Let us take the hypothesis that there is no significant difference in the arithmetic mean of the marks secured by the students of the two groups. Applying t-test :

$$t = \frac{\overline{X}_1 - \overline{X}_2}{S}\sqrt{\frac{n_1 n_2}{n_1 + n_2}}$$

Group I X_1	$(X_1 - \overline{X}_1)$ $\overline{X}_1 = 37$	$(X_1 - \overline{X}_1)^2$	**Group II** X_2	$(X_2 - \overline{X}_2)$ $\overline{X}_2 = 34$	$(X_2 - \overline{X}_2)^2$
18	− 19	361	29	− 5	25
20	− 17	289	28	− 6	36
36	− 1	1	26	− 8	64
50	+ 13	169	35	+ 1	1
49	+ 12	144	30	− 4	16
36	− 1	1	44	+ 10	100
34	− 3	9	46	+ 12	144
49	+ 12	144			
41	+ 4	16			
ΣX_1 = 333	$\Sigma(X_1 - \overline{X}_1)$ = 0	$\Sigma(X_1 - \overline{X}_1)^2$ = 1,134	ΣX_2 = 238	$\Sigma(X_2 - \overline{X}_2)$ = 0	$\Sigma(X_2 - \overline{X}_2)^2$ = 386

$$\overline{X}_1 = \frac{\Sigma X_1}{n_1} = \frac{333}{9} = 37; \ \overline{X}_2 = \frac{\Sigma X_2}{n_2} = \frac{238}{7} = 34$$

$$S = \sqrt{\frac{\Sigma(X_1 - \overline{X}_1)^2 + \Sigma(X_2 - \overline{X}_2)^2}{n_1 + n_2 - 2}} = \sqrt{\frac{1134 + 386}{9 + 7 - 2}} = 10.42$$

$$t = \frac{37 - 34}{10.42}\sqrt{\frac{9 \times 7}{9 + 7}} = \frac{3}{10.42} \times 1.984 = 0.571.$$

$v = n_1 + n_2 - 2 = 9 + 7 - 2 = 14;$

For $v = 14$, $t_{0.05} = 2.14$

The calculated value of t is less than the table value and hence the hypothesis holds true. We, therefore, conclude that the mean marks of the students of the two groups do not differ significantly.

Example 45:

Test the significance of the correlation r = 0.5 from a sample of size 18 against hypothesis correlation r = 0.7.

Solution:

We have to test the hypothesis that correlation in the population is 0.7. Applying Z-transformation

$$Z = \frac{1}{2}\log_e \frac{1+r}{1-r} = 1.1513 \log_{10} \frac{1+0.5}{1-0.5}$$

$$= 1.1513 \log 3 = 1.1513 \times 0.4771 = 0.549$$

$$\xi = \frac{1}{2}\log_e \frac{1+\rho}{1-\rho} = \frac{1}{2}\log_{10} \frac{1+0.7}{1-0.7}$$

$$= 1.1513 \log 5.67 = 1.1513 \times 0.7536 = 0.868.$$

$$S.E._{z} = \frac{1}{\sqrt{n-3}} = \frac{1}{\sqrt{15}} = \frac{1}{3.873} = 0.258$$

$$\frac{\text{Difference}}{\text{S.E.}} = \frac{0.319}{0.258} = 1.236$$

Since the difference is less than 1.96 S.E. (5% level of significance) it could have arisen due to fluctuations of sampling. Hence ρ may very well be 0.7.

Example 46:

The heights of six randomly chosen soldiers are inches : 76, 70, 68, 69, 89, and 68. Those of 6 randomly chosen sailors are 68, 64, 69, 72, 74. Discuss the light the these data throw on the suggestions that soldiers are, on the average, tailor man sailors. Use t-test :

Solution:

Let us take the hypothesis that there is no difference is height of soldiers and sailors. Applying t-test :

$$t = \frac{\overline{X}_1 - \overline{X}_2}{S}\sqrt{\frac{n_1 n_2}{n_1 + n_2}}$$

Height X_1	$(X_1-\overline{X}_1)$	$(X_1-\overline{X}_1)^2$ = 46	Height X_2	$(X_2-\overline{X}_2)$	$(X_2-\overline{X}_2)^2$
76	+6	36	68	+1	1
70	0	0	64	–3	9
68	–2	4	65	–2	4
69	–1	1	69	+2	4
69	–1	1	72	+5	25
68	–2	4	64	–34	9
		$\Sigma(X_1-\overline{X}_1)^2$	$\Sigma X_2 = 402$		$\Sigma(X_2-\overline{X}_2)^2 = 52$

$$\overline{X}_1 = \frac{420}{6} = 70;\ \overline{X}_2 = \frac{402}{6} = 67$$

$$S = \sqrt{\frac{\Sigma(X_1 - \overline{X}_1)^2 + \Sigma(X_2 - \overline{X}_2)^2}{n_1 + n_2 - 2}}$$

$$= \sqrt{\frac{46+52}{6+6-2}} = \sqrt{\frac{98}{10}} = 3.13$$

$$= \frac{70-67}{3.13}\sqrt{\frac{6\times 6}{6+6}} = \frac{3}{3.13}\times 1.732 = 1.66$$

$$v = n_1 + n_2 - 2 = 6 + 6 - 2 = 10$$

For $v = 10,\ t_{0.05} = 2.23.$

The calculated value of t is less than the table value. There is no reason to doubt the hypothesis. Hence the soldiers are not, on an average, taller than sailors.

Example 47:

Out of 20,000 customer's ledger accounts, a sample of 600 was taken to test the accuracy of posting and balancing and 45 mistakes were found. Assign limits which the number of mistakes can be expected at 95% level of confidence.

Solution:

p, i.e., Proportion of mistakes = 45/600 = 0.75

$\therefore \quad q = 1 - .075 = .925$

$$S.E._{p} = \sqrt{\frac{pq}{n}} = \sqrt{\frac{.075 \times .925}{600}} = .011$$

95% Confidence Limits

p ± 1.96 S.E.

.075 ± 1.96 (.011)

.075 ± .022 = .053 to .097.

Number of mistakes = 2000 × 0.053 to 2000 × 0.097

= 1060 to 1940

Hence at 5% level of significance it is expected that the number of mistakes would vary between 5.3 to 9.7 per cent.

Example 48:

10 workers are selected at random from a large number of workers in a factory. The number of items produced by them on a certain day are found to be:

51 52 53 55 56 57 58 59 59 60

In the light of these data, would it be appropriate to suggest that the mean of the number of items produced in the population is 58? (5% value of t for 9 d.f. is 2.262). (M. Com., Raj Univ., 1994)

Solution:

Let us first calculate the sample mean and standard deviation

X	**(X – 56)**	**$(X - 56)^2$**
51	–5	25
52	–4	16
53	–3	9
55	–1	1
56	0	0
57	+1	1
58	+2	4
59	+3	9
59	+3	9
60	+4	16
$\Sigma X = 560$		$\Sigma(X - 56)^2 = 90$

$$\overline{X} = \frac{\Sigma X}{n} = \frac{560}{10} = 56$$

$$S = \sqrt{\frac{\Sigma(X - \overline{X})^2}{n-1}} = \sqrt{\frac{90}{6}} = 3.162$$

Let us take the hypothesis that there is no significant difference in the sample mean and hypothetical population mean. Applying t-test.

$$t = \frac{\overline{X} - \mu}{S}\sqrt{n}$$

X = 56, m = 58, S = 3.162, n = 10

$$t = \frac{|56 - 58|}{3.162} \times \sqrt{10} = \frac{2}{3.162} \times 3.162 = 2$$

v = 10 – 1 = 9. For v 9 $t_{0.05}$ = 2.262

The calculated value of t is less than the table value. The hypothesis is accepted. Hence in the light of the data it is appropriate to suggest that the mean of the number of items produced in the population is 58.

Example 49:

A factory is producing 50,000 pairs of shoes daily. From a sample of 500 pairs, 2% were found to be of sub-standard quality. Estimate the number of pairs that can be reasonably expected to be spoiled in the daily production and assign limits at 95% level of confidence.

Solution:

p = 0.02, q = 0.98

$$S.E._{p} = \sqrt{\frac{pq}{n}} = \sqrt{\frac{0.2 \times .98}{500}} = 0.0063$$

95% confidence limit for percentage of defective pairs of shoes is given by

p ± 1.96 S.E.

0.2 ± 1.96 (.0063)

.02 ± .0123 = .0077 to .0323

The number of pairs that can be reasonably expected to be spoiled in daily production is given by 50,000 × .0077, i.e., 385 and 50,000 × 1615.

Example 50:

In a village 'A' out of a random sample of 1,000 persons 100 were found to be vegetarians while in another 'B' out 1,500 persons 180 were found to be vegetarians. Do you find a significant difference in the food habits of the people of the two villages?

Solution:

Let us take the hypothesis that there is no significant difference in the food habits of the people of two villages. Applying the test of the difference of two proportions; $S.E._{(p_1 - p_2)} = \sqrt{pq\left(\frac{1}{n_1} + \frac{1}{n_2}\right)}$

p_1, i.e., percentage of vegetarians in village 'A' $= \frac{100}{1000} \times 100 = 10$

p_2, i.e., percentage of vegetarians in village 'B' $= \frac{180}{1500} \times 100 = 12$

$$p = \frac{x_1 + x_2}{n_1 + n_2} = \frac{100 + 180}{1000 + 1500} = \frac{280}{2500} = 0.112 \text{ or } 11.2 \text{ per cent}$$

$$q = 100 - 11.2 = 88.8\%$$

$$S.E._{(p_1 - p_2)} = \sqrt{11.2 \times 88.8 \left(\frac{1}{1000} + \frac{1}{1000}\right)} = \sqrt{\frac{994.56 \times 5}{3000}} = 1.288$$

$$\frac{\text{Difference}}{\text{S.E.}} = \frac{12 - 10}{1.288} = 1.55$$

Since the calculated value (1.56) is less than 1.96 at 5% level of significance there is no reason to doubt the hypothesis. Hence is no significant difference in the food habits of the people of two villages 'A' and 'B'.

Example 51:

The following data give samples sizes and correlation coefficients. Test the significance of the difference between two values using Fisher's Z-transformation.

Sample Size	***Value of r***
5	*0.870*
12	*0.560*

Solution:

Let us take the hypothesis that the samples are drawn from the same population. Applying Z-test

$$Z = \frac{Z_1 - Z_2}{\sqrt{\frac{1}{n_1 - 3} + \frac{1}{n_2 - 3}}}$$

$$Z_1 = \log Z_1 = \log_e \frac{1 + r_1}{1 - r_1} \text{ or } 1.1513 \ \log_{10} \frac{1 + r_1}{1 - r_1}$$

Here $Z_1 = 1.1513 \log \frac{1 + 0.87}{1 - 0.87}$

$= 1.1513 \log \frac{1.87}{0.13}$

$= 1.1513 \log 14.385$

$$= 1.1513 \times 1.1579 = 1.333$$

$$Z_2 = \frac{1}{2}\log_e \frac{1+r_2}{1-r_2} \quad \text{or } 1.1513 \ \log_{10} \frac{1+r_2}{1-r_2}$$

$$= 1.1513 \ \log_{10} \frac{1+0.56}{1-0.56}$$

$$= 1.1513 \ \log \frac{1.56}{0.44}$$

$$= 1.1513 \ \log 3.545$$

$$= 1.1513 \times 0.5496 = 0.633$$

$$Z_1 - Z_2 = 1.333 - 0.633 = 0.7$$

$$S.E._{(Z_1 - Z_2)} = \sqrt{\frac{1}{n_1 - 3} + \frac{1}{n_2 - 3}}$$

$$= \sqrt{\frac{1}{5-3} + \frac{1}{12-3}} = 0.782$$

$$\frac{Z_1 - Z_2}{S.E.} = \frac{0.7}{0.782} = 0.895$$

Since the difference is less than 2.58 S.E. (1% level) the experiment provides no evidence against the hypothesis that the samples are drawn from the sample population.

Example 52:

An educator claims that the average I.Q. of American college students is at most 110, and that in a study made to test this claim 150 American college students, selected at random, had an average I.Q. of 111.2 with a standard deviation of 7.2. Use a level of significance of 0.01 to test the claim of the educator.

Solution:

Let us take the hypothesis that there is no significant difference in the claim of the educator and the sample result.

$$\frac{\text{Diff.}}{S.E.\overline{x}} = \frac{111.2 - 110}{0.588} = 2.04$$

Since the difference is less than 2.58 S.E. (1% level of significance), the hypothesis is accepted. Hence the claim of the educator is valid.

Example 53:

A population consists of four numbers 3, 7, 11, 15. Consider all possible samples of size two which can be drawn with replacement from their population.

Find

(i) the population mean,

(ii) the population standard deviation,

(iii) the mean of the sampling distribution of means, and

(iv) the standard deviation of the sampling distribution of means.

Verify (iii) and (iv) directly from (i) and (ii) by use of suitable formulae (which you are to mention). Solve this problem if sampling is without replacement.

Solution:

With 4 numbers 16 samples (i.e. 4 × 4) of size two can be drawn with replacement. The samples with their respective means are shown below:

Sample	Mean	Sample	Mean	Sample	Mean	Sample	Mean
3, 3	3	3, 7	5	3, 11	7	3, 15	9
7, 3	5	7, 7	7	7, 11	9	7, 15	11
11, 3	7	11, 7	9	11, 11	11	11, 15	13
15, 3	9	15, 7	11	15, 11	13	15, 15	15
	24		32		40		48

Mean of the sampling distribution of means :

$$= \frac{24+32+40+48}{16} = 9$$

The mean of the population $= \dfrac{3+7+11+15}{4} = 9$

Hence the mean of the sampling distribution of means is equal to the population mean.

Standard Deviation of Sampling Distribution

Mean	Deviations (d)	d^2	Mean	Deviations (d)	d^2
3	(3 – 9) = – 6	36	7	(7 – 9) = – 2	4
5	(5 – 9) = – 4	16	9	(9 – 9) = 0	0
7	(7 – 9) = – 2	4	11	(11 – 9) = 2	4
9	(9 – 9) = 0	0	13	(13 – 9) = 4	16
5	(5 – 9) = – 4	16	9	(9 – 9) = 0	0
7	(7 – 9) = – 2	4	11	(11 – 9) = 2	4
9	(9 – 9) = 0	0	13	(13 – 9) = 4	16
11	(11 – 9) = 2	4	15	(15 – 9) = 6	36
		$\Sigma d^2 = 80$			$\Sigma d^2 = 80$

S.D. of sampling distribution $= \sqrt{\frac{160}{16}} = 3.162$

Standard Deviation of Population

X	(X – 9) x	x^2
3	– 6	36
7	–2	4
11	+ 2	4
15	+ 6	36
	Σx = 5	Σx^2 = 80

$$\sigma = \sqrt{\frac{\Sigma x^2}{n}} = \sqrt{\frac{80}{4}} = 4.472$$

The standard deviation of sampling distribution of means is equal to standard error of mean, i.e., $\frac{\sigma}{\sqrt{n}} = \frac{4.47}{\sqrt{2}} = 3.16$ (n = size of sample).

If the sampling is without replacement the following six samples can be drawn :

(3, 7) (3, 11), (3, 15), (7, 11), (7, 15), (11, 15).

Their respective means are 5, 7, 9, 9, 11, 13

Mean of sampling distribution of mean

$$= \frac{5 + 7\ 9 + 9 + 11 + 13}{6} = \frac{54}{6} = 9$$

It is the same as population mean.

Variance of sampling distribution of means

$$= \frac{(5-9)^2 + (7-9)^2 + (9-9)^2 + (9-9)^2 + (11-9)^2 + (13-9)^2}{6}$$

$$= \frac{16+4+4+16}{6} = \frac{40}{6} = 6.67$$

$$\sigma = \sqrt{6.67} = 2.58$$

It is equal to

$$\sqrt{\left(\frac{n - n_1}{n-1}\right)\frac{\sigma^2}{\sqrt{n_1}}} = \sqrt{\left(\frac{4-2}{4-1}\right)\frac{20}{3}} = \sqrt{\frac{20}{3}} = \sqrt{6.67} = 2.58.$$

Example 54:

A sample of 100 households in a village was taken and the average income was found to be Rs. 628 per month with a standard deviation of Rs. 60 per month. Find the standard error of mean and determine 99% confidence limits within which the incomes of all the people in this village are expected to lie.

Solution:

$$\text{S.E.}\left(\overline{X}\right) = \frac{\sigma}{\sqrt{n}} = \frac{50}{\sqrt{100}} = \frac{50}{10} = 5$$

99% Confidence Limits

$\overline{X} \pm 2.58$ S.E. $= 628 \pm 2.58(5)$

$= 628 \pm 12.9$ or 615.1 to 640.9

Hence the limits which the incomes of all the people in this village are expected to be are Rs. 615.1 to Rs. 649.9.

Example 55:

A sample of size 25 yielded mean equal 1 to 33 and an estimated variance equal to 100. At the 1% level would we have reasons to doubt the claim that the population mean is not greater than 27?

Solution:

Let us take the hypothesis that there is no significant difference between the sample mean and the hypothetical population mean. Applying t-test.

$$t = \frac{\overline{X} - \mu}{S}\sqrt{n}$$

$\overline{X} = 33, \mu = 27, S = \sqrt{100} = 10, n = 25.$

$$t = \frac{33 - 27}{10} \times \sqrt{25} = 3$$

$v = n - 1 = 25 - 1 = 24$

For $v = 24$, $t_{0.05} = 2.064$.

The calculated value of t is more than the table value. The hypothesis is rejected. Hence we have reasons to doubt the claim that the population mean is not greater than 27.

Example 56:

(a) The mean population of a random sample of 400 villages in Jaipur district was found to be 400 with a standard deviation of 12. The mean population of a random sample of 400 villages in Meerut district was found to be 395 with a standard deviation of 15, is the difference between the two means statistically significant.

$$S.E.\ (\overline{X}_1 - \overline{X}_2) = \sqrt{\frac{\sigma_1^2}{n_1} + \frac{\sigma_2^2}{n_2}}$$

$\sigma_1 = 12,\ n_1 = 400,\ \sigma_2 = 15,\ n_2 = 400,$

$$S.E. = \sqrt{\frac{(12)^2}{400} + \frac{(15)^2}{400}} = \sqrt{\frac{144}{400} + \frac{225}{400}\ 0.96}$$

$$\overline{X}_1 - \overline{X}_2 = 400 - 395 = 5$$

$$\frac{\text{Difference}}{\text{S.E.}}\ \frac{5}{9.6} = 5.21$$

Since the difference is more than 2.58 (1% level of significance), the hypothesis is rejected. Hence the difference between the mean population of the two villages is statistically significant.

(b) Two randomly selected groups of 50 employees each of a very large firm are taught an assembly operation by two different methods and then tested for performance if the first group average 140 points with a standard deviation of 10 points while the second group averaged 135 points with a standard deviation of 8 points, test at 0.05 level whether the difference between their mean scores significant.

Solution:

Let us take the hypothesis that the difference in the mean scores of the two groups of employees is not significant. Applying t-test of difference of means.

$$\text{S.E. } (\overline{X}_1 - \overline{X}_2) = \sqrt{\frac{\sigma_1^2}{n_2} + \frac{\sigma_2^2}{n_2}}$$

$\sigma_1 = 10,\ \sigma_2 = 8,\ n_1 = 50,\ n_2 = 50,\ \overline{X}_1 = 140\ \overline{X}_2 = 135$

$$= \sqrt{\frac{(10)^2}{50} + \frac{(8)^2}{50}} = \sqrt{2 + 128} = 1.811$$

$$\frac{\text{Diff.}}{\text{S.E.}} = \frac{140 - 135}{1.811} = 2.76$$

Since the difference is more than 1.96 S.E. at (5% level) of significance, the hypothesis is rejected. Hence the difference in the mean scores of the two groups of employees is significant.

Example 57:

In measuring reaction time, a psychologist estimates that the standard deviation is 0.05 seconds. How large a sample of measurements must be taken in order to be 95% confident that the error of his estimate will not exceed 0.001 seconds?

Solution:

Since the error of estimate is not to, exceed 0.01 second

$$\frac{1.96\ \sigma}{\sqrt{n}} > 0.01$$

$$1.96 \times 0.05 > 0.01\ \sqrt{n}$$

or $$\sqrt{n} < \frac{1.96 \times 0.05}{0.01} = 9.8$$

$$n \leq (9.8)^2 = 96.04.$$

Hence the size of the sample should be if.

Example 58:

An examination was given to two classes consisting of 40 and 50 students respectively. In the first class the mean mark was 74 with a standard deviation of 8, while in the second class the mean mark was 78 with a standard deviation of 7. Is there a significant difference between the performance of the two classes at a level of significance of 0.05?

Solution:

Let us take the hypothesis that there is no significant difference in the mean marks of the two classes.

$$\text{S.E. } \left(\overline{X}_1 - \overline{X}_2\right) = \sqrt{\frac{\sigma_1^2}{n_1} + \frac{\sigma_2^2}{n_2}}$$

$$\sigma_1 = 8,\ n_1 = 40,\ \sigma_2 = 7,\ n_2 = 50$$

$$\text{S.E. } = \sqrt{\frac{(8)^2}{40} + \frac{(7)^2}{50}} = \sqrt{1.6 + 0.98}\ 1.606$$

$$\frac{\text{Difference}}{\text{S.E.}} = \frac{78 - 74}{1.606} = 2.49$$

Since the difference is more than 1.96 S.E. (5% level) of significance, the hypothesis is rejected. Hence there is a significant difference in the performance of the two classes at 5% level.

Example 59:

You are working as a purchase manager for a company. The following information has been supplied to you by two manufacture of electric bulbs:

	Company A	*Company B*
Mean life in hours	*1300*	*1248*

Standard deviations (in hours)	*82*	*93*
Sample size	*100*	*100*

Which brand of bulbs are you going to purchase if you desire to take a risk of 5%?

Solution:

Let us take the hypothesis that there is no significant difference in the mean life of the two makes of bulbs.

$$\text{S.E.}\left(\overline{X}_1 - \overline{X}_2\right) = \sqrt{\frac{\sigma_1^2}{n_1} + \frac{\sigma_2^2}{n_2}}$$

$\sigma_1 = 82, \sigma_2 = 93, n_1 = 100, n_2 = 100$

$$\text{S.E.}_{(\overline{X}_1 - \overline{X}_2)} = \sqrt{\frac{(82)^2}{100} + \frac{(93)^2}{100}} = \sqrt{\frac{6724}{100} + \frac{8649}{100}}$$

$$\frac{\text{Diff.}}{\text{S.E.}} = \frac{1300 - 1248}{12.399} = 4.19$$

Since the difference is more than 1.96 S.E., the hypothesis is rejected. Hence there is a significant difference in the mean life of the two makes of bulbs. We should prefer to buy the bulbs of Co. A since their average life is more.

Example 60:

(a) 400 labourers were selected at random from a certain district. Their mean income was 140.5 rupees per month, with standard deviation 25.2 rupees. Set up 95% confidence limits whiten which the income of the labour community of district is expected to lie.

(b) Set up 95% confidence limits for the standard deviation of the entire labour community of the district.

Solution:

(a) We are given = 140.5, σ = 25.2 and n = 400

$$E\overline{X} = \frac{\sigma}{\sqrt{n}} = \frac{25.2}{\sqrt{400}} = 1.26$$

95% confidence limits for mean are given by :

$\overline{X} \pm 1.96$ S.E.

$= 140.5 \pm 1.96\ (1.26)$

$= 140.5 \pm 2.47$ or 138.03 to 142.97

(b) $$\text{S.E.}_{\sigma} = \frac{\sigma}{\sqrt{2n}} = \frac{25.2}{\sqrt{2 \times 400}} = 0.89$$

95% confidence limit for standard deviation are given by :

σ + 1.96 S.E.$_\sigma$ or 25.2 ± 1.96 (.89)

25.2 ± 26.944 or 23.456 to 26.944.

Example 61:

An I.O. test was administered to 5 persons before and after they were trained. The results are given below :

Candidates	*I*	*II*	*III*	*IV*	*V*
I.Q. before training	*110*	*120*	*123*	*132*	*125*
I.Q. after training	*120*	*118*	*125*	*136*	*121*

Test whether there is any change I.Q. after the training programme.

[For v = 4, $t_{0.01}$ = 4.6]

Solution:

Let us take the hypothesis that there is no change in I.Q. after the training programme. Applying t-test :

$$t = \frac{\bar{d}\sqrt{n}}{S}$$

I.Q. Before	**I.Q. After**	**d**	**d²**
110	120	+ 10	100
120	118	– 2	4
123	125	+ 2	4
132	136	+ 4	16
125	121	– 4	16
		$\Sigma d = 10$	$\Sigma d^2 = 140$

$$\bar{d} = \frac{\Sigma d}{n} = \frac{10}{5} = 2$$

$$S = \sqrt{\frac{\Sigma d^2 - n(\bar{d})^2}{n-1}} = \sqrt{\frac{140 - 5(2)^2}{5-1}} = 5.477$$

$$t = \frac{2\sqrt{5}}{5.477} = 0.817$$

For v = 4, $t_{0.01}$ = 4.6

The calculated value of t is less than the table value. The hypothesis holds true. Hence there is no change in I.Q. after the training programme.

FORMULAS

Tests of Significance of Attributes :

S.E. of number of success $= \sqrt{npq}$

S.E. of proportion of successes $= \sqrt{\dfrac{pq}{n}}$

S.E. of difference of proportions $= \sqrt{pq\left(\dfrac{1}{n_1}+\dfrac{1}{n_2}\right)}$ where $p = \dfrac{x_1 + x_2}{n_1 + n_2}$

Test of Significance of Large Samples :

$$S.E._{\overline{X}} = \frac{\sigma}{\sqrt{n}}$$

$$S.E._{(\overline{X}_1-\overline{X}_2)} = \sqrt{\frac{\sigma_1^2}{n_1}+\frac{\sigma_2^2}{n_2}}$$

$$S.E._{(\sigma_1-\sigma_2)} = \sqrt{\frac{\sigma_1^2}{n_1}+\frac{\sigma_2^2}{2n_2}}$$

$$S.E._r = \frac{1-r^2}{\sqrt{n}}$$

Tests of Significance of Small Samples :

(1) To test the significance of the mean of a random sample

$$t = \frac{\left(\overline{X}-\mu\right)}{S}\sqrt{n}$$

where $v = n_1 + n_2 = 2$

(2) To test the significance of the difference of the means of two samples

$$t = \frac{\left(\overline{X}_1-\overline{X}_2\right)}{S} \times \sqrt{\frac{n_1 n_2}{n_1+n_2}}$$

where $v = n - 1$

(3) Two test the difference of the means of two samples which are not independent.

$$t = \frac{\overline{d}\sqrt{n}}{S}$$

where $v = n - 1$

(4) To test the significance of an observed correlation coefficient

$$t = \frac{r}{\sqrt{1+r^2}} \times \sqrt{n-2}$$

where v = n – 2

The Variance Ratio Test–F-Test :

$$F = \frac{S_1^2}{S_2^2}$$

where $S_1^2 = \frac{\Sigma\left(X_1 - \overline{X}_1\right)^2}{n_1 - 1}$ and

$$S_2^2 = \frac{\Sigma\left(X_1 - \overline{X}_1\right)^2}{n_2 - 1}.$$

EXERCISES

1. What do you mean by test of significance of a mean or difference between two means ? Define standard error. Explain and mention formula to obtain standard error of mean and standard error of standard deviation both in ungrouped and grouped data.

2. (a) Two different types of drugs A and B were tried on certain patients for increasing weight. Five persons were give drug A and seven persons were given drug B. The increase in weight in pound is given below :

Drug A :	8	12	13	9	3		
Drug B :	10	8	12	15	6	8	11

Do the two drugs differ significantly with regard to their effect in increasing weight? (table value of t for v = 10 at 5% level is 2.23)

(b) A machine puts out 16 imperfect articles in a sample of 500. After the machine is over-handed, it puts out 3 imperfect articles in a batch of 100. Has the machine improved?

3. (a) The breaking strengths of cables produced by a manufacturer have been 1800 lbs-and standard deviation 200 lbs. By a new technique in the manufacturing process, it is claimed that the breaking strength can be increased. To test this claim, a sample of 64 cables is tested and it is found that the mean breaking strength is 1850 lbs. Can we support the claim at (i) 5 per cent (ii) 1 per cent levels of significance?

(b) A sample of 400 persons was taken from a large population. The mean weight and standard deviation of these persons was found to be 68 kg. and 6 kg. Can it reasonably be said that in the population the mean weight would be 67 kgs?

4. A large corporation employs both men and women to do the same kind of work. It is hypothesised that the average hourly output of men is less than that of women. The following information was gathered by the O and M teams of the corporation :

	Men	*Women*
Variance in units	$\sigma 1^2 = 70$	$\sigma 2^2 = 74$
Sample size	$N_1 = 36$	$N_2 = 36$
Sample mean in units	$\overline{X}_1 = 150$	$\overline{X}_2 = 153$

Is the average hourly output of men significantly lower than that of women at 5% level of significance?

[Diff./S.E. = 1.5]

5. (a) In one section of a large city, only 200 out of 250 residents own the dwellings in which they live. In other section, 120 out of 200 residents own their homes. Using a 0.05 level of significance, can we conclude that there is no difference between these two section of the city in terms of house ownership?

(b) A mean buys 200 electric bulbs of each of two well-known makes taken at random from stock for testing purposes. He finds that 'Make A' has a mean life of 2,560 hours with a standard deviation of 90 hours and 'Make B' has mean life of 2,650 hours with a standard deviation of 75 hours. Is there a significant difference in the mean life of these two makes at 5% level of significance?

6. (a) An examination was given to two classes, each consisting of 49 students. In the first class, the mean marks were 74 with a standard deviation of 8, in the second class, the mean marks were 78 with standard deviation 9. Is it reasonable that two classes are having been obtained from two normal populations with equal means? Test your hypothesis at 5% and 10% level of significance.

(b) A large hotel chain is trying to decide whether to convert more of its rooms to non-smoking rooms. In a random sample of 400 guests last year, 16 had requested non-smoking rooms. This year, 205 guests in a sample of 380, preferred the not-smoking rooms. World you recommend that the hotel chain converts more rooms to non-smoking? Support your recommendation by testing the appropriate hypothesis at 0.01 level of significance?

7. (a) State the procedure followed in testing a hypothesis.
   (b) Explain the procedure generally followed in testing of a hypotheses. Point out the difference between one tail and two tail tests.
8. (a) Define null hypothesis, critical region and two sided test used in testing of hypothesis.
   (b) Explain the concept of standard error. How is it useful in testing of hypothesis?
9. Discuss the role of hypothesis in making a research design. What are the characteristics of a good hypothesis?
10. (a) Two independent random samples of size 10 and 15 from two independent normal populations having variance 4 and 8 were found to have means 48 and 46 respectively. Test at 5% significance level whether the population means may be taken to be equal. What assumptions do you make in applying the test? Also find 95% confidence limits for thew difference of the population means.
   (b) Of the two salesmen, X claims that he has made larger sales than Y. For the accounts examined, which were comparable for the two men, the results were:

	X	Y
Number of sales	10	17
Average size	Rs. 6,200	Rs. 5,600
Standard deviation	Rs. 690	Rs. 600

   Do these two 'average size of sales' figures differ significantly? Explain your result. You may use the following extract from the table of Student's t-distribution.
11. The following data show weekly sales of a manufacturer BEFORE and AFTER recognition of the sales organisation, 10 weeks from Sept, to Dec. in two successive years were selected for comparison:

Week No. :	1	2	3	4	5	6	7	8	9	10
sales (before Reorganisation) (in '000 Rs.) :	15	17	12	18	16	13	15	17	1918	
Sales (After Reorganisation) (in '000 Rs.) :	20	19	18	22	20	19	21	23	24	24

   Apply the 't' test to determine whether re-organisation had any effect on the sales.
   [t = 11.86]

12. (a) As a financial manager, you are considering two different television advertisements for promotion of a new financial service. Two test market areas with identical consumer characteristics are selected. Advertisement A is used in one area and B is used in the other area. In a random sample of 60 customers who saw advertisement A, 18 used the service. In another sample of 100 customers who saw advertisement B, 22 tried the new service. Would you say that advertisement A is more effective than advertisement B at 5% level of significance?

    (b) A sample of 900 members is found to have a mean 3.4. Can it be reasonably regarded as a simple sample from a large population with mean 3.25 and standard deviation 2.61?

    [Diff./S.E. = 1.72, Yes]

13. A random sample of 200 villages from Jaipur district was taken and average population per village was found to be 420 with 50 as standard deviation from the same district, another sample of 200 village as 480 and 60 as standard deviation. Is the difference between the average of two samples significant?

14. What do you understand by t-test? Explain. A sample of 11 chickens of two months old was taken which was reared on Hind Lever diet. Weights were found to be 14, 16, 14, 16, 18, 16, 18, 16, 17, 14 and 17 ounces. Another sample of 11 chickens of the same group was taken which was kept on another diet. Their weights were fond to be 12, 14, 13, 14, 16, 17, 16, 17, 16, 14 and 16 ounces. Test whether two diets are significantly different? The table value of t for 20 d.f., at 5% level of significance is 2.086.

15. (a) The wages of 10 workers taken at random from a factory are given below :

    Wages (Rs.): 578, 572, 570, 568, 572,
    578, 570, 572, 596, 584

    Is it possible that the mean wage of all workers of this factory is Rs. 580? (Given for v = 2, t = 2.26

    [t = 1.481; Yes: Yes]

    (b) The annual salary of administrators in the Government averages Rs. 20,000 and has a standard deviation of Rs. 1,500. In the same Government, the salary of doctors averages Rs. 22,500 and has a standard deviation of Rs. 2,250. These data relate to samples of sizes of 25 for each group chosen at random. Test at 5% level of significance whether there is significant difference in the mean salaries of two groups.

16. (a) A random sample of 225 families in a city indicates that they average 150 minutes of television viewing per day with a standard deviation of 30 minutes. What are the 99 per cent confidence limits for the average viewing time of all families in this city?

(b) Suppose you want to estimate the proportion of families in your town which has two or more children. A random sample of 144 families shows that 48 families have two or more children. Construct a 95% confidence interval.

17. (a) 10 tests of two groups, viz., Sawraj and Swadeshi. Give the following results.

Sawaraj	:	mean = 64	S.D. = 8	n = 100
Swadeshi	:	Mean = 61	S.D. = 10	n = 80

(i) Is the difference in mean scores significant?

(ii) Is the difference in standard deviations significant?

[(i) d.f./S.E. = 2.18 (ii) Diff./S.E. = 2.06]

(b) In a sample of 50, there are found to be 27 males and 23 females. Ascertain if the observed proportions are inconsistent with the hypothesis that the male and female should be in equal proportion. (M. Com., Garhwal Univ., 1998)

[Diff./S.E. = 0.56; No]

(c) 250 oranges are taken at random from a basket and 25 are found to be bad. Estimate the proportion of bad oranges in the basket and assign limits within which the percentage most probably lies.

[Diff./S.E. = 5.7]

19. Daily sales figures of 40 shopkeepers were collected and the average sales and standard deviation were fond to be Rs. 528 and Rs. 600 respectively. Is the assertion that daily sale on the average is Rs. 400 only, contradicted at 5% level of significance by the sample?

20. (a) Tests were made at short intervals on spark plugs from two manufacturers. The following table gives the number of hours of service given by plugs from the two sources :

A :	200	210	190	200	190	200	180	200	200	210
B :	190	200	210	190	180	190	200	192		

Do these results indicate a statistically significant difference between the spark plugs os far as mean length of service is concerned?

(b) Intelligence test on two groups of boys and girl gave the following result :

	Mean	*S.D.*	*N*
Girls	75	15	150
Boys	70	20	250

Is there a significant difference in the mean scores obtained by boys and girls?

21. What is a "Point estimator"? Give examples. How will you proceed to construct an interval estimate of the mean of normal population with known variance?

22. (a) Why should there be different test functions for testing the significance of difference in means when samples are (a) small, (b) large?

    (b) Explain the term "Standard Error" and show how standard error is used in large sample test of statistical hypothesis.

23. (a) Explain the terms : (i) Type I and Type II errors; (ii) Critical region; and (iii) Level of significance associated with testing hypothesis.

    (b) Explain the difference between the 'standard error of mean' and 'standard error of proportions'.

24. (a) Explain the concept of sampling distribution in general and also explain the sampling distribution means in particular. What are its properties?

    (b) Explain Type I and II errors associated with testing of hypothesis. How will you proceed to test the difference between the means in two population, based on large samples drawn from the populations?

25. Explain the procedure for testing the hypothesis concerning the difference between two population proportions based on samples taken from each of the two populations.

26. On the basis of their total scores, 200 candidates of a civil service examination are divided into two groups, the upper 30 per cent and the remaining 70 per cent. Consider the first question of this examination. Among the first group, 40 had the correct answer. On the basis of these results, can one conclude that the first question is not good at discriminating ability of the type being examined here?

27. A company is interested in knowing if there is a difference in the average salary received by foremen in two divisions. Accordingly, samples of 12 foremen in the first division and 10 foremen in the second division are selected at random. Based upon experience foremen's salaries are known to be approximately normally distributed, and the standard deviations are about the same.

Sample size	*First Division*	*Second Division*
	12	10
Average monthly salary of foremen (Rs.)	1,050	980
Standard deviation of salaries (Rs.)	68	74

28. (a) A survey of 17 agricultural labourers reveals an income of Rs. 40 per week with a standard deviation of Rs. 8. Find to the limits of mean weekly wages in the population at 5% level of significance.

[Given v = 16, $t_{0.05}$ = 2.12]

(b) It is hypothesized that more than 10 per cent of the families in the United States plan to buy new automobiles this year. A sample of 2,000 families interviewed in a nationwide survey, indicated that 160 planned to purchase new cars this year. Do you accept the hypothesis at 5 per cent level of significance ?

(c) In a sample of 100 respondents it is found that 55 view a T.V. Serial. Find a 99% and 95% confidence interval for the population proportion.

29. (a) Two types of batteries X and Y are tested for their length of life and the following results are obtained :

Battery	*Sample size*	*Mean hours*	*Variance (hours)*
X :	10	1000	100
Y :	12	1020	121

Can you conclude that the two types of batteries are having the same mean life?

[Diff./S.E. = 4.22]

(b) Test the significance of the difference between the means of the sample from the following data :

	Size of sample	Mean	S.D.
Sample A	100	50	4
Sample B	150	51	5

[Dif./S.E. = 1.748]

30. (a) What is t-distribution? Explain some of its applications.

(b) What is t-distribution? Explain its uses in testing the hypothesis.

(c) Explain how Student's 't' test is used to test the significance of the difference between the means of two samples?

31. Differentiate the following pairs of concepts :

(i) Statistics and parameter,

(ii) Critical region of acceptance, and

(iii) Null and alternative hypothesis.

32. Explain how the Student's 't' distribution may be used to :

(i) test the significance of the sample correlation coefficient in a sample drawn from a bivariate normal population.

(ii) test the significance of the difference between the yield of two varieties in agricultural experiment.

(iii) explain Fisher's transformation of the correlation coefficient and indicate its use in tests of significance.

(iv) explain briefly different applications of the t-test.

33. (a) Difference between standard deviation and standard error of estimate.

(b) Explain the following :

(i) Null hypothesis.

(ii) Statistical hypothesis.

(c) Explain briefly the need for sampling in the context of estimation of unknown population parameters and statistical testing.

(d) Explain the concept of degrees of freedom.

(e) Write short notes on :

(i) Type I and Type II errors

(ii) Steps involved in testing of a hypothesis.

34. (a) What is null hypothesis?

(b) Distinguish between large sample and small sample.

(c) State the limitations of tests of significance.

(d) Write a note on the sampling distribution of the Sample Mean.

35. (a) Define student's t-statistic and write its probability density function.

(b) Explain the procedure that is followed i testing of a statistical hypothesis.

(c) What do you mean by estimation? Discuss the desirable properties of an estimator.

36. (a) Explain the terms Sampling distribution and Standard error of a statistics.

(b) Show that mean of the Sampling distribution of the Sample mean is equal to the population mean.

37. A certain stimulus administered to each of 12 patients resulted in the following changes in blood pressure :

5, 2, 8, –1, 3, 0, –2, 15, 0, 4, 6.

Can it be concluded that the stimulus will, in general, be accompanied by an increase in blood pressure?

38. Eleven school boys were given a test in science. They were given a month's special coaching and a second test was held at the end of it. Do the marks give evidence that the students have benefited by the extra coaching?

Marks in first test	:	46	40	38	42	36	40
Marks in Second test	:	48	38	44	36	40	44
Marks in first test	:	30	34	46	32	38	
Marks in second test	:	40	40	46	40	32	

[For v = 10, the table value of t at 5% level of significance = 2.23]

39. The incomes of a random sample of engineers in Factory 1 are Rs. 630, 650, 680, 690, 710 and 720 per month. The incomes of a similar sample in Factory II are Rs. 610, 620, 660, 690, 700, 710, 720 and 730 per month. Discuss the validity of the opinion that Factory 1 pays its engineers much better than Factory II.

40. (a) A sample of 400 items is taken from a population whose standard deviation is 1.5. The mean of the sample is 2.5. Test whether the sample has come from a population with mean 26.8. Also calculate the 98% confidence limits of the population mean.
[Diff./S.E. = 2.4; 23, 25 to 26.76]

(b) Random samples of 200 bolts manufactured by machine A and 100 bolts manufactured by machine B showed 19 and 5 defective bolts respectively. Is there a significant difference between the performance of the two machines?
[Diff./S.E. = 1.36]

41. (a) A coin is tossed 10,000 times and the head appeared 5,195 times. Would you consider the coin biased?

(b) A random sample of 100 mill workers at Kanpur showed their mean wages to be Rs. 350 per week with a standard deviation of Rs. 28. A sample of 150 mill workers in Bombay showed the mean wage to be Rs. 390 per week with a standard deviation of Rs. 40. On the basis of the data, would you say the mean wages of mill workers in Bombay are higher that those at Kanpur?
[(a) Diff./S.E. – 3.9 (b) Diff./S.E. = 9.4]

42. (a) A correlation coefficient based on a sample of size 51 was computed to be 0.32. Use 't' test to find whether it is significant. The table value of 't' for 49 degrees of freedom at 5% level of significance is 2.01.

(b) The management of a company claims that the average weekly income of their employees is Rs. 900. The trade union disputes this claim stressing that it is rather less. An independent survey of 150 randomly selected employees estimated the average to be Rs. 856 and the Standard Deviation to be Rs. 364.26. Would you accept the view of the management or the trade union? [Diff./S.E. = 1.48]

43. (a) Sky Packets guarantee that 90 per cent of their deliveries are on time. In a recent week, 81 deliveries were made of which 6 were late. Sky packets, Managing Director says with 95% confidence that there has been a significant improvement in deliveries. Should managing Director's statement be accepted? [Diff./S.E. = 0.78]

(b) the mean height obtained from a random sample of size 100 is 64 inches. The standard deviation of the height distribution is known to be 3 inches. Set up 95% limits of the mean height of the population.

44. (a) A die is thrown 9,000 times and a throw of 3 or 4 is reckoned as a success. Suppose that 3,240 throws of a 3 or 4 have been made out. Do the data indicate an unbiased die?

(b) A random sample of size 7 from a normal population gave a mean 977.51 and a standard deviation 4.42. Find a 95% confidence interval for the population mean.

(c) Intelligence test of two groups of boys and girls gave the following results :

Girls	12	84	10
Boys	8	81	12

Examine if the difference of means is significant.
[t = 0.607]

45. (a) A random sample of 400 members is found to have a mean of 4.45 cm. Can it be regarded a sample from large population whose mean is 5 cm. and whose variance is 4?
[Diff./S.E. = 5.5]

(b) 500 units from a factory are inspected and 12 are found to be defective. 800 units from another factory are inspected and 12 are found to be defective. Can it be concluded at 5% level of significance that production at second factory is better that in first factory?
[1.184]

46. (a) A departmental store executive reads in a trade journal that a large departmental store has made a survey indicating that over 50% of the persons who enter the store as apparently potential customers,, actually figure and decide to try to see what the percentage of non-buying visitors is for his store. Plans are made calling for discrete surveillance of a random sample consisting of 100 persons, entering and leaving the store. It is found that 41 of these persons made no purchases Test at 5% level of significance, the hypothesis that the percentage of non-buyers is 50.

[Diff./S.E. = 1.8]

(b) The following table presents data on the value of a harvested crop stored in the open and inside a godown:

Sample	size	Mean	$\sum(\overline{X}-X)^2$
Outside	40	117	8,865
Inside	100	132	27,315

Assuming that the two samples are random and they have been drawn from normal population with equal variances, examine if the mean value of the harvested crop is affected by weather conditions.

[Diff./S.E. = 0.342]

47. (a) Below are given the gains in weights (lb.) of lions on two diets X and Y : Gain in weight (lb.)

Diet X :	25	32	30	32	24	14	32			
Diet Y :	24	34	22	30	42	31	40	30	32	35

Test, at 5% level, whether the two diets differ significantly with regard to increase in weight.

(Table value of 't' for 15 degrees of freedom at 5% = 1.573)

[t = 1.62]

(b) The average hourly wage of a sample of 150 workers in a plant 'A' was Rs. 2.56 with a standard deviation of Rs. 1.08. The average wage of a sample of 200 workers in plant 'B' was Rs. 2.87 with a standard deviation of Rs. 1.28 Can an applicant safely assume that the hourly wages paid by plant 'B' are higher than those paid by plant 'A'?

[Diff./S.E. 2.46]

48. Tick the correct answer :

(a) Standard error of number of successes is given by :

(i) $\sqrt{\frac{pq}{n}}$ (ii) $\sqrt{npq}$ (iii) npq (iv) $\sqrt{\frac{np}{q}}$ (v) n_2p_2

(b) 95% fiducial limits of population mean are :

(i) $\overline{X} \pm 3 S.E.$ (ii) $\overline{X} \pm 2.51 S.E.$ (iii) $\overline{X} + 2.58 S.E.$

(iv) $\overline{X} \pm 1.96 S.E.$ (v) $\overline{X} \pm 1.95 S.E.$

(c) 99% fiducial limits of population mean are:

(i) $\overline{X} \pm 2.58 S.E.$ (ii) $\overline{X} \pm 1.96 S.E.$ (iii) $\overline{X} \pm 3 S.E.$

(iv) $\overline{X} \pm 2 S.E.$ (v) none of these.

(d) Large sample theory is applicable when:

(i) N > 30 (ii) N < 30 (iii) N = 30 (iv) N is at least 100 (v) N is at least 1,000.

(e) Student's 't' Distribution was discovered by :

(i) Karl Pearson (ii) Laplace (iii) Fisher (iv) Gosset (v) None of these.

(f) The difference of two means in case of small samples is tested by the formula :

(i) $t = \frac{\overline{X}_1 - \overline{X}_2}{S}$ (ii) $t = \frac{\overline{X}_1 - \overline{X}_2}{S}\sqrt{\frac{n_1 + n_2}{n_1 n_2}}$

(iii) $t = \frac{\overline{X}_1 - \overline{X}_2}{S}\sqrt{\frac{n_1 n_2}{n_1 + n_2}}$ (iv) $t = \sqrt{\frac{n_1 n_2}{n_1 + n_2}}$

(v) $t = \frac{\overline{X}_1 - \overline{X}_2}{S}\sqrt{\frac{n_1 - n_2}{n_1 n_2}}$

(g) While testing significance of the difference of two sample means in case of small samples, the degree of freedom is calculated by:

(i) $v = n_1 + n_2$ (ii) $v = n_1 - 1$ (iii) $v = n_1 + n_2 - 1$ (iv) $v = n_1 - n_2 + 2$ (v) $v = n_1 + n_2 - 2$

[a (ii), b (v), c (i), d (i), e (iv), f (iii), g (v)]

49. Intelligence test given to two groups of boys and girls gave the following information:

	Mean Score	*S.D.*	*Number*
Girls	75	10	50

Boys	70	12	100

Is the difference in the mean scores of boys and girls statistically significant?

50. 500 articles from a factory are examined and found to be 2 per cent defective. 800 similar articles from a second factory are found to have 1.5 per cent defectives. Can it reasonably be concluded that the products of the first factory are inferior to those of the second?

	Sample size	*Sample mean*	*Sample variance*
Yarn A	4	50	42
Yarn B	9	42	56

The strengths are expressed in pounds. Is the difference in mean strengths significant of real difference in the mean strengths of the sources from which the samples are drawn?

51. (a) Distinguish between the one tailed tests and two tailed tests, clearly bringing out the difference between the alternative hypothesis under the two types of tests.

    (b) How does the standard error of the mean measure sampling error? Is the amount of sampling error in the sample mean affected by the amount of variability in the universe? Explain. Why is it important to consider sampling error?

52. Fill in the blanks :

    (i) The null hypothesis asserts that there is no true difference in the .... and the .... in the particular matter under consideration.

    (ii) Type I error is committed when the hypothesis is true but our test .... it.

    (iii) Type II errors are made when we accept a null hypothesis which is ....

    (iv) In .... tail test rejection region is located in one tail.

    (v) The standard deviation of sampling distribution is called ....

    (vi) The distribution formed of all possible values of a statistics is called the ....

    (vii) The mean of sampling distribution of means is equal to the ....

    (viii) Standard error provides an idea about the .... of sample.

    (ix) An estimator is said to be .... if it covers as much information as possible about the parameter which is contained in the sample.

    (x) An .... estimate of a population parameter provides two values, between which it is estimated that the parameter lies.

[(i) Sample, population (ii) rejects (iii) not true (iv) one (v) standard error (vi) sampling distribution (vii) population mean (viii) unreliability (ix) sufficient (x) interval]

53. Which of the following statements are True or False:

(i) H_a represents the null hypothesis and H_0 alternative hypothesis.

(ii) When the hypothesis is true and our test accepts it. This is called Type I error.

(iii) The reciprocal of standard error is a measure of reliability of precision of the sample.

(iv) An estimator is said to be biased. If its expected value is identical with the population parameter being estimated.

(v) A point estimate is a single number which is used as an estimate of the unknown population parameter.

(vi) An interval estimate should generally be preferred to point estimate.

(vii) The term standard error of the mean is used to refer to the standard deviation of the distribution of sample means.

(viii) The sampling distribution has a standard deviation equal to the population standard deviation divided by the sample size.

(ix) The sample mean $\overline{X}$ is the best estimator of the population mean μ.

(x) A t-distribution is lower at the mean and higher at the tail than a normal distribution.

[(i) F (ii) F (iii) T (iv) F (v) T (vi) T (vii) T (viii) F (ix) T]

54. (a) What is the major purpose of hypothesis testing? Explain the various steps involved in hypothesis. What is sampling distribution? Discuss its utility in significance tests.

(b) Explain the concept standard error. How is it useful testing of hypothesis?

55. (a) What is 'test of significance'? Explain the concept of standard and discuss its role in large sample theory.

(b) State the meaning and utility of standard error. Describe briefly the procedure for carrying out a hypothesis testing.

(c) What is sampling distribution of the Mean? State its important properties.

(d) Distinguish between Estimation and Estimator. State the desirable properties of a good estimator.

(e) Describe the various steps involved in testing of a hypothesis. What is the role of standard error in testing hypothesis?

56. (a) Why is testing of hypothesis at all necessary?
(b) Define type I and type II error.
(c) Discuss the following in proper details :
(i) Maximum likelihood method.
(ii) Estimation of population variance.
(iii) Point vs, interval estimates.
(d) (i) Explain the Central Limit Theorem and its usefulness.
(ii) Explain some important properties of estimators.

57. (a) Eleven sales executive trainees are assigned selling jobs right after their recruitment. After a fortnight they are withdrawn from their field duties and given a month's training for executive sales. Sales executed by them in thousands of rupees before and after the training in the same period, are listed below.

Sales ('000Rs.) (Before training)	: 23	20	19	21	18	20	18	17	23	16	19
Sales ('000Rs.) (Before training)	: 24	19	21	18	20	22	20	20	23	20	27

Do these data indicate that the training has contributed to their performance?

(b) A marketing research analyst collects data for a random sample of 100 customers out of the 400 who purchased a particular coupon special'. The 100 people spent an average of Rs. 2457 in the store with a standard deviation of Rs. 660. Using 95% confidence interval estimate (i) the mean purchase amount for all 400 customers and (ii) the total amount (in Rs.) of purchases by the 400 customers.

58. In a sample of 600 students of a certain college, 400 are found to use dot pens. Test whether the two colleges are significantly different with respect to the habit of using dot pens.

59. Two types of batteries A and B are tested for their length of life and the following results are obtained :

Battery	*Sample size*	*Mean (hrs.)*	*Variance (hrs.)*
A	10	500	110
B	12	560	121

Is there a significant difference in the two sample means?

60. Name the test of significance which is most appropriate in following situations :

(i) For judging the significance of differences of more than two samples means at one and the same time.

(ii) For comparing the sample variance to a specified variance of the population.

(iii) For comparing the mean of a random sample to some hypothesised mean for the population when 15 and population variance is known.

(iv) For comparing the variances of two independent random samples.

61. What do you understand by-test? Indicate some practical application of t-test.

62. (a) What is it that which the standard error of estimate measures?

(b) What are Type I and Type II errors in tests of hypothesis. How is a test of hypothesis constructed?

63. (a) Discuss some important application of t-test in making business decisions.

(b) Explain the role of standard error in large sample tests.

(c) What are the main characteristics of normal probability curve? Distinguish between standard normal distribution and t-distribution.

64. (a) A random sample of 16 values from a normal population showed a mean of 48.5 and sum of squares of deviations from the mean equal to 135. Can it be assumed that the mean of the population is 43.5?

(b) A random sample of 12 pairs of observations from a normal population gives a coefficient of correlation of 0.54. Is this value significant of correlation in the population?
[t=1.597]

65. (a) The mean life time of 400 bulbs produced by a company is found to be 2.570 hours with a standard deviation of 120 hours. Test the hypothesis that the alternative hypothesis is greater than 2,6000 hours at the 5 per cent level of significance.
[(a) Diff./S.E.=2.73]

(b) In an examination in Psychology. 12 students in one class had a mean grade of 78 with a standard deviation of 6, while 15 students in another class had a mean grade of 74 with a standard deviation of 8. Is there a significant difference between the means of the two groups?

66. (a) In a random sample of 600 males in Jaipur, 400 were found to be smokers while in another random sample of 900 females in

Delhi, 450 were found to be smokers. Discuss the question whether the data reveal a significant difference in Jaipur to Delhi so far as the proportion of smokers is concerned.

[Diff./S.E.=6.42]

(b) I.Q. Test on two groups of boys and girls gave the following results:

Girls	$\overline{X}=78$	S.D.=10	N=50
Boys	$\overline{X}=73$	S.D.=15	N=100

Is there a significant difference in the mean scores of boys and girls?

[Diff./S.E.=2.43]

67. (a) A company produces two makes of bulbs, A and B, 200 bulbs of each make were tested and it was fond that make A has mean life of 2,560 hours and S.D. 90 hours: whereas make B has 2,650 hours of mean life with S.D. 75 hours. Is there a significant difference between the mean life of two makes?

[Diff/S.E.=8.284 : Yes]

(b) An automatic machine was designed to pack exactly 5 kilograms of oil. A random sample of 15 tins was examined to test the machine. The average weight was fond to be 4.94 kg. with standard deviation 0.10. Assuming normal distribution for the weights of tine packed, test at 5 per cent level of significance, whether the automatic machine is working properly.

[t = 2.32, No]

(c) A stenographer claims that she can take dictation at the rate of 120 words per minute. Can we reject her claim on the basis of 100 trails in which she demonstrate a mean of 116 words with a standard deviation of 15 words. Use 5 per cent level of significance.

[Diff./S.E.=2.67; Yes]

68. You are given the following information relating to purchase of bulbs from two manufacturers A and B:

Manufacture	*No. of Bulbs bought*	*Mean live*	*S.D.*
A	100	2,950 hours	100 hrs.
B	100	2,970 hours	90 hrs.

Is there a significant difference in the mean life of two makes of bulbs?

[Diff./S.E.=1.487; No]

69. (a) A random sample from 200 villages was taken from Meerut district and the average population per village was found to be 250 with S.D. of 50. Another random sample of 200 villages from the same district gave an average population of 580 per village with S.D. of 60. Is the difference between the averages of the two samples statistically significant.

    (b) A sample of 10 measurements of the diameter of a sphere gave a mean $\bar{X}$ = 4.38 inches and a standard deviation 0.06 inches. Find (a) 95%, and (b) 99% confidence limits for the actual diameter.

    [(a) 40.343 to 4,417: (b) 4.331 to 4.22]

70. (a) Ten specimens of copper wires drawn from two large lots have the following breaking strength (in kg. wt.):

    578, 572, 570, 568, 512, 578, 570, 572, 569,548

    Test whether the mean breaking strength of the lot may be taken to be 578 kg. wt.

    (b) Two groups of students are given an intelligence test (X) and an arithmetic test (Y).

    n_1 = 45; r = 0.45

    n_2 = 39; r = 0.08

    Is the difference between the value of r significant?

    [t = 0.3]

72. Rate of oxygen consumption in 15 groups of fishes of macrograthus was studied as 62. 64, 67, 68, 61, 72, 71, 69, 67, 64, 62, 60, 62, 64 and 67 kg/hour/I OOc.c. Obtain standard erforofmean and standard errorof standard deviation both in ungrouped and grouped series.

73. What do you mean by student's 't' test ? Mention formula to obtain 't' ratio.

74. What is degree of freedom ? Mention formula to obtain d.f. in single and paired data while computing 't' ratio. If we get calculated 't' value higher than the table't' value at a certain d.f. and level then the experiment is reliable or not.

75. In a pig form of military camp at Gaya a group of 12 pigs were fed with one variety of food A while another group of 12 pigs of same age, sex and stock were given another variety of food B. After one month weight of both groups was recorded as follows :
    First group on food A : 31, 34, 34, 29, 26, 32, 35, 38, 34, 30.
    Second group on food B : 26, 24, 28, 29, 30, 29, 32, 26, 35, 29.